T0309105

Current Progress in Protein Kinase Research

Current Progress in Protein Kinase Research

Editor: Bailey Graham

MURPHY & MOORE
www.murphy-moorepublishing.com

www.murphy-moorepublishing.com

ⓂMURPHY & MOORE

Cataloging-in-Publication Data

Current progress in protein kinase research / edited by Bailey Graham.
  p. cm.
Includes bibliographical references and index.
ISBN 978-1-63987-733-1
1. Protein kinases. 2. Protein kinases--Research. 3. Biochemistry. I. Graham, Bailey.
QK898.P79 C87 2023
572.62--dc23

© Murphy & Moore Publishing, 2023

Murphy & Moore Publishing
1 Rockefeller Plaza,
New York City,
NY 10020, USA

ISBN 978-1-63987-733-1

Contents

Preface

In my initial years as a student, I used to run to the library at every possible instance to grab a book and learn something new. Books were my primary source of knowledge and I would not have come such a long way without all that I learnt from them. Thus, when I was approached to edit this book; I became understandably nostalgic. It was an absolute honor to be considered worthy of guiding the current generation as well as those to come. I put all my knowledge and hard work into making this book most beneficial for its readers.

Kinase refers to an enzyme that catalyzes the transfer of phosphate groups from high-energy, phosphate-donating molecules to specified molecules or substrates. Kinases may modify lipids, carbohydrates, proteins, or other molecules. Protein modifying kinases are called protein kinases. Kinase signaling is essential for regulating various metabolic events. AMP-activated protein kinase or AMPK is an enzyme, which is activated when the intracellular ATP levels are lower. It is one of the major regulators of cellular and organismal metabolism in eukaryotes, which is necessary for all cells to maintain a balance between ATP consumption and ATP generation. It is also known to play an essential role in regulating growth. AMPK can regulate lipid and glucose metabolism, therefore it is studied extensively as a suitable therapeutic target for treating obesity. This book unfolds the innovative aspects of protein kinase and kinase signaling. With state-of-the-art inputs by acclaimed experts of this field, this book targets students and professionals.

I wish to thank my publisher for supporting me at every step. I would also like to thank all the authors who have contributed their researches in this book. I hope this book will be a valuable contribution to the progress of the field.

Editor

1

Advances in Understanding TKS4 and TKS5: Molecular Scaffolds Regulating Cellular Processes from Podosome and Invadopodium Formation to Differentiation and Tissue Homeostasis

Gyöngyi Kudlik [1], Tamás Takács [1], László Radnai [1,2,3], Anita Kurilla [1], Bálint Szeder [1], Kitti Koprivanacz [1], Balázs L. Merő [1], László Buday [1,4] and Virag Vas [1,*]

1 Institute of Enzymology, Research Centre for Natural Sciences, 1117 Budapest, Hungary; kudlik.gyongyi@ttk.hu (G.K.); takacs.tamas@ttk.hu (T.T.); lradnai@scripps.edu (L.R.); kurilla.anita@abc.naik.hu (A.K.); szeder.balint@ttk.mta.hu (B.S.); koprivanacz.kitti@ttk.hu (K.K.); mero.balazs@ttk.mta.hu (B.L.M.); buday.laszlo@ttk.mta.hu (L.B.)
2 Department of Molecular Medicine, The Scripps Research Institute, Jupiter, FL 33458, USA
3 Department of Neuroscience, The Scripps Research Institute, Jupiter, FL 33458, USA
4 Department of Medical Chemistry, Semmelweis University Medical School, 1085 Budapest, Hungary
* Correspondence: vas.virag@ttk.hu

Abstract: Scaffold proteins are typically thought of as multi-domain "bridging molecules." They serve as crucial regulators of key signaling events by simultaneously binding multiple participants involved in specific signaling pathways. In the case of epidermal growth factor (EGF)-epidermal growth factor receptor (EGFR) binding, the activated EGFR contacts cytosolic SRC tyrosine-kinase, which then becomes activated. This process leads to the phosphorylation of SRC-substrates, including the tyrosine kinase substrates (TKS) scaffold proteins. The TKS proteins serve as a platform for the recruitment of key players in EGFR signal transduction, promoting cell spreading and migration. The TKS4 and the TKS5 scaffold proteins are tyrosine kinase substrates with four or five SH3 domains, respectively. Their structural features allow them to recruit and bind a variety of signaling proteins and to anchor them to the cytoplasmic surface of the cell membrane. Until recently, TKS4 and TKS5 had been recognized for their involvement in cellular motility, reactive oxygen species-dependent processes, and embryonic development, among others. However, a number of novel functions have been discovered for these molecules in recent years. In this review, we attempt to cover the diverse nature of the TKS molecules by discussing their structure, regulation by SRC kinase, relevant signaling pathways, and interaction partners, as well as their involvement in cellular processes, including migration, invasion, differentiation, and adipose tissue and bone homeostasis. We also describe related pathologies and the established mouse models.

Keywords: scaffold protein; tyrosine kinase substrates; TKS4; TKS5; invasion; mesenchymal stem cells; adipose tissue; bone homeostasis; epithelial–mesenchymal transition

---

1. Introduction

Scaffold proteins modulate intracellular signaling by bringing regulatory proteins, enzymes, or cytoskeletal structures in close proximity [1]. TKS molecules are large scaffold proteins earning their name from the early observation that they serve as tyrosine kinase substrates of SRC kinase [2–4]. TKS4 and TKS5 contain one Phox Homology (PX) domain, conserved linear motifs, e.g., several proline-rich motifs (PRMs), and four or five SRC Homology 3 (SH3) domains, respectively. Other names for TKS5 are SH3 and PX domain-containing protein 2A (SH3PXD2A) and Five SH3 domains (FISH), while TKS4 is also known as SH3 and PX domain-containing protein 2B (SH3PXD2B), Homolog of FISH (HOFI), and a factor of adipocyte differentiation 49 (Fad49), reflecting some of their known characteristics [3,5]. The main function of the PX domain is to link the TKS scaffold proteins to the cell membrane via phosphoinositide binding [2,6]. The SH3 domains serve as docking sites for signaling molecules and mediate protein-protein interactions [7]. It is likely that the PRMs of the TKS proteins represent contact sites for SH3 domain-containing molecules (Figure 1). The TKS proteins are phylogenetically related and are expressed in vertebrates, and TKS-like genes are widely present in invertebrates [8]. TKS scaffold proteins are broadly expressed in tissues except for the testis for TKS4, and the spleen and testis for TKS5 [2,3]. They are also expressed in several transformed cell lines [2,9].

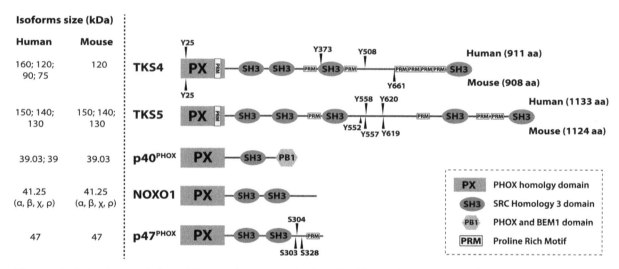

Figure 1. Members of the p47 organizer protein family. The p47 organizer family consists of five structurally similar adaptor/scaffold proteins containing an N-terminal PX domain followed by several SH3 domains, namely p40[phox], NOXO1, p47[phox], TKS4, and TKS5. Experimentally confirmed SRC kinase tyrosine phosphorylation sites ("Y") in the human and mouse TKS proteins are shown above and below the depicted domain architecture, respectively.

In this review, we summarize the current knowledge of the properties of TKS4 and TKS5 and their involvement in specific cellular processes, including growth factor signaling, formation of actin-rich membrane protrusions, and generation of reactive oxygen species by regulating NADPH oxidase (NOX) transmembrane enzyme complexes. We also provide an overview of the function of TKS proteins in cancer metastasis and genetic diseases.

2. The Regulated Localization of TKS Proteins Determines Their Signal Recruiting Function

Both TKS4 and TKS5 have a cytoplasmic, inactive state, and a membrane-bound, active state in cells. The transition between these states is most likely regulated by phosphorylation [2,10,11]. Direct evidence for these conformational states is still limited [12]. However, this hypothesis is supported by the auto-inhibitory intramolecular interactions known to regulate p47[phox], a homologous protein that is structurally highly similar to the N-terminal regions of TKS4 and TKS5. Based on the similarities in their domain architecture, all of these proteins belong to the p47-related organizer protein family [13] (Figure 1). In mammals, the p47-related organizer family consists of five members: p40[phox], NOXO1 and p47[phox], TKS4, and TKS5 [1]. These proteins share many functional and conformational similarities [4,14]. A common feature is the presence of an N-terminal PX domain, followed by one or more SH3 domains (Figure 1). An intramolecular regulatory mechanism was first described for p47[phox]. In the auto-inhibited state, its first and second SH3 domains ("tandem SH3") bind a specific proline-rich motif within the C-terminal region. The assembly of this auto-inhibitory organization makes the PX domain inaccessible for phosphatidylinositol phosphates. Therefore, p47[phox] remains in the cytoplasm [15–17]. The autoinhibitory interaction is disrupted by phosphorylation of several C-terminal serine residues located in close proximity to the proline-rich motif. Consequently, the tandem SH3 domains release the proline-rich motif and become available to bind an interacting partner, p22[phox]. In association with these events, the locked PX domain is released and anchors the protein to the membrane via phospholipid binding [18].

The intramolecular elements known to be necessary for the autoinhibited state of p47[phox] (the tandem SH3 domains and the proline-rich autoinhibitory region) are conserved and are present in both TKS4 and TKS5 [14]. Therefore, the similarities between the structures of the p47[phox] and TKS proteins support the idea that similar intramolecular interactions could regulate their functional states. A study by Abram and colleagues supports the existence of intramolecular regulation in TKS5 [6]. Based on their results, they speculated that, when TKS5 is in its auto-inhibited conformation, its PX lipid recognition module is masked, and the molecule is distributed diffusely in the cell. Upon SRC phosphorylation on tyrosine residues, the PX domain is released, thus becoming available to bind to membrane lipids, resulting in translocation of TKS5 to the cell membrane [6]. Simultaneously, the SH3 domains can bind to signaling proteins to recruit them to the cell membrane, allowing intracellular signal transduction [6,12,19].

3. TKS4 and TKS5 Affect Multiple Biological Processes from Growth Factor Receptor Signaling to Metastasis to Tissue Homeostasis

3.1. EGFR Signaling via TKS4 and TKS5

Receptor tyrosine kinases (RTK) are transmembrane proteins that control several cellular processes, ranging from proliferation to differentiation and cell migration. Following the binding of their extracellular ligands, RTKs dimerize, undergo auto-phosphorylation on multiple tyrosine residues in their cytoplasmic region, and associate with intracellular signaling molecules. Diverse molecular cascades transmit the signal from RTKs to their final effector molecules, ultimately leading to the modulation of distinct biological processes within the cell [20].

Epidermal growth factor receptor (EGFR) is one of the most well-studied RTKS. Upon activation, it initiates several signal transduction cascades, including the RAS-RAF-MEK, phosphatidylinositol 3 (PI3)-kinase-AKT, PLCγ, and JAK-STAT pathways [21]. Moreover, active EGFR binds cytosolic SRC tyrosine-kinase, which then becomes activated [22–25]. This process leads to the phosphorylation of SRC-substrates, including the TKS scaffold proteins, which are known to be involved in EGFR signaling [11,26,27]. The TKS proteins serve as a platform for the recruitment of key players in EGFR signal transduction (Figure 2, Table 1), promoting cell spreading and migration [9,11,28–30]. In response to EGFR activation, PI3 kinases are activated, and lipids are phosphorylated in the plasma membrane. For example, phosphatidylinositol (4,5)-bisphosphate (PI(4,5)P$_2$) is converted to phosphatidylinositol (3,4,5)-trisphosphate (PI(3,4,5)P$_3$) [31]. According to a model proposed by Bögel et al., the phosphorylated lipid residues anchor the PX domain of TKS4 and translocate the scaffold protein from the cytoplasm to the plasma membrane [11]. On the other arm of the signaling pathway, SRC kinase is also activated by binding to the intracellular tail of EGFR [22–25] subsequently phosphorylating tyrosine residues on TKS4 (i.e., Tyr25, Tyr373, Tyr508) (Figure 1) [2]. Phosphorylated TKS4 can bind activated SRC by interacting with both its SH2 and SH3 domains. In this complex, SRC remains active for a prolonged period of time and may phosphorylate multiple downstream molecules/partners [32]. This direct interaction between TKS4 with SRC was shown to involve the proline-rich region PSRPLPDAP (residues 466–474) and the tyrosine-phosphorylated pYEEI motif (residues 508–511) of TKS4 (both located between the third and fourth SH3 domains) and the SH3 and SH2 domains of SRC, respectively [32]. Upon PI3 kinase activation, TKS5 also translocates to the plasma membrane in epidermal growth factor (EGF)-stimulated cells [26]. The PX domain of both TKS4 and 5 was found to be essential for the participation of the molecules in EGFR signaling and for the phosphorylation of TKS4 and 5 by activated SRC [11,26]. TKS4 forms a complex with EGFR in which either SRC or a yet unidentified protein may serve as a bridge between the two molecules [11,32].

For example, growth factor receptor binding protein 2 (GRB2) has been identified as a binding partner of both EGFR and TKS4 [28]. No strong interaction between TKS5 and EGFR or SRC has been detected so far, suggesting that, despite their structural similarities, there is only a partial overlap between the regulation of TKS4 and TKS5 in EGF signaling [11,26].

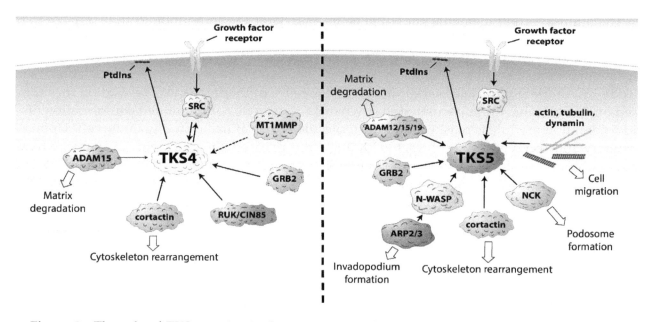

Figure 2. The role of TKS proteins in the recruitment of signaling molecules. ADAM12/15/19–a disintegrin and metalloprotease 12/15/19, ARP2/3–actin-related protein 2/3, GRB2–growth factor receptor binding protein 2, MT1MMP–membrane type 1 matrix metalloprotease, NCK–non-catalytic region of tyrosine kinase adaptor protein, N-WASP–neural Wiskott-Aldrich syndrome protein, PtdIns–phosphatidylinositol, RUK/CIN85–regulator of ubiquitous kinase/Cbl-interacting protein of 85 kDa, SRC–proto-oncogene tyrosine-protein kinase Src.

Table 1. Known protein binding partners of TKS4 and TKS5. The known binding partners of (**a**) TKS4 and (**b**) TKS5 are shown with the methods of detection and the binding sites within the TKS molecules. Some of the well-described functions of the binding partners are also listed. ECM–extracellular matrix, EMT–epithelial-mesenchymal transition, ITC–isothermal titration calorimetry, NOX1–NADPH oxidase 1, PRR–proline-rich region, ROS–reactive oxygen species, RTK–receptor tyrosine kinase. * The first and second SH3 domains cooperate to form a common "super SH3 platform" and allow the binding of the proline-rich region of the partner protein [33].

(a)			
TKS4			
Partner	**Method**	**TKS4-Interacting Site**	**Function**
ADAM15 [34]	GST pull-down assay	4th SH3 domain	Ectodomain shedding, cell adhesion, and signaling [35]
Cortactin [9]	Co-immunoprecipitation, GST pull-down assay, immunofluorescence co-localization	Unknown	Regulation of actin cytoskeleton [36]
CR16 [37]	GST pull-down assay	Weak interaction with the 2nd, 3rd, and 4th SH3 domains	Reorganization of actin cytoskeleton [38]
DNM2 [37]	GST pull-down assay	3rd SH3 domain	Endo-/exocytosis [37]
FasL (CD178) [39]	Phage display screening	3rd and 4th SH3 domains	Apoptosis induction [40]

Table 1. *Cont.*

(a)			
TKS4			
Partner	**Method**	**TKS4-Interacting Site**	**Function**
GRB2 [28]	Affinity purification–selected reaction monitoring mass spectrometry	Unknown	Adaptor protein involved in the regulation of RTK signaling, cycle progression, actin-based cell motility, podosome formation [41]
NOXA1 [42,43]	Co-immunoprecipitation, GST pull-down assay	Unknown	ROS generation through NOX1 activation [44]
N-WASP [37]	GST pull-down assay	2nd SH3 domain	A scaffold protein regulating actin cytoskeleton reorganization, and actin polymerization during cell motility and invasion [45]
OPHN1 [37]	GST pull-down assay	3rd SH3 domain	Endo-/exocytosis [37]
RUK/CIN85 [46]	GST pull-down assay	Unknown	Adaptor protein that recruits endocytotic regulatory proteins, and regulates RTK internalization, trafficking, and degradation [47]
SRC [9,32]	Co-immunoprecipitation; GST pull-down and fluorescence-polazization assays, Duolink proximity ligation assay	PRR (aa: 466–474); P-Tyr motif (aa: 508–511)	Regulation of cell growth, differentiation, proliferation, survival, adhesion, migration, and motility [9,32]
SYNJ1 [37]	GST pull-down assay	3rd SH3 domain and weak interaction with the 4th SH3 domain	Endo-/exocytosis [37]
(b)			
TKS5			
Partner	**Method**	**TKS5-Interacting Site**	**Function**
ADAM12 [6]	Co-immunoprecipitation, immunofluorescence co-localization	5th SH3 domain	Cell adhesion and fusion, extracellular matrix restructuring, reorganization of actin cytoskeleton, regulation of ectodomain shedding [48]
ADAM15 [6]	Co-immunoprecipitation	5th SH3 domain	Cell adhesion, degradation of ECM components, ectodomain shedding of membrane-bound growth factors [35]
ADAM19 [6]	Phage display screen, co-immunoprecipitation	5th SH3 domain	Extracellular matrix breakdown and reconstruction, ectodomain shedding, role in embryogenesis, cardiovascular system development, obesity, and insulin resistance [49]
β-dystroglycan [50]	Phage display screen, GST pull-down assay, co-immunoprecipitation, immunofluorescence co-localization	3rd SH3 domain	Links the extracellular matrix to the intracellular actin cytoskeleton [50]

Table 1. *Cont.*

(b)			
TKS5			
Partner	**Method**	**TKS5-Interacting Site**	**Function**
CircSKA3 [51]	Co-immunoprecipitation, pull-down assay	Not specified	Circular RNA, an inducer of invadopodium formation [51]
Drebrin [52]	Co-immunoprecipitation	Unknown	An actin-binding protein involved in the regulation of actin filament organization, role in cell migration, cell process formation, intercellular communication, metastasis, and brain development [53]
Dynamin [29,33]	Peptide spot membrane assay, GST pull-down assay, ITC, immunofluorescence co-localization, GST pull-down assay, mass spectrometry/Western blotting	1st and 2nd SH3 domains; 1st and 5th SH3 domains	Regulation of actin cytoskeleton, podosome/invadopodium formation, role in endocytosis [54]
F-actin [29]	GST pull-down assay, and mass spectrometry	5th SH3 domain	Component of cytoskeleton [55]
FasL (CD178) [39]	Phage display screening	5th SH3 domain	Apoptosis induction [40]
FGD1 [56]	Co-immunoprecipitation and mass spectrometry, GST pull-down assay, immunofluorescence co-localization	4th and 5th SH3 domains	A guanine nucleotide exchange factor for the Rho-GTPase CDC42, assembly of podosomes and invadopodia, control of secretory membrane-trafficking, and cell cycle [56,57]
Girdin [58]	Co-immunoprecipitation, immunofluorescence co-localization	Unknown	actin-binding protein regulating actin remodeling and cell polarity, collective migration of neuroblasts, epithelial and cancer cells [59]
GRB2 [28,29]	Co-immunoprecipitation	Polyproline sequences	An adaptor protein involved in cell cycle progression and actin-based cell motility, podosome formation [41]
IRTKS [60]	GST pull-down assay	First binding site located in the segment comprising the 1st and 2nd SH3 domains, second binding site located in the segment comprising the 3rd and 4th SH3 domains	Regulation of plasma membrane dynamics, actin cytoskeleton remodeling, cell migration and polarization, insulin signaling [61]
MT4-MMP [62]	Co-immunoprecipitation	Unknown	Induction of invadopodia and amoeboid movement, degradation of ECM components, role in hypoxia-mediated metastasis [62]

Table 1. *Cont.*

	(b)		
	TKS5		
Partner	**Method**	**TKS5-Interacting Site**	**Function**
NCK [52]	Co-immunoprecipitation, fluorescence co-localization	Linker region between the 3rd and 4th SH3 domains containing pY557	Adaptor protein involved in cytoskeletal remodeling, invadopodium formation, cell proliferation [63]
Nogo-B [29]	GST pull-down assay and mass spectrometry	5th SH3 domain	Roles in vascular remodeling, cell migration and proliferation, and EMT [64]
NOXA1 [42,43]	Co-immunoprecipitation, GST pull-down assay	One or more of the five SH3 domains	ROS generation through NOX1 activation [44]
N-WASP [29]	GST pull-down assay and mass spectrometry/Western blotting, co-immunoprecipitation	All five SH3 domains	A scaffold protein regulating actin cytoskeleton reorganization, and actin polymerization during cell motility and invasion [45]
p22^{phox} [65]	Co-immunoprecipitation	1st and 2nd SH3 domains	Subunit of NADPH oxidases involved in ROS generation through NOX activity [66]
Rab40b [67]	GST pull-down assay, co-immunoprecipitation,	PX-domain: sites 14-KRR-19 and Y24 in 23-YVYI-28	A GTPase required for the sorting and secretion of MMP2 and MMP9, promotion of migration, invasion, and metastasis of cancer cells [67,68]
RET [69]	Co-immunoprecipitation, immunofluorescence co-localization	Unknown	A receptor tyrosine kinase mediating stress fiber formation, cell polarization, directional migration and invasion, enhancement of proteolytic activity [69]
SOS1 [33]	Immunofluorescence co-localization, peptide spot membrane assay, GST pull-down assay, isothermal titration calorimetry	1st and 2nd SH3 domains *	A guanine nucleotide exchange factor promoting Ras and Rac activation downstream of a variety of receptors such as RTKs [70]
Tubulin [29]	GST pull-down assay and mass spectrometry	3rd SH3 domain	Component of microtubules, affects cell division, differentiation, intracellular transport, motility [71]
WIP [29]	GST pull-down assay and mass spectrometry	3rd and 5th SH3 domains	Regulation of actin cytoskeleton assembly and remodeling [72]
XB130 [73]	Yeast two-hybrid screening, co-immunoprecipitation, GST pull-down assay, immunofluorescence co-localization	5th SH3 domain	A scaffold protein influencing cell growth, survival, and migration [73]
Zyxin [29]	GST pull-down assay and mass spectrometry	3rd and 5th SH3 domains	A focal adhesion protein involved in actin cytoskeleton assembly [74]

In recent years, more possible interaction partners of TKS4 have been identified (Table 1a). One possible partner is cortactin [9], a well-known substrate of SRC localized to cortical actin structures within cells. Cortactin can bind the actin-related protein-2/3 (ARP2/3) and neural Wiskott-Aldrich syndrome protein (N-WASP) proteins, and it mediates actin polymerization [75–77]. Therefore, TKS4 was expected to be involved in EGFR signaling-mediated actin cytoskeleton assembly and rearrangement. This proposed mechanism was confirmed by Lányi et al. [9]. They found that, in response to EGF stimulation, TKS4 associates with cellular motility-associated membrane ruffles. They also showed that, when constitutively active SRC is present, TKS4 accumulates in podosomes (actin-rich membrane protrusions involved in cell motility, see below) while forming a complex with SRC and cortactin [9]. TKS5 has also been reported to bind cortactin and other proteins important in the regulation of actin cytoskeleton assembly, including N-WASP, non-catalytic region of tyrosine kinase adaptor protein (NCK), and GRB2 (for a full list, see Table 1b) [10,29,52,78,79].

3.2. Molecular Organizers of Podosome and Invadopodium Assembly

Podosomes and invadopodia are dynamically formed actin-rich protrusions formed on the ventral surface of cells facing the extracellular matrix (ECM). Both structures share the function of motility promotion and pericellular proteolytic activity [80]. The migration of normal cells is driven by podosomes, specialized structures that allow cells to adhere to and enter into their surroundings [81]. Podosomes are formed by a variety of different cells under normal circumstances, including endothelial cells, smooth muscle cells, osteoclasts, macrophages, and dendritic cells [82,83]. Invadopodia, by contrast, are used by cancer cells to break ECM barriers and metastasize. Via a coordinated stepwise process at the site of these protruding membrane structures, matrix metalloproteinases accumulate and are secreted into the extracellular space, leading to ECM degradation to facilitate invasion [84].

Both TKS proteins have been implicated in regulating podosome/invadopodium formation and function [2,3,27,85,86]. The processes involved in the assembly of these structures share many similarities. Depending on the cell type, the sequential process of podosome/invadopodium assembly can be initiated by cell adhesion via integrins [87], vascular endothelial growth factor (VEGF) [88], platelet-derived growth factor (PDGF) [89], transforming growth factor beta (TGFβ) [90], keratinocyte growth factor (KGF) [91], colony-stimulating factor-1 (CSF-1) [92], or EGF-derived [93] signals. Upon receptor tyrosine kinase activation, the non-receptor c-SRC kinase becomes active and phosphorylates several substrates, including the TKS proteins, cortactin, N-WASP, focal adhesion kinase (FAK), and other signaling molecules [94,95]. TKS5 has a key role at this point in podosome precursor formation [29]. The PX domain of TKS5 is responsible for docking the molecule to membranes by binding PtdIns(3)P or PtdIns(3,4)P_2 [6]. During podosome formation, TKS5 is recruited to the plasma membrane by PtdIns(3,4)P_2 and the GRB2 adaptor while binding to PtdIns(3,4)P_2 via its PX domain [29]. In this way, the TKS5/GRB2 complex cooperates in the recruitment of proteins necessary for the final podosome-maturation steps. Among the recruited proteins, N-WASP and TKS5 binding has been analyzed in detail and it was experimentally proven that the strong protein interaction is mediated by all five SH3 domains of TKS5, stimulating robust N-WASP accumulation at the adhesion site (Table 1b) [29]. N-WASP is a multi-functional protein containing several different protein subunits that interact with the small GTPase CDC42, PtdIns(4,5)P_2, and the actin regulatory complex ARP2/3. Upon N-WASP activation and ARP2/3 association, new actin filament polymerization can begin [96,97]. TKS5 functions as a platform to recruit the ARP2/3 complex and ultimately facilitates actin cytoskeleton rearrangement. Finally, the newly formed actin branches allow the cells to change shape and protrude podosomes (Figure 2) [98]. Meanwhile, the other TKS protein TKS4 is phosphorylated by SRC kinase and is recruited to the site of the developing podosome, where it is directly anchored to phosphoinositide residues via its PX domain. The suggested function of TKS4 in podosome maturation is to recruit matrix metalloproteases (MMPs) to the protruding membrane edge and to specifically allow membrane type 1 matrix metalloprotease (MT1-MMP) activation [2]. Buschman et al. showed that loss of TKS4 results in incomplete podosome assembly in which most of the known podosome proteins colocalize at the site of podosome formation but fail to associate with the filamentous actin. Furthermore, MT1-MMP failed to localize to the sites of the pre-podosome structures, resulting in decreased ECM degradation. TKS5 overexpression could rescue podosome formation in the absence of TKS4. However, it could not restore ECM degradation. Thus, the two TKS proteins seem to have overlapping functions in filamentous actin formation, and the upregulation of TKS5 expression can substitute for TKS4 in this process [2]. TKS5 has also been implicated in regulating proper MT1-MMP cell surface expression by controlling its exocytosis [85]. At the end of the podosome formation process (and depending on the cell type), single, clustered, ring-like, belt-like, or rosette-like podosomes form with localized MMPs, imparting the cells with motility and ECM-remodeling activities [99].

Invadopodium assembly is very similar to podosome assembly [80]. However, there are a few differences. While podosomes usually contain an actin-rich core surrounded by a vinculin/paxilin adhesion ring, this organized ring structure is absent in invadopodia [100]. NCK1 and GRB2, two TKS-interacting partners (Table 1), have been shown to be present in podosomes and invadopodia. However, their localization patterns are different in the two structures [101]. According to Oser et al., the adaptor protein NCK1 is specifically restricted to invadopodia while GRB2 functions mainly in podosome-like structures and not in the invadopodia of metastatic cells [101]. It is likely that the two degradative cell compartments use distinct mediators for N-WASP recruitment to the site of action [93,96]. For example, in invadopodia, TKS5 might recruit N-WASP via NCK1, while TKS5 interacts with GRB2 during podosome assembly [29,52]. This feature might explain how the different invasive structures can both assemble using the same scaffold molecule.

Besides endowing the cell with an invading phenotype and coordinating cell motility, podosomes might also play a role in cell-cell communication. It was demonstrated that the fifth SH3 domain of TKS5 associates with the intracellular tail of certain ADAM (a disintegrin and metalloprotease) protein family members (Table 1b). TKS5 and ADAM 12, 15, and 19 co-immuno-precipitate and co-localize at podosome sites in SRC-transformed fibroblasts [6]. Although only in a GST pull-down experiment, TKS4 was also reported to bind ADAM15 with its fourth SH3 domain (Table 1a) [34]. An interesting feature of these membrane-localized proteases is that they can act as sheddases [102]. In fact, ADAM proteins are involved in the growth factor or ligand activation by cleaving the inactive, membrane-anchored forms of these molecules to release the active forms, which has been demonstrated in the case of an insulin-like growth factor-binding protein (IGF-BP) [103], Delta-like ligand 1 (DLL1) [104], E-cadherin [105], amphiregulin [106], heparin-binding EGF-like growth factor (HB-EGF) [107], transforming growth factor alpha (TGFα), EGF [108], or tumor necrosis factor alpha (TNFα) [109]. After cleavage by ADAMs, the released cytokines can act on the same, adjacent, or distant cells to allow the "signal sender" cell to communicate with "receiver" cells.

Based on these observations, we propose that TKS proteins might be involved in diverse cell fate-determining mechanisms via podosome organization.

3.3. Significance of TKS5 in Invasiveness

TKS5 has been described in several studies as an invadopodium (and podosome) marker [110–118], as it is not found in other types of protrusions and adhesive motility structures [80]. Elevated expression levels of the protein have been reported in a number of cancer types [119,120]. As already discussed, TKS5 is involved in the regulation of invadopodium formation and is also known as a key player in metastasis-related processes [85,121]. So far, elevated TKS5 expression has been demonstrated in lung adenocarcinoma [122], glioblastoma cells [119], breast cancer, melanoma cells [120], and keratocystic odontogenic tumor samples [123], where it is primarily correlated with the invasive phenotypes. In addition, upregulated TKS5 expression has been linked to a tissue-invasive, hypermobile pro-inflammatory T cell phenotype in a rheumatoid arthritis model [124]. Analogous to metastasis formation in vivo, it has been shown that altered TKS5 expression can influence the invasive properties of tumor cell lines in vitro [86]. In lung adenocarcinoma, it was also demonstrated that the tumor cell invadopodium activity and metastatic behavior depended on the TKS5 isoform type present [122]. The TKS5 protein has three isoforms (molecular weights of ~150 kDa, ~140 kDa, ~130 kDa) generated via alternative promoter usage as a result of intron 5 retention [125]. Li et al. found that the expression level of the long TKS5 isoform was elevated in metastasis-derived cells when compared with its level in non-metastatic tumor cells [122]. Moreover, a higher TKS5 long/TKS5 short isoform ratio induces invadopodium formation and mediates the development of an invasive phenotype. The functional differences between the isoforms depend on the fact that both transcripts encode five SH3 domains, while the PX domain is missing from the short TKS5 isoform. This short TKS5 cannot organize invadopodium assembly as effectively as the long isoform due to the lack of proper PX domain-dependent membrane localization [122]. Although TKS4 also has several isoforms (~75 kDa, ~90 kDa, ~120 kDa, and ~160 kDa) [46], such a biased isotype preference in invadopodia formation has not been reported.

Metastatic cancer cells cross the basement membrane using MMPs enriched in the invadopodium machinery [126]. In addition to degrading the ECM, the proteolytic activity of ADAM proteases might also facilitate the release of ECM-bound tumor-supporting factors (e.g., EGF and TNFα) and maintain the invasive ability of cancer cells [127,128]. In the context of this special type of tumor cell-extracellular environment communication, invadopodium development and exosome formation were described as connected processes [129].

Exosomes are secreted membrane vesicles that contain a cargo of proteins, mRNAs, and miRNAs highly specific to the "exosome-sender" cell [130]. In terms of cancer, exosomes secreted by tumor cells help establish a tumor-promoting niche via the release of angiogenic and survival factors. It has been demonstrated that TKS5 inhibition in the context of invadopodium formation also greatly decreases exosome formation in a carcinoma cell line [129]. Consistent with this observation, the existence of invadopodia might be an important determining step in exosome formation, representing a newly described cell communication method involved in cancer progression.

3.4. The Possible Role of TKS4 and TKS5 in the Compartmentalization of Oxidative Processes

In general, reactive oxygen species (ROS), including hydrogen peroxide (H_2O_2), superoxide anion (O^{2-}), and hydroxyl radical ($\cdot OH$), are generated as by-products of normal metabolic processes [131]. These molecules act as secondary messengers in normal and pathological cells in which they orchestrate various biological processes [131,132]. However, at high concentrations, they can potentially damage vital signaling molecules and the genome [133]. At specific ROS concentrations and under biological control [134], several ROS-dependent processes and signaling pathways exist, including angiogenesis [135], Notch and Wnt stem cell fate determining signaling [136], the anti-microbial function of phagocytes [137], and pain processing within the nociceptive system [138]. But, how can molecules as simple as ROS modulate such diverse pathways? The best way to answer this question is to briefly summarize the regulation of NADPH oxidase (NOX) transmembrane enzyme complexes (i.e., NOX1, NOX2, NOX3, NOX4, NOX5, DUOX1, and DUOX2) as natural ROS sources [139]. At the molecular level, ROS-generating enzyme-complexes comprise a catalytic subunit (one of the NOX family oxidases), an activity providing subunit (p22[phox]), and several regulatory cytosolic proteins (e.g., members of the p47[phox] protein family) (Figure 3) [140,141].

Each mammalian NADPH oxidase has a distinct tissue-specific expression pattern [142]. To precisely channel ROS production to the intended targets and to achieve ROS-pathway specificity, NADPH oxidase activity must be compartmentalized within the cells and restricted to spatial cellular microdomains [138,143].

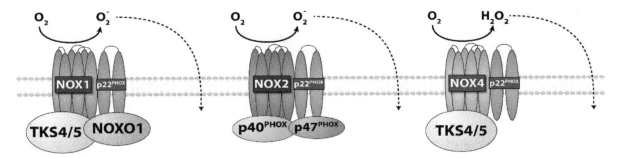

Figure 3. Assembly of NADPH oxidase multi-protein enzyme complexes with the P47 organizer family members. The core of the intracellular enzyme is formed via an interaction between a specific NOX oxidase and p22^{phox} accompanied by activity-modulating organizers such as PX-SH3-structured proteins (NOXO1, TKS4/5, p47^{phox}, p40^{phox}). The extra-membrane-produced O_2 is rapidly transformed into H_2O_2 and can passively diffuse through the cell-membrane. (Reviewed in Reference [144]).

Three members of the p47 organizer protein family (i.e., p47^{phox}, p40^{phox}, and NOXO1) have well-known functions in the regulation of oxidative processes [145]. It was suspected that TKS proteins, which are members of the same family, can also play regulatory roles in channeling and localizing the NOX enzymes to the podosome/invadopodium membrane to locally increase the ROS concentration [65,146]. TKS5 has been shown to colocalize with ROS [147]. Furthermore, Diaz et al. showed that TKS5 associates with p22^{phox} and the TKS proteins can be involved in processes involving both the NOX1-based and NOX4-based enzyme complexes [65]. In an accompanying paper, Gianni et al. demonstrated that, in a colon cancer cell line, TKS4 recruits the NOX1 NADPH oxidase to the sites of invadopodia and allows ECM degradation [42]. These results raised the possibility that the modulation of ROS levels plays a regulatory role in a subcellular compartment of invadopodia [4,43,146].

It has already been reported that, in the presence of ROS, redox-sensitive cysteine residues in several proteins become oxidized, demonstrating that the conformation of certain enzymes can be changed in an ROS-dependent manner [148]. This remodelling can lead to altered three-dimensional structures in the target proteins that might also alter their catalytic activity [149]. The most highly studied ROS-targeted enzymes are phosphatases (PTPases). When PTPases are inhibited by ROS-dependent cysteine modifications at a site in the catalytic domain, they cannot dephosphorylate their substrate proteins. For example, SRC kinase dephosphorylation in invadopodia is known to be regulated by this mechanism [150]. In an interesting model of invadopodium turnover, the NADPH oxidase regulated by the TKS scaffold proteins produces ROS near the cell-membrane, leading to PTPase inactivation via cysteine modification. The inactivated PTPase then primes the sustained activation of phosphorylated SRC kinase [65]. This process might lead to extended activation of all SRC substrates, including the TKS proteins. At this point in invadopodium development, the TKS scaffold proteins can recruit actin-organizing complexes to the sites of membrane protrusion to stabilize the invadopodium. It is tempting to speculate that the concerted action of the TKS proteins is central in controlling invadopodium turnover via ROS-dependent phosphatase inactivation and coordination of distinct intracellular signaling.

To avoid the harmful side-effects of free radicals, cells control ROS levels by maintaining a balance between ROS production and elimination [151,152]. TKS proteins, as regulators of NOX localization, participate in this process by regulating ROS compartmentalization. Moreover, via their SH3 domains, they facilitate the recruitment of actin cytoskeleton modifiers and ECM degrading machinery in invadopodia [133]. Future studies are needed to determine whether cooperation between the TKS molecules and the NOX complex is also involved in podosome and invadopodium formation in vivo.

3.5. Absence of TKS4 Induces Epithelial-Mesenchymal Transition (EMT) and Promotes Invasive Behavior

A novel function of TKS4 in EMT-like processes has been recently discovered by Szeder and colleagues [153]. During EMT, epithelial cells lose epithelial features and functions. The expression levels of E-cadherin, claudins, occludins, and α6β4 integrins are reduced. Thus, cells lose apicobasal polarity and cell-cell attachment ability. This loss of epithelial characteristics is concomitant with the gain of mesenchymal-like features, including N-cadherin, vimentin, fibronectin, β1 and β3 integrins, and MMP expression, as well as the acquisition of motility and invasive properties. These changes are governed by the EMT-inducing transcription factors zinc finger E-box-binding homeobox (ZEB), Snail, and Twist, which inhibit expression of genes responsible for epithelial characteristics and activate expression of mesenchymal-associated genes [154,155]. EMT occurs at certain stages of developmental processes and wound-healing and is an important mechanism in cancer progression from tumor initiation to metastasis and colonization [154,156]. According to Szeder et al., TKS4 knockout (KO) HCT116 colon cancer cells showed a mesenchymal morphology with increased motility and decreased cell-cell adhesion. Loss of E-cadherin and apicobasal polarity was observed together with increased fibronectin and Snail2 transcription factor expression, indicating a shift from an epithelial to mesenchymal-like phenotype. Furthermore, decreased spheroid forming capacity and increased invasiveness in collagen matrix were also observed [153]. The exact mechanism for how the absence of TKS4 may induce EMT in these cells remains unknown. However, two cellular processes known to be influenced by TKS4, i.e., EGFR signaling [157,158] and ROS balance [159,160], have been implicated in affecting EMT in various cancer model systems (see above). EGFR signaling is a known inducer of EMT, causing increased Snail2/Slug and ZEB1 levels and decreased E-cadherin levels [157,158]. Thus, changes in EGFR or ROS signaling in the absence of TKS4 might cause these cancer cells to shift into EMT. Szeder et al. hypothesized that a temporary loss of TKS4 could negatively affect podosome-related cell migration (as described earlier), while the prolonged effect of the absence of the molecule could be increased invasiveness by activating an EMT-like program. They also concluded that the molecule might act differently in cells of epithelial origin (like HCT116) than in those of mesenchymal origin [153]. Despite these interesting findings on the role of TKS4 in EMT processes, more studies are needed to confirm the results of Szeder et al. and to elucidate the exact mechanisms behind this phenomenon.

EMT, along with the reverse process known as mesenchymal-epithelial transition (MET), are crucial processes involved in embryonic development during which they facilitate body formation and tissue and organ differentiation. EMT is involved in such central developmental processes as gastrulation and neural crest cell formation, somitogenesis, cardiac morphogenesis, and trophoblast invasion, which affects placental development, among others [161,162]. A disturbance in EMT (or in MET) can, therefore, have diverse developmental consequences that could also explain the phenotype observed in patients [163–165] and mouse models [34,164,166] that lack a functional TKS4 protein.

3.6. Cell Differentiation Modulated by TKS Molecules

Cellular differentiation is a fundamental process throughout the lifetime of an organism. During embryogenesis, the spectrum of cells comprising the tissues and organs, which perform a vast number of functions, are derived from a single zygote via differentiation and proper localization in the developing organism via migration [167,168]. During the lifetime of tissues and organs, cells lost through injury or normal cell turnover must be continuously replaced via differentiation from tissue-resident stem and progenitor cells [169,170]. The role of TKS proteins in such contexts will be discussed in this section.

The first observation regarding the role of the TKS proteins in cell fate determination came from Hishida et al. They reported that TKS4 expression is necessary in the early phase of adipogenesis for the expansion and commitment toward adipocytes [12]. By using the 3T3-L1 mouse cell line as an in vitro adipocyte differentiation model in conjunction with a TKS4-silenced derivative, they found that TKS4 down-regulation impaired adipocyte differentiation. (This study was performed before the detailed description of TKS4. Therefore, the original name of the protein, Fad49, was based on its newly identified function, i.e., factor for adipocyte differentiation 49.)

A recent analysis of bone marrow-derived mesenchymal stem/stromal cells (MSCs) revealed a central role of TKS4 in the adipogenesis and osteogenesis of MSCs [166]. These cells serve as common precursors of adipocytes and bone-forming osteoblasts (among others) [171]. During adipogenic or osteogenic induction of mouse MSC cultures, differentiating TKS4 KO cells failed to accumulate lipid droplets or deposit calcium-containing minerals, respectively. Analysis of the expression levels of lipid-regulated genes during adipogenic induction revealed reduced or delayed levels of adipogenic transcription factors, genes driving lipid droplet formation, and sterol and fatty acid metabolism in TKS4 KO cultures [166]. Furthermore, PPARγ2, which is a key transcription factor and regulator of adipose tissue expansion [172], showed no detectable expression at the protein level in TKS4 KO MSC cultures [166]. Upon osteogenic induction, the key osteogenesis-driving transcription factors RUNX2 and osterix showed highly reduced expression levels in TKS4 KO cultures accompanied by reduced bone-forming capacity when compared to wild-type MSCs [166]. Related to this topic, Vas et al. published interesting findings by studying the adipogenic potential of adipose tissue-derived stromal vascular fraction cells. These cell populations also contain MSCs. However, no difference was found between the adipogenic and osteogenic differentiation potential of cells isolated from TKS4 KO and wild-type mice [173]. Perhaps the complexity of the adipose tissue microenvironment (MSCs, preadipocytes, vascular endothelial cells, pericytes) together with the ECM can rescue the differentiation defects of TKS4 KO MSCs or preadipocytes (e.g., 3T3-L1 cells), which fail to properly differentiate in pure in vitro cultures.

It was hypothesized by Oikawa et al. that TKS5 might have an effect on cell-cell fusion. In two studies, TKS5 expression was found to be induced during the course of osteoclast development [19,60]. Osteoclast precursor cells developed podosome-like structures characterized by TKS5 enrichment during the multinucleation process. RNAi-mediated knockdown of TKS5 markedly reduced podosome formation in maturing monocytes (the precursors of osteoclasts) and abolished cell-fusion. The authors also suggested that TKS5, as a master regulator of invadopodium formation, might mediate the fusion-competent protrusion generation necessary for bone metastasis.

TKS scaffold proteins have an instructive effect not only on cell specialization in adult organisms but also during embryonic development. The morphogenic effects of TKS5 were studied by Murphy et al. and Cejudo-Martin et al. in zebrafish [174] and mouse [125] embryos, respectively. TKS5 was found to be necessary in neural crest patterning in zebrafish. TKS5-morpholino zebrafish have cardiac failure, abnormal craniofacial structures, and melanophores with decreased pigmentation. These morphological defects might be explained by reduced neural cell migration during embryonic development due to abnormal podosome-like structure formation in neural stem cells [174]. TKS5 gene-trapped mice are born with a complete cleft of the secondary palate and die shortly after birth [125]. Since trophoblast podosome formation is important in trophoblast function and implantation, and because TKS4 and TKS5 have been implicated in influencing podosome formation, the question arose whether the absence of TKS proteins causes lethality before or after implantation. By genotyping E3.5 pre-implantation blastocysts from TKS4-TKS5 double heterozygous intercross matings, Cejudo-Martin et al. found adequate amounts of double-null blastocysts, suggesting a post-implantation role for TKS5 in mammalian development [125]. These results reveal that the influence of TKS proteins seems to extend to several steps and processes involved in differentiation and embryonic development.

3.7. Role of TKS4 in Tissue Homeostasis

As described in the previous section, TKS proteins seem to have determining roles in the cell specialization processes of tissues. A role of TKS4 has also been implicated in the homeostasis of mature adipose and bone tissue. A genome-wide association scan on obesity in a large US Caucasian population found a strong link between body mass index and the chromosomal region of 5q35 with the *Sh3pxd2b* gene [175], supporting the idea that TKS4 has a role in adipose tissue development and/or regulation. No similar association was found by Vogel et al. based on a dataset from children and adolescents [176]. These contradictory results most likely reflect the multifactorial nature of obesity. The development of obesity is dependent on alterations in the composition of the adipose depots [177].

A recent study revealed that TKS4 KO mice had a disturbed adipose depot phenotype involving the beige-ing/browning of white adipose tissue (WAT) depots and a concomitant "whitening" of brown adipose tissue (BAT) [173]. WAT was found to be enriched with smaller and more multi-locular adipocytes and showed higher expression of uncoupling protein 1 (UCP1) at both the RNA and protein levels in tissue samples of TKS4 KO mice compared with the same features in wild-type samples [173]. UCP1 is a marker of brown and beige adipocytes, which are responsible for uncoupling the mitochondrial respiratory chain to stimulate heat production instead of ATP generation [178]. UCP1 showed increased expression in TKS4 KO WAT, indicating white adipocyte beige-ing/browning [173]. The reverse trend was true for the BAT of TKS4 KO mice, which showed an increased adipocyte size with fewer multilocular cells and reduced UCP1 expression, indicating BAT "whitening" or impaired BAT function. A more detailed analysis of one WAT depot showed a shift in the expression patterns of PPARγ-regulated adipogenesis-related genes (e.g., downregulation of the PPARγ target genes *Cebpd*, *Lpl*, *Lipe*, and *Adipoq* and upregulation of the beige transcription factors *Prdm16* and *Ppargc1a*), favoring beige-ing and highlighting PPARγ as a central regulator through which TKS4 can exert its effect on adipocyte homeostasis [173]. Based on these results, TKS4 emerged as an organizer molecule of adipocyte homeostasis-regulating signaling networks.

Studies of the bone structure of a patient with a defective TKS4 gene and TKS4 KO mice revealed altered trabecular systems with increased trabecular separation and porosity resembling an osteoporotic phenotype [179]. Osteoporosis arises from dysregulated bone tissue remodeling when the fine-tuned balance between bone formation and bone resorption is disturbed [180], even though the exact mechanisms of osteoporotic processes are still under investigation. Vas et al. demonstrated that the osteoporotic-like phenotype in TKS4 KO mice did not arise due to an increased osteoclast number or activity and that it likely arose instead due to defective osteoblast differentiation and activity [179]. As it has been mentioned above, TKS4 was found to be indispensable in the differentiation of bone marrow MSCs into functioning osteoblasts [166]. The higher TKS4 expression levels in the immature cell type-enriched fraction of the bone marrow and its presence throughout the differentiation process of osteoblasts as they arise from their precursors (bone-marrow MSCs) highlight the importance of the molecule in bone differentiation [179]. TKS4 was also shown to affect the levels of bone formation markers, i.e., decreased RUNX2 expression in KO bone tissue and reduced osteocalcin levels in TKS4 KO bone marrow, suggesting a role of TKS4 in osteoporotic processes and bone homeostasis [179].

4. TKS4- and TKS5-Related Pathological Conditions and Mouse Models

4.1. Pathological Conditions Related to TKS Protein Dysfunction

Frank-ter Haar syndrome (FTHS, OMIM:249420) is a rare autosomal recessive disease described and named by two groups in the 1970s [181,182]. Most families affected by FTHS have documented consanguinity, and most affected individuals carry a homozygous mutation in the TKS4 gene (*Sh3pxd2b*) on chromosome 5q35.1 [164,165]. FTHS is characterized by craniofacial abnormalities, including a wide anterior fontanel, prominent eyes, and dental anomalies. Other skeletal malformations, including bowing and shortened long bones and kyphosis, are often associated with FTHS, and the most fatal consequences of the disease are cardiac anomalies caused by valve or septal defects. Genome analysis of FTHS patient samples uncovered several major mutations in the TKS4 region, including mutations in the PX domain and between the second and third SH3 domains as well as an extensive deletion from exon 13 that leads to a truncated TKS4 protein/gene product with only two SH3 domains (Figure 4a) [164,165,183,184]. Early stop codon-introducing homozygous mutations (c.147insT or F49X) or a deletion (c.969delG), which lead to the expression of truncated $TKS4^{1-48}$ and $TKS4^{1-341}$ mutant proteins, respectively, were detected in some FTHS-affected families (Figure 4a) [164,185]. In transfected cells, the truncated mutant $TKS4^{1-48}$ protein showed no expression, while $TKS4^{1-341}$ abnormally accumulated in the nucleus, suggesting that these mutations result in dysfunctional TKS4 proteins that could lead to FTHS [5]. Recently, another two mutations in the TKS4-encoding gene (in intron 5 and exon 13) have been linked to an FTHS-related phenotype (Figure 4a) called Borrone dermato-cardio-skeletal syndrome (BDCSS), which causes symptoms such as a coarse face, broad forehead, broad nasal bridge, hypertelorism, megalo-cornea, glaucoma, osteopenia, kyphoscoliosis, and mitral valve prolapse [186]. Despite having an intact TKS4 gene, individuals presenting typical FTHS clinical symptoms are thought to have mis-regulated TKS4 expression at the protein level [164]. The exact mechanism by which mutant TKS4 proteins cause the FTHS symptoms is not known. One suspected mechanism for FTHS development is based on observations of Bögel et al. [11]. In their study, an R43W substitution (c.129C>T) was introduced into the wild type TKS4 protein. This mutation is present in one of the affected FTHS families, and it is located in the conserved region within the PX domain (which is involved in lipid-binding) of the p47 organizer family members [6,187]. The R43W-mutant TKS4 failed to localize to the plasma membrane and was presumably misfolded [11]. Ádám et al. demonstrated that the accumulation of the R43W-mutant TKS4 in aggresomes (at the juxtanuclear region of cells via the microtubule network) is associated with loss of function [5]. These results suggest that the R43W-mutant TKS4 protein might also show similar functional defects in FTHS patients.

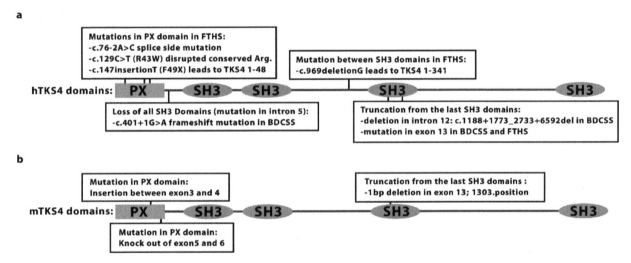

Figure 4. Summary of the location of already documented mutations in human TKS4 gene of patients and mouse TKS4 gene of TKS4 KO strains. (**a**) Mutations along the hTKS4 protein in Frank-ter Haar syndrome (FTHS) and in FTHS-like Borrone dermato-cardio-skeletal syndrome (BDCSS) individuals. The exact sequence location of the mutations in hTKS4 is numbered and the nucleotide deletions, substitutions, or insertions are depicted in black boxes. Iqbal et al. and Bendon et al. published the five mutations connected to FTHS [164,165]. The two BDCSS-associated mutations are described by Wilson et al. [186]. (**b**) Mutations along the mTKS4 protein in the three existing TKS4 KO mouse strains are depicted in black boxes [34,164,166]. PX–Phox homology domain, SH3–SRC homology 3 domain.

Another pathological condition that might be related to a TKS4-dependent process is glaucoma [188]. Although glaucoma is not considered a major diagnostic criteria for FTHS, several FTHS patients suffer from it [163]. This observation led to the hypothesis that TKS4 also has a role in eye development. Analysis of 178 patients with three different forms of glaucoma revealed that the TKS4 gene might harbor rare variants that could affect the pathophysiology of glaucoma. Moreover, TKS4 was present in several ocular cell types important in disease development, reinforcing the possible pathogenic role of the TKS4 variants [189].

In addition to the well-defined function of TKS5 in tumor progression, TKS5 has also been implicated in Alzheimer's disease-related amyloid-β (Aβ) peptide-mediated neurotoxicity [190]. Malinin et al. demonstrated that ADAM12, a TKS5 binding partner, shows reduced expression in diseased brain samples. Furthermore, they provided evidence that TKS5 is phosphorylated in cultured human neuronal cells exposed to the toxic Aβ protein, resulting in a TKS5-ADAM12 interaction and, ultimately, ADAM12 self-cleavage [191]. Another recent study described a potential role for TKS5 in another pathological condition known as pre-eclampsia [192]. Analysis of affected and healthy placentas led to the identification of two upregulated factors, i.e., leptin and TKS5. Although the methylation pattern of the TKS5 promoter region was not significantly altered, one CpG island in the gene body showed higher methylation. The authors propose that TKS5, which is a major organizer of podosome formation, might be involved in trophoblast cell migration in the placenta. As a consequence, impaired TKS5-dependent pathways in trophoblast cells might lead to pre-eclampsia [193].

4.2. Knockout Mouse Models

The first TKS4 KO mouse line arose spontaneously in the 51st generation of a B10.A-H2h4/(4R)SgDvEg mouse strain [34]. The mutant mice were called "nee mice" based on a distinct phenotype, i.e., "nose-ear-eye" deformities. A genome analysis showed that the nee mice carry a 1-bp deletion (which introduces a frameshift mutation) in exon 13 of the $Sh3pxd2b$ gene (Figure 4b) [34]. Iqbal et al. has generated another mouse line with a mixed genetic background carrying an insertion between exons 3 and 4 of the $Sh3pxd2b$ gene on chromosome 11 [164] (Figure 4b). Recently, an FTHS mouse model has been reported in a C57Bl/6 background by Dülk et al. In these TKS4$^{-/-}$ mice, TKS4 was knocked out by introducing an insertion between exons 5 and 6 of the $Sh3pxd2b$ gene (Figure 4b) [166]. All of these TKS4-mutant strains show very similar phenotypes that are reminiscent of the clinical symptoms of FTHS. These phenotypes include a shorter nasal bone, an overall decreased size, and a tendency to develop early onset glaucoma in their enlarged/prominent eyes [34,164,166,179]. Moreover, cardiac examination of the artificially generated TKS4 KO mice revealed variable deficiencies, including septal and mitral valve defects remarkably similar to those of FTHS patients [164]. Skeletal abnormalities, e.g., kyphosis [34,164,166], and reduced bone mineral density [34,179], which are both reminiscent of FTHS features, have also been described in the TKS4-mutant mice. Lipodystrophy is a general feature characterized by highly reduced visceral WAT mass in the case of TKS4 null mice [34,173], even though a low amount of subcutaneous fat tissue has only been reported in one diseased patient [164].

Taken together, the abnormalities observed in TKS4 KO mouse lines support the hypothesis that the presence of a mutant TKS4 gene has a role in FTHS. The exact detailed mechanism by which mutations in TKS4 cause such a diverse range of phenotypes in patients is still unknown. A few putative mechanisms are summarized in Figure 5. TKS4 has its highest expression levels in embryonic tissues [2], suggesting that the most notable phenotype-determining effects of the molecule are exerted during embryonic development. Podosome formation and migration of patterning immature cells are tightly linked processes during healthy embryonic development in response to instructive signals [81]. Since TKS4 and TKS5 are key players in functional podosome formation [2,86], mis-regulated podosome assembly early in development in FTHS patients might have a causative role in the manifestation of the related symptoms. Defective endothelial cell motility in complex three-dimensional ECM environments and diminished vessel sprout growth have also been reported in the absence of TKS4, possibly leading to negative effects on tubular heart formation and cardiac development [194]. If the TKS4 scaffold function is absent, EGF-induced cell migration might also be defective [11]. Therefore, TKS4 KO cells might fail to migrate in response to growth factor signal stimulation in general. The involvement of TKS4 in the EMT processes [153] may also affect embryonic development at several developmental stages, including endocardium formation, which is a process involving three consecutive EMT/MET events [161]. Disturbed EMT function could explain why patients and mouse models show defective cardiac development and functionality, expanding the list of processes by which the absence of TKS4 could cause such characteristic gross phenotypes.

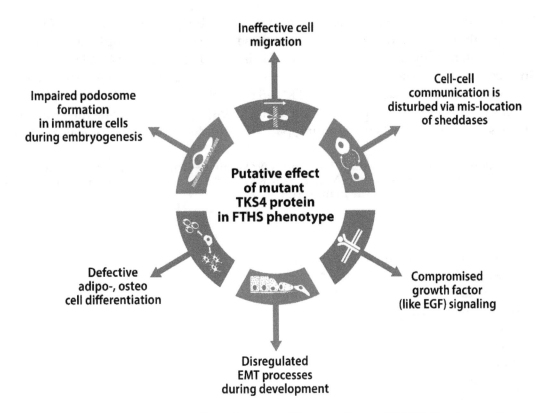

Figure 5. Putative role of mutated TKS4 in Frank-ter Haar syndrome (FTHS). During embryogenesis, differentiating cells must react properly to growth factor signals, migrate via podosome machinery to the determined position in the embryo, degrade the ECM during their migration, send extracellular signaling molecules to other cells, and occasionally undergo EMT. In each step, TKS4 has an experimentally described role. However, there are insufficient evidence at the organismal, tissue, cellular, and even biochemical levels to have a clear understanding of how TKS4 participates in disease manifestation. ECM–extracellular matrix, EMT–epithelial-mesenchymal transition.

Homozygous loss of TKS5 is usually lethal in early neonatal life [125]. Cejudo-Martin et al. created TKS5 KO mice by inserting a trapping vector VICTR 37 between exons 1 and 2 on chromosome 19 in a C57Bl/6Jx129/SvJ mixed background and in a C57Bl/6J pure background. TKS5 gene-trapped (TKS5$^{trap/trap}$) mutant mice of both genetic backgrounds are born at Mendelian ratios, but have a reduced lifespan. In the C57Bl/6Jx129/SvJ background, 50% of the mutant mice have a complete cleft of the secondary palate and die within 24 h after birth, and only 20% reach adulthood with no visible phenotypic defects. Furthermore, 30% of the neonates die between day one after birth and weaning despite having a normal palate. By contrast, in the C57Bl/6J pure background, neonatal mortality and cleft palate incidence rises to 90%. The remaining 10% of animals show no cleft palate but die shortly after birth of unknown causes. The authors concluded that strain purity is a determining factor in the phenotypic manifestation of TKS5 loss. No TKS4$^{-/-}$ and TKS5$^{trap/trap}$ double null mice are born from crosses of heterozygous TKS4 and TKS5 mutant parents, suggesting that the functions of the two molecules can overlap or complement each other during development [125]. Neither TKS4 nor TKS5 mutant animals show gross phenotypic alterations when they are heterozygous for the mutation [34,125,164].

5. Conclusions and Future Perspectives

Besides mediating the assembly of signaling components, the scaffold proteins TKS4 and TKS5 have an emerging role as regulated, active facilitators of the crosstalk between multiple signal transduction pathways modulating diverse and complex molecular networks. Future research focusing on the discovery of novel binding partners is expected to reveal not only the detailed molecular level mechanisms resulting in TKS-related phenotypes in FTHS patients and TKS KO animal models, but also, to shed light on yet unknown functions of these proteins. For example, while both TKS4 and 5 were found to be potential binding partners of the Fas ligand (FasL, CD178) (Table 1), which is a known cell death inducer [39,40], possible functions for TKS4/5 in the regulation of cell death have not yet been investigated. A detailed understanding of the TKS4/5 interactome, bolstered by an in-depth description of the molecular modifications of the TKS4/5 proteins during signaling may facilitate the development of new therapies to correct defects in signaling pathways underlying pathological conditions.

Author Contributions: Conceptualization, L.B. and V.V. Writing—original draft preparation, V.V. and G.K. Writing—reviewing, L.R. Writing—editing, A.K., B.S., K.K., and B.L.M. Visualization, T.T. All authors have read and agreed to the published version of the manuscript.

Abbreviations

ADAM	a disintegrin and metalloproteinase
BAT	brown adipose tissue
BDCSS	Borrone dermato-cardio-skeletal syndrome
ECM	extracellular matrix
EGF	epidermal growth factor
EMT	epithelial-mesenchymal transition
FTHS	Frank-ter Haar syndrome
GST	glutathione S-transferase
H_2O_2	hydrogen peroxide
ITC	isothermal titration calorimetry
KO	knockout
MET	mesenchymal-epithelial transition
miRNA	microRNA
MMP	matrix metalloproteases
mRNA	messenger RNA
MSCs	mesenchymal stem/stromal cells
nee mice	"nose-ear-eye" deformities mice
NOX	NADPH oxidase
O^{2-}	superoxide anion
·OH	hydroxyl radical
PI(4,5)P2	phosphatidylinositol (4,5)-bisphosphate
PI(3,4,5)P3	phosphatidylinositol (3,4,5)-trisphosphate
PPARγ	peroxisome proliferator-activated receptor gamma
PRR	proline-rich region
PTPases	phosphatases
PX domain	Phox homology domain
ROS	reactive oxygen species
RTK	receptor tyrosine kinase
SH3	SRC homology 3 domain
TKS	tyrosine kinase substrate
UCP1	uncoupling protein 1
WAT	white adipose tissue

References

1. Buday, L.; Tompa, P. Functional classification of scaffold proteins and related molecules. *FEBS J.* **2010**, *277*, 4348–4355. [CrossRef]
2. Buschman, M.D.; Bromann, P.A.; Cejudo-Martin, P.; Wen, F.; Pass, I.; Courtneidge, S.A. The Novel Adaptor Protein Tks4 (SH3PXD2B) is Required for Functional Podosome Formation. *Mol. Biol. Cell* **2009**, *20*, 1302–1311. [CrossRef] [PubMed]

3. Lock, P.; Abram, C.L.; Gibson, T.; Courtneidge, S.A. A new method for isolating tyrosine kinase substrates used to identify Fish, an SH3 and PX domain-containing protein, and Src substrate. *EMBO J.* **1998**, *17*, 4346–4357. [CrossRef] [PubMed]

4. Weaver, A.M. Regulation of Cancer Invasion by Reactive Oxygen Species and Tks Family Scaffold Proteins. *Sci. Signal.* **2009**, *2*, pe56-pe56. [CrossRef] [PubMed]

5. Ádám, C.; Fekete, A.; Bőgel, G.; Németh, Z.; Tőkési, N.; Ovádi, J.; Liliom, K.; Pesti, S.; Geiszt, M.; Buday, L. Accumulation of the PX domain mutant Frank-ter Haar syndrome protein Tks4 in aggresomes. *Cell Commun. Signal.* **2015**, *13*, 33. [CrossRef] [PubMed]

6. Abram, C.L.; Seals, D.F.; Pass, I.; Salinsky, D.; Maurer, L.; Roth, T.M.; Courtneidge, S.A. The Adaptor Protein Fish Associates with Members of the ADAMs Family and Localizes to Podosomes of Src-transformed Cells. *J. Biol. Chem.* **2003**, *278*, 16844–16851. [CrossRef] [PubMed]

7. Kurochkina, N.; Guha, U. SH3 domains: Modules of protein–protein interactions. *Biophys. Rev.* **2013**, *5*, 29–39. [CrossRef] [PubMed]

8. Kawahara, T.; Lambeth, J.D. Molecular evolution of Phox-related regulatory subunits for NADPH oxidase enzymes. *BMC Evol. Biol.* **2007**, *7*, 178. [CrossRef]

9. Lányi, Á.; Baráth, M.; Péterfi, Z.; Bőgel, G.; Orient, A.; Simon, T.; Petrovszki, E.; Kis-Tóth, K.; Sirokmány, G.; Rajnavölgyi, É.; et al. The Homolog of the Five SH3-Domain Protein (HOFI/SH3PXD2B) Regulates Lamellipodia Formation and Cell Spreading. *PLoS ONE* **2011**, *6*, e23653. [CrossRef]

10. Daly, C.; Logan, B.; Breeyear, J.; Whitaker, K.; Ahmed, M.; Seals, D.F. Tks5 SH3 domains exhibit differential effects on invadopodia development. *PLoS ONE* **2020**, *15*, e0227855. [CrossRef]

11. Bögel, G.; Gujdár, A.; Geiszt, M.; Lányi, Á.; Fekete, A.; Sipeki, S.; Downward, J.; Buday, L. Frank-ter Haar Syndrome Protein Tks4 Regulates Epidermal Growth Factor-dependent Cell Migration. *J. Biol. Chem.* **2012**, *287*, 31321–31329. [CrossRef]

12. Hishida, T.; Eguchi, T.; Osada, S.; Nishizuka, M.; Imagawa, M. A novel gene, fad49, plays a crucial role in the immediate early stage of adipocyte differentiation via involvement in mitotic clonal expansion. *FEBS J.* **2008**, *275*, 5576–5588. [CrossRef] [PubMed]

13. Teasdale, R.D.; Collins, B.M. Insights into the PX (phox-homology) domain and SNX (sorting nexin) protein families: Structures, functions and roles in disease. *Biochem. J.* **2012**, *441*, 39–59. [CrossRef] [PubMed]

14. Yaffe, M.B. The p47phox PX Domain: Two Heads Are Better Than One! *Structure* **2002**, *10*, 1288–1290. [CrossRef]

15. Groemping, Y.; Lapouge, K.; Smerdon, S.J.; Rittinger, K. Molecular Basis of Phosphorylation-Induced Activation of the NADPH Oxidase. *Cell* **2003**, *113*, 343–355. [CrossRef]

16. Belambri, S.A.; Rolas, L.; Raad, H.; Hurtado-Nedelec, M.; Dang, P.M.-C.; El-Benna, J. NADPH oxidase activation in neutrophils: Role of the phosphorylation of its subunits. *Eur. J. Clin. Investig.* **2018**, *48*, e12951. [CrossRef]

17. El-Benna, J.; Dang, P.M.-C.; Gougerot-Pocidalo, M.-A.; Marie, J.-C.; Braut-Boucher, F. p47phox, the phagocyte NADPH oxidase/NOX2 organizer: Structure, phosphorylation and implication in diseases. *Exp. Mol. Med.* **2009**, *41*, 217–225. [CrossRef]

18. Meijles, D.N.; Fan, L.M.; Howlin, B.J.; Li, J.-M. Molecular Insights of p47phox Phosphorylation Dynamics in the Regulation of NADPH Oxidase Activation and Superoxide Production. *J. Biol. Chem.* **2014**, *289*, 22759–22770. [CrossRef]

19. Oikawa, T.; Oyama, M.; Kozuka-Hata, H.; Uehara, S.; Udagawa, N.; Saya, H.; Matsuo, K. Tks5-dependent formation of circumferential podosomes/invadopodia mediates cell–cell fusion. *J. Cell Biol.* **2012**, *197*, 553–568. [CrossRef] [PubMed]

20. Casaletto, J.B.; McClatchey, A.I. Spatial regulation of receptor tyrosine kinases in development and cancer. *Nat. Rev. Cancer* **2012**, *12*, 387–400. [CrossRef]

21. Wee, P.; Wang, Z. Epidermal Growth Factor Receptor Cell Proliferation Signaling Pathways. *Cancers* **2017**, *9*, 52. [CrossRef]

22. Belsches, A.P.; Haskell, M.D.; Parsons, S.J. Role of c-Src tyrosine kinase in EGF-induced mitogenesis. *Front. Biosci.* **1997**, *2*, d501–d518. [CrossRef]

23. Alonso, G.; Koegl, M.; Mazurenko, N.; Courtneidge, S.A. Sequence Requirements for Binding of Src Family Tyrosine Kinases to Activated Growth Factor Receptors. *J. Biol. Chem.* **1995**, *270*, 9840–9848. [CrossRef] [PubMed]

24. Bromann, P.A.; Korkaya, H.; Courtneidge, S.A. The interplay between Src family kinases and receptor tyrosine kinases. *Oncogene* **2004**, *23*, 7957–7968. [CrossRef] [PubMed]

25. Sierke, S.L.; Longo, G.M.; Koland, J.G. Structural Basis of Interactions Between Epidermal Growth Factor Receptor and SH2 Domain Proteins. *Biochem. Biophys. Res. Commun.* **1993**, *191*, 45–54. [CrossRef]

26. Fekete, A.; Bőgel, G.; Pesti, S.; Péterfi, Z.; Geiszt, M.; Buday, L. EGF regulates tyrosine phosphorylation and membrane-translocation of the scaffold protein Tks5. *J. Mol. Signal.* **2013**, *8*, 8. [CrossRef] [PubMed]

27. Gianni, D.; Taulet, N.; DerMardirossian, C.; Bokoch, G.M. c-Src–Mediated Phosphorylation of NoxA1 and Tks4 Induces the Reactive Oxygen Species (ROS)–Dependent Formation of Functional Invadopodia in Human Colon Cancer Cells. *Mol. Biol. Cell* **2010**, *21*, 4287–4298. [CrossRef]

28. Bisson, N.; James, D.A.; Ivosev, G.; Tate, S.A.; Bonner, R.; Taylor, L.; Pawson, T. Selected reaction monitoring mass spectrometry reveals the dynamics of signaling through the GRB2 adaptor. *Nat. Biotechnol.* **2011**, *29*, 653–658. [CrossRef]

29. Oikawa, T.; Itoh, T.; Takenawa, T. Sequential signals toward podosome formation in NIH-src cells. *J. Cell Biol.* **2008**, *182*, 157–169. [CrossRef]

30. Crimaldi, L.; Courtneidge, S.A.; Gimona, M. Tks5 recruits AFAP-110, p190RhoGAP, and cortactin for podosome formation. *Exp. Cell Res.* **2009**, *315*, 2581–2592. [CrossRef]

31. Thapa, N.; Tan, X.; Choi, S.; Lambert, P.F.; Rapraeger, A.C.; Anderson, R.A. The Hidden Conundrum of Phosphoinositide Signaling in Cancer. *Trends Cancer* **2016**, *2*, 378–390. [CrossRef]

32. Dülk, M.; Szeder, B.; Glatz, G.; Merő, B.L.; Koprivanacz, K.; Kudlik, G.; Vas, V.; Sipeki, S.; Cserkaszky, A.; Radnai, L.; et al. EGF Regulates the Interaction of Tks4 with Src through Its SH2 and SH3 Domains. *Biochemistry* **2018**, *57*, 4186–4196. [CrossRef]

33. Rufer, A.C.; Rumpf, J.; von Holleben, M.; Beer, S.; Rittinger, K.; Groemping, Y. Isoform-Selective Interaction of the Adaptor Protein Tks5/FISH with Sos1 and Dynamins. *J. Mol. Biol.* **2009**, *390*, 939–950. [CrossRef] [PubMed]

34. Mao, M.; Thedens, D.R.; Chang, B.; Harris, B.S.; Zheng, Q.Y.; Johnson, K.R.; Donahue, L.R.; Anderson, M.G. The podosomal-adaptor protein SH3PXD2B is essential for normal postnatal development. *Mamm. Genome* **2009**, *20*, 462. [CrossRef] [PubMed]

35. Mattern, J.; Roghi, C.S.; Hurtz, M.; Knäuper, V.; Edwards, D.R.; Poghosyan, Z. ADAM15 mediates upregulation of Claudin-1 expression in breast cancer cells. *Sci. Rep.* **2019**, *9*, 12540. [CrossRef]

36. Sharafutdinov, I.; Backert, S.; Tegtmeyer, N. Cortactin: A Major Cellular Target of the Gastric Carcinogen Helicobacter pylori. *Cancers* **2020**, *12*, 159. [CrossRef] [PubMed]

37. Kropyvko, S.V. New partners of TKS4 scaffold protein. *Biopolym. Cell* **2015**, *31*, 395–401. [CrossRef]

38. Kropyvko, S.; Gryaznova, T.; Morderer, D.; Rynditch, A. Mammalian verprolin CR16 acts as a modulator of ITSN scaffold proteins association with actin. *Biochem. Biophys. Res. Commun.* **2017**, *484*, 813–819. [CrossRef]

39. Voss, M.; Lettau, M.; Janssen, O. Identification of SH3 domain interaction partners of human FasL (CD178) by phage display screening. *BMC Immunol.* **2009**, *10*, 53. [CrossRef]

40. Glukhova, X.A.; Trizna, J.A.; Proussakova, O.V.; Gogvadze, V.; Beletsky, I.P. Impairment of Fas-ligand–caveolin-1 interaction inhibits Fas-ligand translocation to rafts and Fas-ligand-induced cell death. *Cell Death Dis.* **2018**, *9*, 1–12. [CrossRef]

41. Giubellino, A.; Burke, J.T.R.; Bottaro, D.P. Grb2 signaling in cell motility and cancer. *Expert Opin. Ther. Targets* **2008**, *12*, 1021–1033. [CrossRef]

42. Gianni, D.; Diaz, B.; Taulet, N.; Fowler, B.; Courtneidge, S.A.; Bokoch, G.M. Novel p47phox-Related Organizers Regulate Localized NADPH Oxidase 1 (Nox1) Activity. *Sci. Signal.* **2009**, *2*, ra54. [CrossRef]

43. Gianni, D.; DerMardirossian, C.; Bokoch, G.M. Direct interaction between Tks proteins and the N-terminal proline-rich region (PRR) of NoxA1 mediates Nox1-dependent ROS generation. *Eur. J. Cell Biol.* **2011**, *90*, 164–171. [CrossRef]

44. Schröder, K.; Weissmann, N.; Brandes, R.P. Organizers and activators: Cytosolic Nox proteins impacting on vascular function. *Free Radic. Biol. Med.* **2017**, *109*, 22–32. [CrossRef] [PubMed]

45. Hou, J.; Yang, H.; Huang, X.; Leng, X.; Zhou, F.; Xie, C.; Zhou, Y.; Xu, Y. N-WASP promotes invasion and migration of cervical cancer cells through regulating p38 MAPKs signaling pathway. *Am. J. Transl. Res.* **2017**, *9*, 403–415. [PubMed]

46. Bazalii, A.V.; Samoylenko, A.A.; Petukhov, D.M.; Rynditch, A.V.; Redowicz, M.-J.; Drobot, L.B. Interaction between adaptor proteins Ruk/CIN85 and Tks4 in normal and tumor cells of different tissue origins. *Biopolym. Cell* **2014**, *30*, 37–41. [CrossRef]

47. Kong, M.S.; Hashimoto-Tane, A.; Kawashima, Y.; Sakuma, M.; Yokosuka, T.; Kometani, K.; Onishi, R.; Carpino, N.; Ohara, O.; Kurosaki, T.; et al. Inhibition of T cell activation and function by the adaptor protein CIN85. *Sci. Signal.* **2019**, *12*. [CrossRef] [PubMed]

48. Nyren-Erickson, E.K.; Jones, J.M.; Srivastava, D.K.; Mallik, S. A disintegrin and metalloproteinase-12 (ADAM12): Function, roles in disease progression, and clinical implications. *Biochim. Biophys. Acta BBA Gen. Subj.* **2013**, *1830*, 4445–4455. [CrossRef] [PubMed]

49. Weerasekera, L.; Rudnicka, C.; Sang, Q.-X.; Curran, J.E.; Johnson, M.P.; Moses, E.K.; Göring, H.H.H.; Blangero, J.; Hricova, J.; Schlaich, M.; et al. ADAM19: A Novel Target for Metabolic Syndrome in Humans and Mice. *Mediat. Inflamm.* **2017**, *2017*. [CrossRef]

50. Thompson, O.; Kleino, I.; Crimaldi, L.; Gimona, M.; Saksela, K.; Winder, S.J. Dystroglycan, Tks5 and Src Mediated Assembly of Podosomes in Myoblasts. *PLoS ONE* **2008**, *3*, e3638. [CrossRef]

51. Du, W.W.; Yang, W.; Li, X.; Fang, L.; Wu, N.; Li, F.; Chen, Y.; He, Q.; Liu, E.; Yang, Z.; et al. The Circular RNA circSKA3 Binds Integrin β1 to Induce Invadopodium Formation Enhancing Breast Cancer Invasion. *Mol. Ther.* **2020**, *28*, 1287–1298. [CrossRef] [PubMed]

52. Stylli, S.S.; Stacey, T.T.I.; Verhagen, A.M.; Xu, S.S.; Pass, I.; Courtneidge, S.A.; Lock, P. Nck adaptor proteins link Tks5 to invadopodia actin regulation and ECM degradation. *J. Cell Sci.* **2009**, *122*, 2727–2740. [CrossRef] [PubMed]

53. Shirao, T.; Hanamura, K.; Koganezawa, N.; Ishizuka, Y.; Yamazaki, H.; Sekino, Y. The role of drebrin in neurons. *J. Neurochem.* **2017**, *141*, 819–834. [CrossRef] [PubMed]

54. Zhang, R.; Lee, D.M.; Jimah, J.R.; Gerassimov, N.; Yang, C.; Kim, S.; Luvsanjav, D.; Winkelman, J.; Mettlen, M.; Abrams, M.E.; et al. Dynamin regulates the dynamics and mechanical strength of the actin cytoskeleton as a multifilament actin-bundling protein. *Nat. Cell Biol.* **2020**, *22*, 674–688. [CrossRef] [PubMed]

55. Grintsevich, E.E.; Yesilyurt, H.G.; Rich, S.K.; Hung, R.-J.; Terman, J.R.; Reisler, E. F-actin dismantling through a redox-driven synergy between Mical and cofilin. *Nat. Cell Biol.* **2016**, *18*, 876–885. [CrossRef]

56. Zagryazhskaya-Masson, A.; Monteiro, P.; Macé, A.-S.; Castagnino, A.; Ferrari, R.; Infante, E.; Duperray-Susini, A.; Dingli, F.; Lanyi, A.; Loew, D.; et al. Intersection of TKS5 and FGD1/CDC42 signaling cascades directs the formation of invadopodia. *J. Cell Biol.* **2020**, *219*. [CrossRef]

57. Genot, E.; Daubon, T.; Sorrentino, V.; Buccione, R. FGD1 as a central regulator of extracellular matrix remodelling—Lessons from faciogenital dysplasia. *J. Cell Sci.* **2012**, *125*, 3265–3270. [CrossRef]

58. Ke, Y.; Bao, T.; Zhou, Q.; Wang, Y.; Ge, J.; Fu, B.; Wu, X.; Tang, H.; Shi, Z.; Lei, X.; et al. Discs large homolog 5 decreases formation and function of invadopodia in human hepatocellular carcinoma via Girdin and Tks5. *Int. J. Cancer* **2017**, *141*, 364–376. [CrossRef]

59. Wang, X.; Enomoto, A.; Weng, L.; Mizutani, Y.; Abudureyimu, S.; Esaki, N.; Tsuyuki, Y.; Chen, C.; Mii, S.; Asai, N.; et al. Girdin/GIV regulates collective cancer cell migration by controlling cell adhesion and cytoskeletal organization. *Cancer Sci.* **2018**, *109*, 3643–3656. [CrossRef]

60. Oikawa, T.; Matsuo, K. Possible role of IRTKS in Tks5-driven osteoclast fusion. *Commun. Integr. Biol.* **2012**, *5*, 511–515. [CrossRef]

61. Li, L.; Liu, H.; Baxter, S.S.; Gu, N.; Ji, M.; Zhan, X. The SH3 domain distinguishes the role of I-BAR proteins IRTKS and MIM in chemotactic response to serum. *Biochem. Biophys. Res. Commun.* **2016**, *479*, 787–792. [CrossRef]

62. Yan, X.; Cao, N.; Chen, Y.; Lan, H.-Y.; Cha, J.-H.; Yang, W.-H.; Yang, M.-H. MT4-MMP promotes invadopodia formation and cell motility in FaDu head and neck cancer cells. *Biochem. Biophys. Res. Commun.* **2020**, *522*, 1009–1014. [CrossRef] [PubMed]

63. Chaki, S.P.; Barhoumi, R.; Rivera, G.M. Nck adapter proteins promote podosome biogenesis facilitating extracellular matrix degradation and cancer invasion. *Cancer Med.* **2019**, *8*, 7385–7398. [CrossRef]

64. Zhu, B.; Chen, S.; Hu, X.; Jin, X.; Le, Y.; Cao, L.; Yuan, Z.; Lin, Z.; Jiang, S.; Sun, L.; et al. Knockout of the Nogo-B Gene Attenuates Tumor Growth and Metastasis in Hepatocellular Carcinoma. *Neoplasia* **2017**, *19*, 583–593. [CrossRef] [PubMed]

65. Diaz, B.; Shani, G.; Pass, I.; Anderson, D.; Quintavalle, M.; Courtneidge, S.A. Tks5-Dependent, Nox-Mediated Generation of Reactive Oxygen Species Is Necessary for Invadopodia Formation. *Sci. Signal.* **2009**, *2*, ra53. [CrossRef]

66. Nagaraj, C.; Tabeling, C.; Nagy, B.M.; Jain, P.P.; Marsh, L.M.; Papp, R.; Pienn, M.; Witzenrath, M.; Ghanim, B.; Klepetko, W.; et al. Hypoxic vascular response and ventilation/perfusion matching in end-stage COPD may depend on p22phox. *Eur. Respir. J.* **2017**, *50*. [CrossRef]

67. Jacob, A.; Linklater, E.; Bayless, B.A.; Lyons, T.; Prekeris, R. The role and regulation of Rab40b–Tks5 complex during invadopodia formation and cancer cell invasion. *J. Cell Sci.* **2016**, *129*, 4341–4353. [CrossRef]

68. Li, Y.; Jia, Q.; Wang, Y.; Li, F.; Jia, Z.; Wan, Y. Rab40b upregulation correlates with the prognosis of gastric cancer by promoting migration, invasion, and metastasis. *Med. Oncol.* **2015**, *32*, 126. [CrossRef] [PubMed]

69. Lian, E.Y.; Hyndman, B.D.; Moodley, S.; Maritan, S.M.; Mulligan, L.M. RET isoforms contribute differentially to invasive processes in pancreatic ductal adenocarcinoma. *Oncogene* **2020**, 1–18. [CrossRef]

70. Gerboth, S.; Frittoli, E.; Palamidessi, A.; Baltanas, F.C.; Salek, M.; Rappsilber, J.; Giuliani, C.; Troglio, F.; Rolland, Y.; Pruneri, G.; et al. Phosphorylation of SOS1 on tyrosine 1196 promotes its RAC GEF activity and contributes to BCR-ABL leukemogenesis. *Leukemia* **2018**, *32*, 820–827. [CrossRef]

71. Prassanawar, S.S.; Panda, D. Tubulin heterogeneity regulates functions and dynamics of microtubules and plays a role in the development of drug resistance in cancer. *Biochem. J.* **2019**, *476*, 1359–1376. [CrossRef] [PubMed]

72. Sokolik, C.G.; Qassem, N.; Chill, J.H. The Disordered Cellular Multi-Tasker WIP and Its Protein–Protein Interactions: A Structural View. *Biomolecules* **2020**, *10*, 1084. [CrossRef] [PubMed]

73. Moodley, S.; Hui Bai, X.; Kapus, A.; Yang, B.; Liu, M. XB130/Tks5 scaffold protein interaction regulates Src-mediated cell proliferation and survival. *Mol. Biol. Cell* **2015**, *26*, 4492–4502. [CrossRef] [PubMed]

74. Kotb, A.; Hyndman, M.E.; Patel, T.R. The role of zyxin in regulation of malignancies. *Heliyon* **2018**, *4*, e00695. [CrossRef]

75. Buday, L.; Downward, J. Roles of cortactin in tumor pathogenesis. *Biochim. Biophys. Acta BBA Rev. Cancer* **2007**, *1775*, 263–273. [CrossRef] [PubMed]

76. MacGrath, S.M.; Koleske, A.J. Cortactin in cell migration and cancer at a glance. *J. Cell Sci.* **2012**, *125*, 1621–1626. [CrossRef] [PubMed]

77. Mader, C.C.; Oser, M.; Magalhaes, M.A.O.; Bravo-Cordero, J.J.; Condeelis, J.; Koleske, A.J.; Gil-Henn, H. An EGFR–Src–Arg–Cortactin Pathway Mediates Functional Maturation of Invadopodia and Breast Cancer Cell Invasion. *Cancer Res.* **2011**, *71*, 1730–1741. [CrossRef]

78. Chen, Y.-C.; Baik, M.; Byers, J.T.; Chen, K.T.; French, S.W.; Díaz, B. Experimental supporting data on TKS5 and Cortactin expression and localization in human pancreatic cancer cells and tumors. *Data Brief* **2019**, *22*, 132–136. [CrossRef]

79. Thuault, S.; Mamelonet, C.; Salameh, J.; Ostacolo, K.; Chanez, B.; Salaün, D.; Baudelet, E.; Audebert, S.; Camoin, L.; Badache, A. A proximity-labeling proteomic approach to investigate invadopodia molecular landscape in breast cancer cells. *Sci. Rep.* **2020**, *10*, 1–14. [CrossRef]

80. Paterson, E.K.; Courtneidge, S.A. Invadosomes are coming: New insights into function and disease relevance. *FEBS J.* **2018**, *285*, 8–27. [CrossRef]

81. Alonso, F.; Spuul, P.; Daubon, T.; Kramer, I.; Génot, E. Variations on the theme of podosomes: A matter of context. *Biochim. Biophys. Acta BBA Mol. Cell Res.* **2019**, *1866*, 545–553. [CrossRef] [PubMed]

82. Linder, S.; Wiesner, C. Tools of the trade: Podosomes as multipurpose organelles of monocytic cells. *Cell. Mol. Life Sci.* **2015**, *72*, 121–135. [CrossRef]

83. Alonso, F.; Spuul, P.; Génot, E. Podosomes in endothelial cell–microenvironment interactions. *Curr. Opin. Hematol.* **2020**, *27*, 197–205. [CrossRef]

84. Jacob, A.; Prekeris, R. The regulation of MMP targeting to invadopodia during cancer metastasis. *Front. Cell Dev. Biol.* **2015**, *3*. [CrossRef]

85. Iizuka, S.; Abdullah, C.; Buschman, M.D.; Diaz, B.; Courtneidge, S.A. The role of Tks adaptor proteins in invadopodia formation, growth and metastasis of melanoma. *Oncotarget* **2016**, *7*, 78473–78486. [CrossRef]

86. Seals, D.F.; Azucena, E.F.; Pass, I.; Tesfay, L.; Gordon, R.; Woodrow, M.; Resau, J.H.; Courtneidge, S.A. The adaptor protein Tks5/Fish is required for podosome formation and function, and for the protease-driven invasion of cancer cells. *Cancer Cell* **2005**, *7*, 155–165. [CrossRef]

87. Destaing, O.; Planus, E.; Bouvard, D.; Oddou, C.; Badowski, C.; Bossy, V.; Raducanu, A.; Fourcade, B.; Albiges-Rizo, C.; Block, M.R. β1A Integrin Is a Master Regulator of Invadosome Organization and Function. *Mol. Biol. Cell* **2010**, *21*, 4108–4119. [CrossRef] [PubMed]

88. Daubon, T.; Spuul, P.; Alonso, F.; Fremaux, I.; Génot, E. VEGF-A stimulates podosome-mediated collagen-IV proteolysis in microvascular endothelial cells. *J. Cell Sci.* **2016**, *129*, 2586–2598. [CrossRef]

89. Quintavalle, M.; Elia, L.; Condorelli, G.; Courtneidge, S.A. MicroRNA control of podosome formation in vascular smooth muscle cells in vivo and in vitro. *J. Cell Biol.* **2010**, *189*, 13–22. [CrossRef]

90. Varon, C.; Tatin, F.; Moreau, V.; Obberghen-Schilling, E.V.; Fernandez-Sauze, S.; Reuzeau, E.; Kramer, I.; Génot, E. Transforming Growth Factor β Induces Rosettes of Podosomes in Primary Aortic Endothelial Cells. *Mol. Cell. Biol.* **2006**, *26*, 3582–3594. [CrossRef] [PubMed]

91. Sa, G.; Liu, Z.; Ren, J.; Wan, Q.; Xiong, X.; Yu, Z.; Chen, H.; Zhao, Y.; He, S. Keratinocyte growth factor (KGF) induces podosome formation via integrin-Erk1/2 signaling in human immortalized oral epithelial cells. *Cell. Signal.* **2019**, *61*, 39–47. [CrossRef] [PubMed]

92. Yamaguchi, H.; Pixley, F.; Condeelis, J. Invadopodia and podosomes in tumor invasion. *Eur. J. Cell Biol.* **2006**, *85*, 213–218. [CrossRef]

93. Yamaguchi, H.; Lorenz, M.; Kempiak, S.; Sarmiento, C.; Coniglio, S.; Symons, M.; Segall, J.; Eddy, R.; Miki, H.; Takenawa, T.; et al. Molecular mechanisms of invadopodium formation the role of the N-WASP–Arp2/3 complex pathway and cofilin. *J. Cell Biol.* **2005**, *168*, 441–452. [CrossRef] [PubMed]

94. Boateng, L.R.; Huttenlocher, A. Spatiotemporal regulation of Src and its substrates at invadosomes. *Eur. J. Cell Biol.* **2012**, *91*, 878–888. [CrossRef] [PubMed]

95. Pan, Y.-R.; Chen, C.-L.; Chen, H.-C. FAK is required for the assembly of podosome rosettes. *J. Cell Biol.* **2011**, *195*, 113–129. [CrossRef] [PubMed]

96. García, E.; Jones, G.E.; Machesky, L.M.; Antón, I.M. WIP: WASP-interacting proteins at invadopodia and podosomes. *Eur. J. Cell Biol.* **2012**, *91*, 869–877. [CrossRef]

97. Bompard, G.; Caron, E. Regulation of WASP/WAVE proteins making a long story short. *J. Cell Biol.* **2004**, *166*, 957–962. [CrossRef]

98. Murphy, D.A.; Courtneidge, S.A. The "ins" and "outs" of podosomes and invadopodia: Characteristics, formation and function. *Nat. Rev. Mol. Cell Biol.* **2011**, *12*, 413–426. [CrossRef]

99. Schachtner, H.; Calaminus, S.D.J.; Thomas, S.G.; Machesky, L.M. Podosomes in adhesion, migration, mechanosensing and matrix remodeling. *Cytoskeleton* **2013**, *70*, 572–589. [CrossRef]

100. Linder, S. Invadosomes at a glance. *J. Cell Sci.* **2009**, *122*, 3009–3013. [CrossRef]

101. Oser, M.; Dovas, A.; Cox, D.; Condeelis, J. Nck1 and Grb2 localization patterns can distinguish invadopodia from podosomes. *Eur. J. Cell Biol.* **2011**, *90*, 181–188. [CrossRef] [PubMed]

102. Augoff, K.; Hryniewicz-Jankowska, A.; Tabola, R. Invadopodia: Clearing the way for cancer cell invasion. *Ann. Transl. Med.* **2020**, *8*, 902. [CrossRef]

103. Walkiewicz, K.; Nowakowska-Zajdel, E.; Kozieł, P.; Muc-Wierzgoń, M. The role of some ADAM-proteins and activation of the insulin growth factor-related pathway in colorectal cancer. *Central Eur. J. Immunol.* **2018**, *43*, 109–113. [CrossRef]

104. Sinderen, M.V.; Oyanedel, J.; Menkhorst, E.; Cuman, C.; Rainczuk, K.; Winship, A.; Salamonsen, L.; Edgell, T.; Dimitriadis, E. Soluble Delta-like ligand 1 alters human endometrial epithelial cell adhesive capacity. *Reprod. Fertil. Dev.* **2017**, *29*, 694–702. [CrossRef] [PubMed]

105. Seike, S.; Takehara, M.; Kobayashi, K.; Nagahama, M. Clostridium perfringens Delta-Toxin Damages the Mouse Small Intestine. *Toxins* **2019**, *11*, 232. [CrossRef]

106. Dong, W.; Liu, L.; Dou, Y.; Xu, M.; Liu, T.; Wang, S.; Zhang, Y.; Deng, B.; Wang, B.; Cao, H. Deoxycholic acid activates epidermal growth factor receptor and promotes intestinal carcinogenesis by ADAM17-dependent ligand release. *J. Cell. Mol. Med.* **2018**, *22*, 4263–4273. [CrossRef]

107. Miller, M.A.; Moss, M.L.; Powell, G.; Petrovich, R.; Edwards, L.; Meyer, A.S.; Griffith, L.G.; Lauffenburger, D.A. Targeting autocrine HB-EGF signaling with specific ADAM12 inhibition using recombinant ADAM12 prodomain. *Sci. Rep.* **2015**, *5*, 15150. [CrossRef] [PubMed]

108. Giebeler, N.; Zigrino, P. A Disintegrin and Metalloprotease (ADAM): Historical Overview of Their Functions. *Toxins* **2016**, *8*, 122. [CrossRef]

109. Feng, Y.; Tsai, Y.-H.; Xiao, W.; Ralls, M.W.; Stoeck, A.; Wilson, C.L.; Raines, E.W.; Teitelbaum, D.H.; Dempsey, P.J. Loss of ADAM17-Mediated Tumor Necrosis Factor Alpha Signaling in Intestinal Cells Attenuates Mucosal Atrophy in a Mouse Model of Parenteral Nutrition. *Mol. Cell. Biol.* **2015**, *35*, 3604–3621. [CrossRef] [PubMed]

110. Hoffmann, C.; Mao, X.; Brown-Clay, J.; Moreau, F.; Al Absi, A.; Wurzer, H.; Sousa, B.; Schmitt, F.; Berchem, G.; Janji, B.; et al. Hypoxia promotes breast cancer cell invasion through HIF-1α-mediated up-regulation of the invadopodial actin bundling protein CSRP2. *Sci. Rep.* **2018**, *8*, 10191. [CrossRef]

111. Zacharias, M.; Brcic, L.; Eidenhammer, S.; Popper, H. Bulk tumour cell migration in lung carcinomas might be more common than epithelial-mesenchymal transition and be differently regulated. *BMC Cancer* **2018**, *18*, 717. [CrossRef] [PubMed]

112. Baik, M.; French, B.; Chen, Y.-C.; Byers, J.T.; Chen, K.T.; French, S.W.; Díaz, B. Identification of invadopodia by TKS5 staining in human cancer lines and patient tumor samples. *MethodsX* **2019**, *6*, 718–726. [CrossRef]

113. Ren, X.L.; Qiao, Y.D.; Li, J.Y.; Li, X.M.; Zhang, D.; Zhang, X.J.; Zhu, X.H.; Zhou, W.J.; Shi, J.; Wang, W.; et al. Cortactin recruits FMNL2 to promote actin polymerization and endosome motility in invadopodia formation. *Cancer Lett.* **2018**, *419*, 245–256. [CrossRef]

114. Kedziora, K.M.; Leyton-Puig, D.; Argenzio, E.; Boumeester, A.J.; van Butselaar, B.; Yin, T.; Wu, Y.I.; van Leeuwen, F.N.; Innocenti, M.; Jalink, K.; et al. Rapid Remodeling of Invadosomes by Gi-coupled Receptors: Dissecting the Role of Rho GTPases. *J. Biol. Chem.* **2016**, *291*, 4323–4333. [CrossRef] [PubMed]

115. Weidmann, M.D.; Surve, C.R.; Eddy, R.J.; Chen, X.; Gertler, F.B.; Sharma, V.P.; Condeelis, J.S. MenaINV dysregulates cortactin phosphorylation to promote invadopodium maturation. *Sci. Rep.* **2016**, *6*, 36142. [CrossRef]

116. Chen, Y.-C.; Baik, M.; Byers, J.T.; Chen, K.T.; French, S.W.; Díaz, B. TKS5-positive invadopodia-like structures in human tumor surgical specimens. *Exp. Mol. Pathol.* **2019**, *106*, 17–26. [CrossRef]

117. Al Haddad, M.; El-Rif, R.; Hanna, S.; Jaafar, L.; Dennaoui, R.; Abdellatef, S.; Miskolci, V.; Cox, D.; Hodgson, L.; El-Sibai, M. Differential regulation of rho GTPases during lung adenocarcinoma migration and invasion reveals a novel role of the tumor suppressor StarD13 in invadopodia regulation. *Cell Commun. Signal.* **2020**, *18*, 144. [CrossRef] [PubMed]

118. Ngan, E.; Stoletov, K.; Smith, H.W.; Common, J.; Muller, W.J.; Lewis, J.D.; Siegel, P.M. LPP is a Src substrate required for invadopodia formation and efficient breast cancer lung metastasis. *Nat. Commun.* **2017**, *8*, 15059. [CrossRef] [PubMed]

119. Mao, L.; Whitehead, C.A.; Paradiso, L.; Kaye, A.H.; Morokoff, A.P.; Luwor, R.B.; Stylli, S.S. Enhancement of invadopodia activity in glioma cells by sublethal doses of irradiation and temozolomide. *J. Neurosurg.* **2017**, *129*, 598–610. [CrossRef]

120. Blouw, B.; Patel, M.; Iizuka, S.; Abdullah, C.; You, W.K.; Huang, X.; Li, J.-L.; Diaz, B.; Stallcup, W.B.; Courtneidge, S.A. The Invadopodia Scaffold Protein Tks5 Is Required for the Growth of Human Breast Cancer Cells In Vitro and In Vivo. *PLoS ONE* **2015**, *10*, e0121003. [CrossRef]

121. Bayarmagnai, B.; Perrin, L.; Pourfarhangi, K.E.; Graña, X.; Tüzel, E.; Gligorijevic, B. Invadopodia-mediated ECM degradation is enhanced in the G1 phase of the cell cycle. *J. Cell Sci.* **2019**, *132*. [CrossRef] [PubMed]

122. Li, C.M.-C.; Chen, G.; Dayton, T.L.; Kim-Kiselak, C.; Hoersch, S.; Whittaker, C.A.; Bronson, R.T.; Beer, D.G.; Winslow, M.M.; Jacks, T. Differential Tks5 isoform expression contributes to metastatic invasion of lung adenocarcinoma. *Genes Dev.* **2013**, *27*, 1557–1567. [CrossRef] [PubMed]

123. Ribeiro, A.L.R.; da Costa, N.M.M.; de Siqueira, A.S.; Brasil da Silva, W.; da Silva Kataoka, M.S.; Jaeger, R.G.; de Melo Alves-Junior, S.; Smith, A.M.; de Jesus Viana Pinheiro, J. Keratocystic odontogenic tumor overexpresses invadopodia-related proteins, suggesting invadopodia formation. *Oral Surg. Oral Med. Oral Pathol. Oral Radiol.* **2016**, *122*, 500–508. [CrossRef] [PubMed]

124. Shen, Y.; Wen, Z.; Li, Y.; Matteson, E.L.; Hong, J.; Goronzy, J.J.; Weyand, C.M. Metabolic control of the scaffold protein TKS5 in tissue-invasive, proinflammatory T cells. *Nat. Immunol.* **2017**, *18*, 1025–1034. [CrossRef]

125. Cejudo-Martin, P.; Yuen, A.; Vlahovich, N.; Lock, P.; Courtneidge, S.A.; Díaz, B. Genetic Disruption of the Sh3pxd2a Gene Reveals an Essential Role in Mouse Development and the Existence of a Novel Isoform of Tks5. *PLoS ONE* **2014**, *9*, e107674. [CrossRef]

126. Peláez, R.; Morales, X.; Salvo, E.; Garasa, S.; de Solórzano, C.O.; Martínez, A.; Larrayoz, I.M.; Rouzaut, A. β3 integrin expression is required for invadopodia-mediated ECM degradation in lung carcinoma cells. *PLoS ONE* **2017**, *12*, e0181579. [CrossRef]

127. Courtneidge, S.A. Cell migration and invasion in human disease: The Tks adaptor proteins. *Biochem. Soc. Trans.* **2012**, *40*, 129–132. [CrossRef]

128. Gonzalez-Avila, G.; Sommer, B.; Mendoza-Posada, D.A.; Ramos, C.; Garcia-Hernandez, A.A.; Falfan-Valencia, R. Matrix metalloproteinases participation in the metastatic process and their diagnostic and therapeutic applications in cancer. *Crit. Rev. Oncol. Hematol.* **2019**, *137*, 57–83. [CrossRef]

129. Hoshino, D.; Kirkbride, K.C.; Costello, K.; Clark, E.S.; Sinha, S.; Grega-Larson, N.; Tyska, M.J.; Weaver, A.M. Exosome Secretion Is Enhanced by Invadopodia and Drives Invasive Behavior. *Cell Rep.* **2013**, *5*, 1159–1168. [CrossRef]

130. Pant, S.; Hilton, H.; Burczynski, M.E. The multifaceted exosome: Biogenesis, role in normal and aberrant cellular function, and frontiers for pharmacological and biomarker opportunities. *Biochem. Pharmacol.* **2012**, *83*, 1484–1494. [CrossRef]

131. Yang, H.; Villani, R.M.; Wang, H.; Simpson, M.J.; Roberts, M.S.; Tang, M.; Liang, X. The role of cellular reactive oxygen species in cancer chemotherapy. *J. Exp. Clin. Cancer Res. CR* **2018**, *37*, 266. [CrossRef] [PubMed]

132. Chen, J.; Wang, Y.; Zhang, W.; Zhao, D.; Zhang, L.; Fan, J.; Li, J.; Zhan, Q. Membranous NOX5-derived ROS oxidizes and activates local Src to promote malignancy of tumor cells. *Signal Transduct. Target. Ther.* **2020**, *5*, 1–12. [CrossRef]

133. Block, K.; Gorin, Y. Aiding and abetting roles of NOX oxidases in cellular transformation. *Nat. Rev. Cancer* **2012**, *12*, 627–637. [CrossRef]

134. Blaser, H.; Dostert, C.; Mak, T.W.; Brenner, D. TNF and ROS Crosstalk in Inflammation. *Trends Cell Biol.* **2016**, *26*, 249–261. [CrossRef] [PubMed]

135. Singh, A.; Kukreti, R.; Saso, L.; Kukreti, S. Oxidative Stress: A Key Modulator in Neurodegenerative Diseases. *Molecules* **2019**, *24*, 1583. [CrossRef]

136. Coant, N.; Mkaddem, S.B.; Pedruzzi, E.; Guichard, C.; Tréton, X.; Ducroc, R.; Freund, J.-N.; Cazals-Hatem, D.; Bouhnik, Y.; Woerther, P.-L.; et al. NADPH Oxidase 1 Modulates WNT and NOTCH1 Signaling To Control the Fate of Proliferative Progenitor Cells in the Colon. *Mol. Cell. Biol.* **2010**, *30*, 2636–2650. [CrossRef]

137. Smallwood, M.J.; Nissim, A.; Knight, A.R.; Whiteman, M.; Haigh, R.; Winyard, P.G. Oxidative stress in autoimmune rheumatic diseases. *Free Radic. Biol. Med.* **2018**, *125*, 3–14. [CrossRef]

138. Kallenborn-Gerhardt, W.; Schröder, K.; Geisslinger, G.; Schmidtko, A. NOXious signaling in pain processing. *Pharmacol. Ther.* **2013**, *137*, 309–317. [CrossRef] [PubMed]

139. Koju, N.; Taleb, A.; Zhou, J.; Lv, G.; Yang, J.; Cao, X.; Lei, H.; Ding, Q. Pharmacological strategies to lower crosstalk between nicotinamide adenine dinucleotide phosphate (NADPH) oxidase and mitochondria. *Biomed. Pharmacother.* **2019**, *111*, 1478–1498. [CrossRef] [PubMed]

140. Leto, T.L.; Morand, S.; Hurt, D.; Ueyama, T. Targeting and Regulation of Reactive Oxygen Species Generation by Nox Family NADPH Oxidases. *Antioxid. Redox Signal.* **2009**, *11*, 2607–2619. [CrossRef] [PubMed]

141. Altenhöfer, S.; Kleikers, P.W.M.; Radermacher, K.A.; Scheurer, P.; Rob Hermans, J.J.; Schiffers, P.; Ho, H.; Wingler, K.; Schmidt, H.H.H.W. The NOX toolbox: Validating the role of NADPH oxidases in physiology and disease. *Cell. Mol. Life Sci.* **2012**, *69*, 2327–2343. [CrossRef]

142. Lambeth, J.D.; Kawahara, T.; Diebold, B. Regulation of Nox and Duox enzymatic activity and expression. *Free Radic. Biol. Med.* **2007**, *43*, 319–331. [CrossRef] [PubMed]

143. Petry, A.; Weitnauer, M.; Görlach, A. Receptor Activation of NADPH Oxidases. *Antioxid. Redox Signal.* **2009**, *13*, 467–487. [CrossRef]

144. Mortezaee, K. Nicotinamide adenine dinucleotide phosphate (NADPH) oxidase (NOX) and liver fibrosis: A review. *Cell Biochem. Funct.* **2018**, *36*, 292–302. [CrossRef] [PubMed]

145. Drummond, G.R.; Selemidis, S.; Griendling, K.K.; Sobey, C.G. Combating oxidative stress in vascular disease: NADPH oxidases as therapeutic targets. *Nat. Rev. Drug Discov.* **2011**, *10*, 453–471. [CrossRef] [PubMed]

146. Pani, G.; Galeotti, T.; Chiarugi, P. Metastasis: Cancer cell's escape from oxidative stress. *Cancer Metastasis Rev.* **2010**, *29*, 351–378. [CrossRef] [PubMed]

147. Caires-Dos-Santos, L.; da Silva, S.V.; Smuczek, B.; de Siqueira, A.S.; Cruz, K.S.P.; Barbuto, J.A.M.; Augusto, T.M.; Freitas, V.M.; Carvalho, H.F.; Jaeger, R.G. Laminin-derived peptide C16 regulates Tks expression and reactive oxygen species generation in human prostate cancer cells. *J. Cell. Physiol.* **2020**, *235*, 587–598. [CrossRef]

148. Petushkova, A.I.; Zamyatnin, A.A. Redox-Mediated Post-Translational Modifications of Proteolytic Enzymes and Their Role in Protease Functioning. *Biomolecules* **2020**, *10*, 650. [CrossRef]

149. Finkel, T. Signal transduction by reactive oxygen species. *J. Cell Biol.* **2011**, *194*, 7–15. [CrossRef]

150. Miki, H.; Funato, Y. Regulation of intracellular signalling through cysteine oxidation by reactive oxygen species. *J. Biochem.* **2012**, *151*, 255–261. [CrossRef]

151. Hurd, T.R.; DeGennaro, M.; Lehmann, R. Redox regulation of cell migration and adhesion. *Trends Cell Biol.* **2012**, *22*, 107–115. [CrossRef] [PubMed]

152. Milkovic, L.; Cipak Gasparovic, A.; Cindric, M.; Mouthuy, P.-A.; Zarkovic, N. Short Overview of ROS as Cell Function Regulators and Their Implications in Therapy Concepts. *Cells* **2019**, *8*, 793. [CrossRef] [PubMed]

153. Szeder, B.; Tárnoki-Zách, J.; Lakatos, D.; Vas, V.; Kudlik, G.; Merő, B.; Koprivanacz, K.; Bányai, L.; Hámori, L.; Róna, G.; et al. Absence of the Tks4 Scaffold Protein Induces Epithelial-Mesenchymal Transition-Like Changes in Human Colon Cancer Cells. *Cells* **2019**, *8*, 1343. [CrossRef]

154. Dongre, A.; Weinberg, R.A. New insights into the mechanisms of epithelial–mesenchymal transition and implications for cancer. *Nat. Rev. Mol. Cell Biol.* **2019**, *20*, 69–84. [CrossRef]

155. Horejs, C.-M. Basement membrane fragments in the context of the epithelial-to-mesenchymal transition. *Eur. J. Cell Biol.* **2016**, *95*, 427–440. [CrossRef] [PubMed]

156. Song, J.; Wang, W.; Wang, Y.; Qin, Y.; Wang, Y.; Zhou, J.; Wang, X.; Zhang, Y.; Wang, Q. Epithelial-mesenchymal transition markers screened in a cell-based model and validated in lung adenocarcinoma. *BMC Cancer* **2019**, *19*, 680. [CrossRef] [PubMed]

157. Clapéron, A.; Mergey, M.; Nguyen Ho-Bouldoires, T.H.; Vignjevic, D.; Wendum, D.; Chrétien, Y.; Merabtene, F.; Frazao, A.; Paradis, V.; Housset, C.; et al. EGF/EGFR axis contributes to the progression of cholangiocarcinoma through the induction of an epithelial-mesenchymal transition. *J. Hepatol.* **2014**, *61*, 325–332. [CrossRef]

158. Cheng, J.-C.; Auersperg, N.; Leung, P.C.K. EGF-Induced EMT and Invasiveness in Serous Borderline Ovarian Tumor Cells: A Possible Step in the Transition to Low-Grade Serous Carcinoma Cells? *PLoS ONE* **2012**, *7*, e34071. [CrossRef]

159. Jiao, L.; Li, D.-D.; Yang, C.-L.; Peng, R.-Q.; Guo, Y.-Q.; Zhang, X.-S.; Zhu, X.-F. Reactive oxygen species mediate oxaliplatin-induced epithelial-mesenchymal transition and invasive potential in colon cancer. *Tumor Biol.* **2016**, *37*, 8413–8423. [CrossRef]

160. Jung, S.-H.; Kim, S.-M.; Lee, C.-E. Mechanism of suppressors of cytokine signaling 1 inhibition of epithelial-mesenchymal transition signaling through ROS regulation in colon cancer cells: Suppression of Src leading to thioredoxin up-regulation. *Oncotarget* **2016**, *7*, 62559–62571. [CrossRef]

161. Nakaya, Y.; Sheng, G. EMT in developmental morphogenesis. *Cancer Lett.* **2013**, *341*, 9–15. [CrossRef] [PubMed]

162. Kim, D.H.; Xing, T.; Yang, Z.; Dudek, R.; Lu, Q.; Chen, Y.-H. Epithelial Mesenchymal Transition in Embryonic Development, Tissue Repair and Cancer: A Comprehensive Overview. *J. Clin. Med.* **2018**, *7*, 1. [CrossRef] [PubMed]

163. Maas, S.M.; Kayserili, H.; Lam, J.; Apak, M.Y.; Hennekam, R.C.M. Further delineation of Frank–ter Haar syndrome. *Am. J. Med. Genet. Part A* **2004**, *131*, 127–133. [CrossRef] [PubMed]

164. Iqbal, Z.; Cejudo-Martin, P.; de Brouwer, A.; van der Zwaag, B.; Ruiz-Lozano, P.; Scimia, M.C.; Lindsey, J.D.; Weinreb, R.; Albrecht, B.; Megarbane, A.; et al. Disruption of the Podosome Adaptor Protein TKS4 (SH3PXD2B) Causes the Skeletal Dysplasia, Eye, and Cardiac Abnormalities of Frank-Ter Haar Syndrome. *Am. J. Hum. Genet.* **2010**, *86*, 254–261. [CrossRef]

165. Bendon, C.L.; Fenwick, A.L.; Hurst, J.A.; Nürnberg, G.; Nürnberg, P.; Wall, S.A.; Wilkie, A.O.; Johnson, D. Frank-ter Haar syndrome associated with sagittal craniosynostosis and raised intracranial pressure. *BMC Med. Genet.* **2012**, *13*, 104. [CrossRef]

166. Dülk, M.; Kudlik, G.; Fekete, A.; Ernszt, D.; Kvell, K.; Pongrácz, J.E.; Merő, B.L.; Szeder, B.; Radnai, L.; Geiszt, M.; et al. The scaffold protein Tks4 is required for the differentiation of mesenchymal stromal cells (MSCs) into adipogenic and osteogenic lineages. *Sci. Rep.* **2016**, *6*, 34280. [CrossRef]

167. Dushnik-Levinson, M.; Benvenisty, N. Embryogenesis in vitro: Study of Differentiation of Embryonic Stem Cells. *Neonatology* **1995**, *67*, 77–83. [CrossRef]

168. Kurosaka, S.; Kashina, A. Cell biology of embryonic migration. *Birth Defects Res. Part C Embryo Today Rev.* **2008**, *84*, 102–122. [CrossRef]

169. Pellettieri, J.; Alvarado, A.S. Cell Turnover and Adult Tissue Homeostasis: From Humans to Planarians. *Annu. Rev. Genet.* **2007**, *41*, 83–105. [CrossRef]

170. Post, Y.; Clevers, H. Defining Adult Stem Cell Function at Its Simplest: The Ability to Replace Lost Cells through Mitosis. *Cell Stem Cell* **2019**, *25*, 174–183. [CrossRef]

171. Chen, Q.; Shou, P.; Zheng, C.; Jiang, M.; Cao, G.; Yang, Q.; Cao, J.; Xie, N.; Velletri, T.; Zhang, X.; et al. Fate decision of mesenchymal stem cells: Adipocytes or osteoblasts? *Cell Death Differ.* **2016**, *23*, 1128–1139. [CrossRef] [PubMed]

172. Medina-Gomez, G.; Gray, S.L.; Yetukuri, L.; Shimomura, K.; Virtue, S.; Campbell, M.; Curtis, R.K.; Jimenez-Linan, M.; Blount, M.; Yeo, G.S.H.; et al. PPAR gamma 2 Prevents Lipotoxicity by Controlling Adipose Tissue Expandability and Peripheral Lipid Metabolism. *PLoS Genet.* **2007**, *3*, e64. [CrossRef] [PubMed]

173. Vas, V.; Háhner, T.; Kudlik, G.; Ernszt, D.; Kvell, K.; Kuti, D.; Kovács, K.J.; Tóvári, J.; Trexler, M.; Merő, B.L.; et al. Analysis of Tks4 Knockout Mice Suggests a Role for Tks4 in Adipose Tissue Homeostasis in the Context of Beigeing. *Cells* **2019**, *8*, 831. [CrossRef] [PubMed]

174. Murphy, D.A.; Diaz, B.; Bromann, P.A.; Tsai, J.H.; Kawakami, Y.; Maurer, J.; Stewart, R.A.; Izpisúa-Belmonte, J.C.; Courtneidge, S.A. A Src-Tks5 Pathway Is Required for Neural Crest Cell Migration during Embryonic Development. *PLoS ONE* **2011**, *6*, e22499. [CrossRef] [PubMed]

175. Liu, Y.-J.; Liu, X.-G.; Wang, L.; Dina, C.; Yan, H.; Liu, J.-F.; Levy, S.; Papasian, C.J.; Drees, B.M.; Hamilton, J.J.; et al. Genome-wide association scans identified CTNNBL1 as a novel gene for obesity. *Hum. Mol. Genet.* **2008**, *17*, 1803–1813. [CrossRef]

176. Vogel, C.I.; Greene, B.; Scherag, A.; Müller, T.D.; Friedel, S.; Grallert, H.; Heid, I.M.; Illig, T.; Wichmann, H.-E.; Schäfer, H.; et al. Non-replication of an association of CTNNBL1polymorphisms and obesity in a population of Central European ancestry. *BMC Med. Genet.* **2009**, *10*, 14. [CrossRef]

177. Cleal, L.; Aldea, T.; Chau, Y.-Y. Fifty shades of white: Understanding heterogeneity in white adipose stem cells. *Adipocyte* **2017**, *6*, 205–216. [CrossRef]

178. Kajimura, S.; Spiegelman, B.M.; Seale, P. Brown and Beige Fat: Physiological Roles beyond Heat Generation. *Cell Metab.* **2015**, *22*, 546–559. [CrossRef]

179. Vas, V.; Kovács, T.; Körmendi, S.; Bródy, A.; Kudlik, G.; Szeder, B.; Mező, D.; Kállai, D.; Koprivanacz, K.; Merő, B.L.; et al. Significance of the Tks4 scaffold protein in bone tissue homeostasis. *Sci. Rep.* **2019**, *9*, 5781. [CrossRef]

180. Langdahl, B.; Ferrari, S.; Dempster, D.W. Bone modeling and remodeling: Potential as therapeutic targets for the treatment of osteoporosis. *Ther. Adv. Musculoskelet. Dis.* **2016**. [CrossRef]

181. Ter Haar, B.; Hamel, B.; Hendriks, J.; de Jager, J.; Opitz, J.M. Melnick-Needles syndrome: Indication for an autosomal recessive form. *Am. J. Med. Genet.* **1982**, *13*, 469–477. [CrossRef] [PubMed]

182. Frank, Y.; Ziprkowski, M.; Romano, A.; Stein, R.; Katznelson, M.B.; Cohen, B.; Goodman, R.M. Megalocornea associated with multiple skeletal anomalies: A new genetic syndrome? *J. Genet. Hum.* **1973**, *21*, 67–72. [PubMed]

183. Zrhidri, A.; Jaouad, I.C.; Lyahyai, J.; Raymond, L.; Egéa, G.; Taoudi, M.; El Mouatassim, S.; Sefiani, A. Identification of two novel SH3PXD2B gene mutations in Frank-Ter Haar syndrome by exome sequencing: Case report and review of the literature. *Gene* **2017**, *628*, 190–193. [CrossRef] [PubMed]

184. Ratukondla, B.; Prakash, S.; Reddy, S.; Puthuran, G.V.; Kannan, N.B.; Pillai, M.R. A Rare Case Report of Frank Ter Haar Syndrome in a Sibling Pair Presenting With Congenital Glaucoma. *J. Glaucoma* **2020**, *29*, 236–238. [CrossRef] [PubMed]

185. Durand, B.; Stoetzel, C.; Schaefer, E.; Calmels, N.; Scheidecker, S.; Kempf, N.; De Melo, C.; Guilbert, A.-S.; Timbolschi, D.; Donato, L.; et al. A severe case of Frank-ter Haar syndrome and literature review: Further delineation of the phenotypical spectrum. *Eur. J. Med. Genet.* **2020**, *63*, 103857. [CrossRef]

186. Wilson, G.R.; Sunley, J.; Smith, K.R.; Pope, K.; Bromhead, C.J.; Fitzpatrick, E.; Di Rocco, M.; van Steensel, M.; Coman, D.J.; Leventer, R.J.; et al. Mutations in SH3PXD2B cause Borrone dermato-cardio-skeletal syndrome. *Eur. J. Hum. Genet.* **2014**, *22*, 741–747. [CrossRef] [PubMed]

187. Kanai, F.; Liu, H.; Field, S.J.; Akbary, H.; Matsuo, T.; Brown, G.E.; Cantley, L.C.; Yaffe, M.B. The PX domains of p47phox and p40phox bind to lipid products of PI(3)K. *Nat. Cell Biol.* **2001**, *3*, 675–678. [CrossRef]

188. Chang, T.C.; Bauer, M.; Puerta, H.S.; Greenberg, M.B.; Cavuoto, K.M. Ophthalmic findings in Frank-ter Haar syndrome: Report of a sibling pair. *J. Am. Assoc. Pediatr. Ophthalmol. Strabismus* **2017**, *21*, 514–516. [CrossRef]

189. Mao, M.; Solivan-Timpe, F.; Roos, B.R.; Mullins, R.F.; Oetting, T.A.; Kwon, Y.H.; Brzeskiewicz, P.M.; Stone, E.M.; Alward, W.L.M.; Anderson, M.G.; et al. Localization of SH3PXD2B in human eyes and detection of rare variants in patients with anterior segment diseases and glaucoma. *Mol. Vis.* **2012**, *18*, 705–713.

190. Bernstein, H.-G.; Keilhoff, G.; Dobrowolny, H.; Lendeckel, U.; Steiner, J. From putative brain tumor marker to high cognitive abilities: Emerging roles of a disintegrin and metalloprotease (ADAM) 12 in the brain. *J. Chem. Neuroanat.* **2020**, *109*, 101846. [CrossRef]

191. Malinin, N.L.; Wright, S.; Seubert, P.; Schenk, D.; Griswold-Prenner, I. Amyloid-β neurotoxicity is mediated by FISH adapter protein and ADAM12 metalloprotease activity. *Proc. Natl. Acad. Sci. USA* **2005**, *102*, 3058–3063. [CrossRef]

192. Xiang, Y.; Cheng, Y.; Li, X.; Li, Q.; Xu, J.; Zhang, J.; Liu, Y.; Xing, Q.; Wang, L.; He, L.; et al. Up-Regulated Expression and Aberrant DNA Methylation of LEP and SH3PXD2A in Pre-Eclampsia. *PLoS ONE* **2013**, *8*, e59753. [CrossRef] [PubMed]

193. Patel, A.; Dash, P.R. Formation of atypical podosomes in extravillous trophoblasts regulates extracellular matrix degradation. *Eur. J. Cell Biol.* **2012**, *91*, 171–179. [CrossRef] [PubMed]

194. Mehes, E.; Barath, M.; Gulyas, M.; Bugyik, E.; Geiszt, M.; Szoor, A.; Lanyi, A.; Czirok, A. Enhanced endothelial motility and multicellular sprouting is mediated by the scaffold protein TKS4. *Sci. Rep.* **2019**, *9*, 14363. [CrossRef] [PubMed]

2

Insight into the Interactome of Intramitochondrial PKA using Biotinylation-Proximity Labeling

Yasmine Ould Amer [1,2] and **Etienne Hebert-Chatelain** [1,2,*]

1   Department of Biology, University of Moncton, Moncton, NB E1A 3E9, Canada; eyo9935@umoncton.ca
2   Canada Research Chair in Mitochondrial Signaling and Physiopathology, University of Moncton, Moncton, NB E1A 3E9, Canada
*   Correspondence: etienne.hebert.chatelain@umoncton.ca

Abstract: Mitochondria are fully integrated in cell signaling. Reversible phosphorylation is involved in adjusting mitochondrial physiology to the cellular needs. Protein kinase A (PKA) phosphorylates several substrates present at the external surface of mitochondria to maintain cellular homeostasis. However, few targets of PKA located inside the organelle are known. The aim of this work was to characterize the impact and the interactome of PKA located inside mitochondria. Our results show that the overexpression of intramitochondrial PKA decreases cellular respiration and increases superoxide levels. Using proximity-dependent biotinylation, followed by LC-MS/MS analysis and in silico phospho-site prediction, we identified 21 mitochondrial proteins potentially targeted by PKA. We confirmed the interaction of PKA with TIM44 using coimmunoprecipitation and observed that TIM44-S80 is a key residue for the interaction between the protein and the kinase. These findings provide insights into the interactome of intramitochondrial PKA and suggest new potential mechanisms in the regulation of mitochondrial functions.

Keywords: mitochondria; protein kinase A; BioID2; proteomics; serine/threonine phosphoprediction; TIM44

1. Introduction

Mitochondria are crucial in the regulation of cell metabolism, proliferation and survival. These organelles are responsible for generating most of the cellular adenosine triphosphate (ATP) through oxidative phosphorylation (OXPHOS) which links oxidation of metabolic fuels to the production of ATP [1,2]. During this process, electrons are transferred from NADH or $FADH_2$ to O_2 by four electron carriers (enzymatic complexes named complex I to IV). Simultaneously, protons are pumped from the mitochondrial matrix to the intermembrane space establishing an electrochemical gradient across the inner mitochondrial membrane (IMM), which is ultimately used by the ATP synthase to generate ATP. Mitochondria are also involved in intracellular calcium homeostasis, generation of reactive oxygen species (ROS) and apoptosis [3].

Reversible phosphorylation is involved in the maintenance of mitochondrial functions and cellular homeostasis [4–8]. Several serine (S) and threonine (T) kinases translocate to mitochondria in particular conditions to phosphorylate mitochondrial substrates [9]. For instance, AMP-activated protein kinase (AMPK) and pyruvate dehydrogenase kinase (PDK) target cytochrome c and pyruvate dehydrogenase E1-α, respectively, to adapt mitochondrial functions during ischemia/reperfusion-induced injury [10–12].

Protein kinase A (PKA) is involved in numerous physiological processes, including metabolism, gene transcription, cell division, and cell differentiation [13]. Eucaryotic PKA holoenzymes are composed of two regulatory subunits (PKA-R) bound to two catalytic subunits (PKA-C). There are four

PKA-R isoforms (i.e., RI-α, RI-β, RII-α and RII-β) and three PKA-C isoforms (i.e., C-α, C-β and C-γ). Binding of cAMP to the regulatory subunits induces dissociation of the holoenzyme and subsequent phosphorylation of key substrates by the free and active PKA-C [14]. Cyclic AMP is degraded by cAMP phosphodiesterases (PDEs). Phosphorylation of PDEs by PKA reduces cAMP levels and downregulates cAMP signaling in a negative feedback loop [15]. PKA is targeted to specific subcellular compartments by A-kinase anchoring proteins (AKAPs). PKA was one of the first kinase to be associated to phosphorylation of mitochondrial proteins [16], through the observation of cAMP-dependent phosphorylation both at the surface and inside mitochondria [17–20]. AKAPs such as AKAP-1, anchor PKA to the outer mitochondrial membrane (OMM) where it targets several proteins [21–23]. Notably, PKA phosphorylates the glutathione S-transferase alpha 4 protein on S189 to promote its interaction with the chaperone heat shock protein 70 (Hsp70), facilitate its translocation to mitochondria and increase its activity [24]. PKA also targets three subunits of the OMM import machinery, i.e., TOM22, TOM40 and TOM70, to blunt protein import into mitochondria [25–27]. Additionally, phosphorylation of BAD-S112/155, BAX-S60 and BIM-S83 by PKA inhibits apoptosis [28–31]. Phosphorylation of Drp1-S637 by PKA prevents mitochondrial fission [32,33].

Several intramitochondrial proteins were also shown to be phosphorylated by PKA. For instance, PKA phosphorylates COXIV-1-S58, preventing complex IV inhibition by ATP [34], whereas PKA-dependent phosphorylation of the ATPase inhibitory factor 1 (AIF1) on S39 impedes its binding to ATP synthase [35]. PKA also downregulates mtDNA replication and mitochondrial biogenesis by phosphorylating the mitochondrial transcription factor A, which impairs its ability to bind DNA and activate transcription [36]. Whether PKA targets these proteins in the cytosol before their translocation or directly within the organelle is, however, still debated. For instance, PKA can phosphorylate NDUFS4 in the cytosol to modulate its transport inside the organelle and the assembly of complex I [37]. Interestingly, the AKAP sphingosine kinase interacting protein (also named SKIP) tethers PKA in the intermembrane space and in the matrix where it interacts with MIC19 and MIC60 [38]. In addition, indirect evidence suggest that a soluble adenylyl cyclase is localized inside the mitochondrial matrix where it generates cAMP to directly activate intramitochondrial PKA and modulate ATP levels [39–41]. Therefore, it appears important to better describe the role of PKA localized inside mitochondria for which only few targets were identified so far.

The aim of this work was to examine the interactome of PKA localized in mitochondria. To address this, we generated a construct encoding PKA specifically targeted to the interface of the IMM and the matrix (mt-PKA) and examined its impact on mitochondrial activity. Our findings indicate that overexpressed mt-PKA alters mitochondrial functions and phosphorylates several mitochondrial proteins. Using proximity-dependent biotin identification, LC-MS/MS and in silico phospho-site prediction, we identified potential new substrates of mt-PKA including TIM44-S80. We also observed that TIM44-S80 phosphomutants impact OXPHOS and the physical interaction between TIM44 and mt-PKA.

2. Material and Methods

2.1. Cloning and Site-Directed Mutagenesis

PKA-Myc and mt-PKA-Myc constructs were generated as described [42]. Briefly, the mitochondrial leading sequence of COXVIIIa subunit was fused to the Myc-tagged PKA-C-α sequence [42]. The mt-PKA-BioID2-HA plasmid was generated by fusing the human influenza hemagglutinin (HA) tagged BioID2 sequence (Addgene plasmid #74224) to the mt-PKA-C-α sequence [43,44].

pcDNA3-TIM44-V5 was kindly provided by Pr. Elena Bonora (University of Bologna, Bologna, Italy). Site-directed mutagenesis was performed following the Q5® High-Fidelity DNA Polymerase (ref. M0491, New England Biolabs, Ipswich, MA,) PCR protocol, using the following synthetic oligonucleotides, forward: 5'-AATGAAAGAAgcTATAAAAAAATTCCGTGACGAG-3' with reverse 5'-TCTTTGTTTTTGGCTAATTC-3', and forward 5'-AATGAAAGAAgaTATAAAAAAA

TTCCGTGACGAG-3' with reverse 5'-TCTTTGTTTTTGGCTAATTC-3' to generate the phosphomimetic (TIM44-S80D) and the phosphodeficient (TIM44-S80A) mutants of TIM44, respectively.

2.2. Cell Culture and Transient Transfection

HeLa and HEK cells were cultured in high glucose (4,5 g/L Dulbecco's modified Eagle's medium (DMEM) supplemented with 2 mM glutamine, 1 mM pyruvate, 10% (v/v) FBS and penicillin-streptomycin. Cells were kept at 37 °C in 5% CO_2 and 95% humidity. Cells were transiently transfected with polyethylenimine (PolySciences, Warrington, PA, USA) and analyzed 48 h following transfection.

2.3. Subcellular Fractionation

Isolation of subcellular fractions was performed as described previously [45]. Briefly, cells were harvested and resuspended in mitochondrial isolation buffer (250 mM sucrose, 1 mM EDTA, 5 mM HEPES, pH 7.4) supplemented with 1% protease inhibitor cocktail (Bioshop, ON, Canada), 2 mM sodium orthovanadate and 1 mM sodium fluoride. Cells were lysed with 15 strokes using a 25-gauge syringe on ice and centrifuged at 1500× g for 5 min (4 °C). The resulting supernatant (TCL) was centrifuged at 12,500× g for 10 min (4 °C). The obtained supernatant was considered as the cytosolic fraction (cyto), the pellet was resuspended in the mitochondrial buffer and a cycle of centrifugation at 1500× g and 12,500× g was repeated. The final pellet was considered as the mitochondria-enriched fraction (mito). Protein concentration was determined by Bradford assay [46].

2.4. Trypsin Sensitivity Assay

The trypsin sensitivity assay was carried out as described previously [47] with minor modifications. Briefly, isolated mitochondria were suspended in mitochondrial isolation buffer and incubated at 37 °C for 10 min in presence or absence of trypsin (0.5%) and triton X-100 (1%). Reaction was stopped by the addition of 1% of the protease inhibitor cocktail. Mitochondria were then centrifuged at 12,500× g at 4 °C for 10 min. The pellets were processed for SDS-PAGE.

2.5. Mitochondrial Subfractionation

Isolated mitochondria were resuspended in 1M HEPES-KOH buffer (pH 7.4) in the presence of 0.125, 0.25 or 0.5% of digitonin and incubated at room temperature with continuous shaking at 1000 rpm. After 15 min, samples were centrifuged at 12,500× g at 4 °C for 10 min. The resulting pellets and supernatants were processed for SDS-PAGE. The presence of proteins from the different mitochondrial subcompartments in pellets and supernatants was analyzed using immunoblotting.

2.6. SDS-PAGE and BN-PAGE

Samples were analyzed by SDS-PAGE at 200 V using 7, 10 or 12% SDS-polyacrylamide mini-gel containing 0.35% (v/v) of 2,2,2-trichloroethanol for staining loaded proteins. For BN-PAGE, cells were solubilized in 2% digitonin for 10 min on ice and centrifuged during 10 min at 12,500× g. Supernatant was then supplemented with 0.5% Coomassie G-250 and samples were loaded in 4–16% Bis-Tris gels (LifeTechnologies, Pleasanton, CA, USA). Voltage was initially set to 150 V for 45 min and at 300 V during the following 30 min. Proteins were then transferred to polyvinylidene difluoride (PVDF) membranes. The total amount of proteins loaded in SDS-PAGE gels was visualized and quantified using the stain-free method, as described [48]. Briefly, 2,2,2-trichloroethanol interacts with tryptophan in loaded proteins and induce UV light-induced fluorescence which can be visualized on a 300 nm transilluminator. Immunolabeling can then be normalized to the total UV light-induced fluorescence (corresponding to the total protein load) instead of using unique proteins (such as tubulin or actin) as loading control. Membranes were blocked for 1 h in TBS-T (50 mM Tris-Cl, pH 7.6; 150 mM NaCl, 0.1% Tween) containing 5% BSA or 5% skim milk and incubated with primary antibodies overnight at 4 °C.

Protein immunodetection was performed using primary antibodies directed against PKA C-α (Cell Signaling; 4782S), PKA RI-α/β (Cell Signaling; 3927), Phospho-PKA Substrate (Cell Signaling; 9624S), α-Tubulin (Cell Signaling; 2144) SDHa (Abcam; ab14715), SOD2 (Santa Cruz; sc-30080), NDUF9a (Abcam; ab14713), Cytochrome c (Abcam; ab110325), TOM20 (Santa Cruz; (F-10) sc-17764), UQCRC2 (Abcam; ab14745), PAM16 (Abcam; Ab184157), Hsp60 (Santa Cruz; sc-13115), Hsp70 (Santa Cruz; sc-24), V5 (Cell Signaling; 13202S), Myc (Cell Signaling; 2276S), ATP5a (Abcam; ab14730), MFN2 (Cell signaling; 9482S), ERp57 (R&Dsystems; AF8219). Membranes were then incubated for 1 h with peroxidase-conjugated antimouse or antirabbit secondary antibodies. Immunoblots were visualized by chemiluminescence using the ChemiDoc Touch imaging system (Biorad, Irvine, CA, USA).

2.7. Confocal Microscopy Imaging

Cells seeded on 18-mm round glass coverslips, transfected as indicated were placed on the stage of the Olympus FV3000 confocal fluorescence microscope (Tokyo, Japan) and imaged using a 60X oil objective (UPLAN 60× oil, 1.35 NA, Olympus), and appropriate excitation laser and filters. For each experiment, 25 cells were randomly selected and analyzed. Stacks of 30 images separated by 0.2 μm along the Z axis were acquired. Three-dimensional reconstruction and volume rendering of the stacks were carried out with the appropriate plug-in of ImageJ (NIH, Bethesda, MD, USA).

3. Measurement of the Mitochondrial Membrane Potential, Mitochondrial Mass and ROS Levels

Mitochondrial membrane potential was examined using tetramethylrhodamin, methyl ester (TMRM, LifeTechnologies, Pleasanton, CA, USA). Briefly, cells expressing the different constructs were rinsed with PBS and incubated with 100 nM TMRM during 15 min at 37 °C. Cells coincubated with FCCP (6 μM during 15 min) had no TMRM fluorescence (data not shown).

Oxidative stress was evaluated using MitoSox™. Briefly, cells expressing the different constructs were incubated with 5 μM MitoSox™ during 45 min at 37 °C in 5% CO_2 and 95% humidity. Pretreatment of cells with 0.5 μM rotenone and 2.5 μM antimycin A during 15 min significantly increased MitoSox™ labeling (data not shown).

Mitochondrial mass was evaluated using Mitotracker Green™. Cells were incubated with 200 nM Mitotracker Green™ for 30 min at 37 °C.

TMRM, MitoSox™ and Mitotracker Green™ fluorescence were examined using the EVOS FL Auto 2 imaging system with a 40× objective (LPLAN 40×, 0.65NA, EVOS). For each independent experiment, fluorescence intensity was quantified in 25 cells using ImageJ (NIH, MD, USA).

4. Measurement of Oxygen Consumption

Oxygen consumption assays were performed using the high-resolution respirometry system Oroboros™ Oxygraph-2k (Innsbruck, Austria). Cell respiration was measured with 5×10^5 cells mL^{-1} at 37 °C in 2 mL chambers at a stirring rate of 750 rpm. Three different states of endogenous respiration with intact cells were measured: (i) basal respiration representing the endogenous physiological coupled state, (ii) respiration with oligomycin (2 μg.mL^{-1}) representing the non-coupled resting respiration, and (iii) maximal uncoupled respiration induced by FCCP (0.5 μM steps with 2.5 μM final concentration) providing the maximal respiration.

5. Proximity-Dependent Biotinylation and Streptavidin Bead Pulldown Assay

4×10^6 HEK cells expressing pcDNA or mtPKA-BioID2-HA were incubated with 0 or 50 μM biotin during 24 h. Cells were then rinsed twice with PBS and lysed in 500 μL of lysis buffer (50 mM Tris, pH 7.4, 500 mM NaCl, 0.4% SDS and 1 mM dithiothreitol) supplemented with 1% protease inhibitor cocktail, 2 mM sodium orthovanadate and 1 mM sodium fluoride. Lysed cells were sonicated at 60% amplitude three times for 2 s. An equal volume of ice-cold 50 mM Tris, pH 7.4 was then added before centrifugation at 15,000× g for 15 min. 20 μL of the supernatant was processed for SDS-PAGE to verify the efficiency of protein biotinylation. The remaining supernatant was incubated with 25 μL of streptavidin beads at 4 °C overnight under constant agitation. Beads were then washed twice with lysis buffer and centrifuged at 1500× g for 5 min. Supernatant was then removed and beads were resuspended in 100 μL of 50 μM ammonium bicarbonate. 20 μL of beads were processed for immunoblotting to verify the efficiency of protein biotinylation. Proteins on beads were rinsed three times with 50 mM ammonium bicarbonate buffer and kept at −80 °C prior to mass spectrometry analyses.

6. LC-MS/MS Analysis

6.1. Tryptic Digestion

Mass spectrometry was performed at the Proteomics platform of the CHU de Quebec Research Center (Quebec City, QC, Canada). Proteins on beads were suspended in 25 μL of 50 mM ammonium bicarbonate containing 1 μg trypsin and incubated overnight at 37 °C. Trypsin reaction was stopped by acidification with 3% acetonitrile, 1% trifluoroacetic acid, 0.5% acetic acid. Beads were then removed by centrifugation. Peptides were purified from supernatant on stage tip (C18) and vacuum-dried before MS injection. Samples were solubilized into 5 μL of 0.1% formic acid.

6.2. Peptide Separation and MS Detection

Peptide samples were separated by online reversed-phase (RP) nanoscale capillary liquid chromatography (nanoLC) and analyzed by electrospray mass spectrometry (ES MS/MS). The experiments were performed with an Ekspert NanoLC425 coupled to a 5600+ mass spectrometer (Sciex, Framingham, MA, USA) equipped with a nanoelectrospray ion source. Peptide separation took place on a self-packed picofrit column with reprosil 3u, 120A C18, 17 cm × 0.075 mm internal diameter, (Dr Maisch HPLC GmbH, Ammerbuch, Germany). Peptides were eluted with a linear gradient from 5–35% solvent B (acetonitrile, 0.1% formic acid) in 35 min, at 300 nL/min. Mass spectra were acquired using a data dependent acquisition mode using Analyst software version 1.7. Each full scan mass spectrum (400 to 1250 m/z) was followed by collision-induced dissociation of the twenty most intense ions. Dynamic exclusion was set for a period of 12 s and a tolerance of 100 ppm.

6.3. Database Searching

MGF peak list files were created using Protein Pilot version 4.5 software (Sciex, Framingham, MA, USA). MGF sample files were then analyzed using Mascot (Matrix Science, London, UK; version 2.5.1).

6.4. Criteria for Protein Identification

Scaffold (version Scaffold_4.7.1, Proteome Software Inc., Portland, OR, USA) was used to validate MS/MS based peptide and protein identification.

6.5. Coimmunoprecipitation

Whole cell extracts from HEK293 cells transiently transfected with the indicated plasmids were resuspended in lysis buffer (PathScan® Sandwich ELISA 1X, Cell Signaling #7018) supplemented with 1% protease inhibitor cocktail, 2 mM sodium orthovanadate and 1 mM sodium fluoride. Samples were then centrifuged at 12,500× g during 10 min (4 °C). Immunoprecipitation was performed by incubating 2 mg protein with anti-V5 tag antibody (Cell Signaling, D3H8Q Rabbit mAb #13202) under overnight agitation at 4 °C. Then, 20 μL of protein A/G plus-agarose (Santa Cruz, sc-2003) beads were added and incubated for 4 h at 4 °C. Beads were washed three times with lysis buffer and elution was performed with SDS-PAGE sample buffer for 5 min at 95 °C. Samples were then processed for Western blotting.

6.6. Cycloheximide Chase Assay

HeLa cells expressing different constructs were treated with cycloheximide (40 μg/mL) during 0, 10, 20, 30, 60 and 180 min. Cells were then harvested and processed for immunoblotting to analyze the stability of TIM44-V5 protein variants.

6.7. Statistical Analyses

Data are presented as mean ± s.e.m. Statistical analyses were performed using GraphPad Prism v. 8.0 (GraphPad, San Diego, CA, USA). Data were analyzed using one-way or two-way ANOVA, as appropriate. $p < 0.05$ was considered statistically different.

7. Results

7.1. Functional Impact of mt-PKA on Mitochondria

PKA influences mitochondrial functions through phosphorylation of several targets [49–51]. PKA signaling has distinct targets and biological effects inside and outside mitochondria [52]. The role of PKA localized inside mitochondria remains, however, poorly understood. We first examined endogenous expression and subcellular localization of PKA in HeLa cells. Previous works have shown that PKA subunits are present in multiple cellular compartments, including mitochondria [16,53,54]. Similarly, our findings showed that PKA-RI-α/β and PKA-C-α subunits are present in both cytosolic and mitochondria-enriched fractions (Figure 1A). To examine the localization of PKA subunits among the mitochondrial subcompartments, we then performed trypsin sensitivity on mitochondria-enriched fractions. Treatment of mitochondria with trypsin led to strong degradation of the RI-α/β subunits similar to the OMM protein TOM20 (Figure 1B). In contrast, the PKA-C-α subunits resisted to trypsin proteolysis similar to the IMM/matrix protein SDHa (Figure 1B). These results thus suggest that a significant pool of PKA-C-α resides inside mitochondria of HeLa cells.

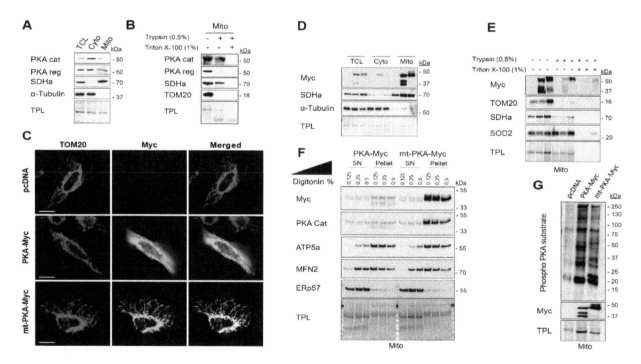

Figure 1. Protein kinase A (PKA) catalytic subunit α is localized within mitochondria of HeLa cells. (**A**) Representative immunoblots ($n = 3$) of PKA catalytic (cat) subunit, PKA regulatory (reg) subunit, the mitochondrial protein SDHa and of the cytosolic protein α-Tubulin with corresponding total protein load (TPL) in total cell lysate (TCL), cytosolic (Cyto) and mitochondrial (Mito) fractions isolated from HeLa cells. (**B**) Representative immunoblots ($n = 3$) of PKA cat, PKA reg, SDHa and TOM20 with corresponding total protein load (TPL) in mitochondrial fractions isolated from HeLa cells and treated as indicated. (**C**) Representative micrographs ($n = 6$) of HeLa cells expressing empty vector (pcDNA), PKA-Myc or mt-PKA-Myc and labelled with anti-Myc (green) and anti-TOM20 (red) as a mitochondrial marker. Scale bars = 5 μm. (**D**) Representative immunoblots ($n = 3$) of Myc, SDHa and α-Tubulin with

corresponding total protein load (TPL) in TCL, cyto and mito fractions isolated from HeLa cells expressing empty vector (pcDNA), PKA-Myc or mt-PKA-Myc. (**E**) Representative immunoblots ($n = 3$) of Myc, TOM20, SDHa and SOD2 with corresponding total protein load (TPL) in mitochondrial enriched fractions isolated from HeLa cells expressing empty vector (pcDNA), PKA-Myc or mt-PKA-Myc and treated as indicated. (**F**) Representative immunoblots ($n = 3$) of Myc, PKA Cat, ATP5a, MFN2 and ERp57 with corresponding TPL in pellet and supernatant (SN) obtained after treatment of mitochondria isolated from HeLa cells expressing PKA-Myc or mt-PKA-Myc and treated as indicated. (**G**) Representative immunoblots ($n = 5$) of phospho-PKA substrates and Myc with corresponding total protein load (TPL) in mitochondrial enriched fractions isolated from HeLa cells expressing empty vector (pcDNA), PKA-Myc or mt-PKA-Myc.

In order to decipher the specific role of PKA catalytic subunit α localized inside mitochondria, we generated two constructs: (i) Myc tagged PKA-C-α (named hereafter PKA-Myc) and (ii) Myc tagged PKA-C-α fused to the mitochondrial leading sequence of COXVIIIa, a protein localized at the interface of the IMM and the matrix [55] (named hereafter mt-PKA-Myc). Immunodetection of Myc and TOM20 by confocal microscopy showed that PKA-Myc spreads throughout the cell whereas mt-PKA-Myc is specifically targeted to mitochondria (Figure 1C). Subcellular fractionation confirmed that both constructs express PKA in specific subcellular compartments (Figure 1D). However, the resolution of conventional confocal microscopes is not sufficient to analyze the submitochondrial distribution of proteins [56]. We thus performed trypsin sensitivity assays on mitochondria-enriched fractions of HeLa cells expressing the different constructs. Similar to TOM20, PKA-Myc was mostly degraded by trypsin, whereas mt-PKA-Myc was only partly degraded by trypsin, similar to SDHa and SOD2 (Figure 1E). Similar findings were obtained when mitochondria-enriched fractions were solubilized by digitonin. As shown in Figure 1F, the OMM protein MFN2 and the IMM protein ATP5a were dose-dependently released in the supernatant after treatment with digitonin. Immunoblotting of Myc in the same conditions further suggest that mt-PKA-Myc is mostly recruited at the mitochondrial matrix and/or IMM (Figure 1F). Considering that mitochondria-enriched fractions are often contaminated by the endoplasmic reticulum (ER), we also examined the release of the ER protein ERp57 after treatment with digitonin. The obtained findings further suggest that mt-PKA is not localized in ER (Figure 1F). Using immunoblotting, we confirmed that expression of both PKA constructs increased phosphorylation of serine and threonine on RRXS/T motifs of mitochondrial proteins in HeLa cells (Figure 1G). Overall, these findings indicate that the mt-PKA-Myc construct is an appropriate tool to characterize the role of PKA-C-α localized at the interface of matrix and IMM.

In order to evaluate the functional impact of mt-PKA, HeLa cells were transfected with pcDNA, PKA-Myc or mt-PKA-Myc and labeled with different probes to examine mitochondrial physiology. Live imaging of cells stained with MitoTracker Green™ and TMRM indicated no difference of mitochondrial mass and membrane potential among cells expressing pcDNA, PKA-Myc or mt-PKA-Myc (Figure 2A,B). MitoSOX™ fluorescence indicated higher levels of superoxide in cells expressing mt-PKA-Myc (Figure 2A,B). Overexpression of mt-PKA-Myc significantly decreased basal and uncoupled respiration rates (Figure 2C). PKA-Myc had a similar impact on respiration although only uncoupled respiration was significantly reduced by this construct (Figure 2C). Overall, these findings suggest that the overexpression of intramitochondrial PKA-C-α alters mitochondrial functions.

7.2. Identification of Potential Substrates of mt-PKA

In order to identify the proteins involved in the regulation of mitochondrial functions by mt-PKA-Myc, we used proximity-dependent biotin identification (BioID) [44]. We first fused mt-PKA to HA tagged promiscuous biotin ligase BioID2. Immunofluorescence confirmed the mitochondrial localization of mt-PKA-BioID2-HA since HA colocalized with the mitochondrial protein TOM20 (Figure 3A). Addition of biotin (50 μM) to the culture medium of cells expressing mt-PKA-BioID2-HA promotes biotinylation of proteins within a ~10 nm radius of the BioID2 (Figure 3B) [43]. Immunoblotting confirmed that biotin treatment triggered protein biotinylation in cells expressing mt-PKA-BioID2-HA (Figure 3C). Biotinylated proteins were then pulled down using streptavidin-coated beads. Biotinylated-purified proteins from cell expressing pcDNA or mt-PKA-BioID2-HA and treated with biotin were submitted to on-bead tryptic digestion and LC/MS-MS analysis. A total of 1898 proteins were first identified including 190 mitochondrial proteins (Table S1). Filters were then applied to increase robustness of MS data and optimize the identification of mitochondrial proteins most likely interacting with PKA. We first selected the 537 proteins for which a minimum of five peptides was detected by LC/MS-MS in at least one sample, using protein and peptide FDR thresholds of 95% (Table S2). To minimize the selection and identification of proteins biotinylated independently of mt-PKA-BioID2-HA, we then selected proteins with a minimum difference of five peptides between cells expressing mt-PKA-BioID2-HA and pcDNA. We thus selected 208 proteins following this criterion (Table S3), including 33 mitochondrial proteins (Table S4). It is possible that the mt-PKA-BioID2-HA biotinylated proteins within a radius of 10 nm not directly interacting with mt-PKA. We thus performed in silico analysis to identify proteins with phospho-PKA binding motifs and potentially phosphorylated by mt-PKA. PKA phosphorylation of NDUFS4-S173 was used as a positive control (Table S5) [37].

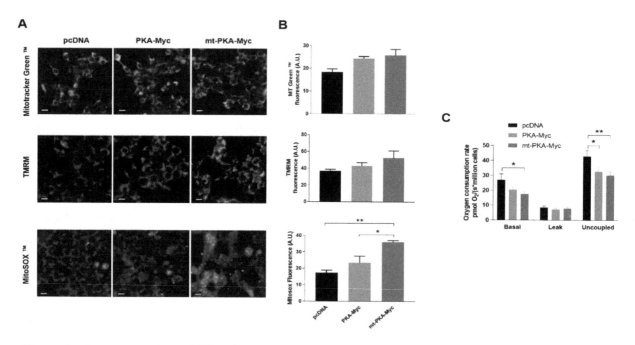

Figure 2. Overexpression of PKA alters mitochondrial activity. (**A**) Representative micrographs of HeLa cells expressing empty vector (pcDNA), PKA-Myc or mt-PKA-Myc and labeled with Mitotracker Green™ (MT Green), TMRM and MitoSOX™. Scale bars = 10 μm. (**B**) Quantification of Mitotracker Green™, TMRM and MitoSOX™ fluorescence intensity ($n = 3$, with 25 random cells per experiment). A.U.: Arbitrary Units. (**C**) Oxygen consumption rates of HeLa cells expressing empty vector (pcDNA), PKA-Myc or mt-PKA-Myc ($n = 3$). Data are presented as mean ± s.e.m. analyzed by one-way ANOVA followed by Tukey's test (* $p < 0.05$), (** $p < 0.01$).

Using the phospho-site prediction software pkaPS (http://mendel.imp.univie.ac.at/sat/pkaPS) [57], NetPhos 3.1 (http://www.cbs.dtu.dk/services/NetPhos/) [58] and GPS 3.0 (http://gps.biocuckoo.org/online.php) [59], we identified 21 mitochondrial proteins for which at least one phospho-site was predicted simultaneously by the three software (Table 1 and Table S5).

Overall, our approach identified 21 mitochondrial proteins potentially targeted by mt-PKA, involved in various molecular functions ranging from metabolite exchange, protein synthesis and degradation, to tricarboxylic (TCA) cycle, mitochondrial dynamics, protein import and cell survival (Table 1).

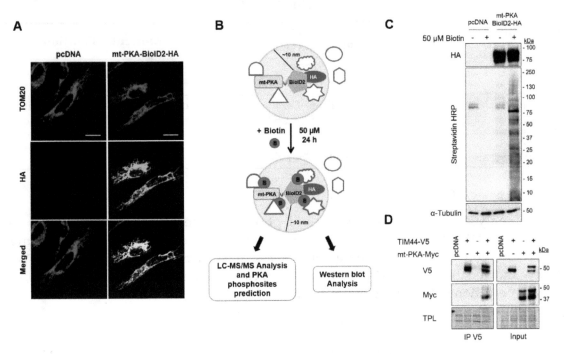

Figure 3. Biotinylation-based proximity labeling of potential interactors of mt-PKA. (**A**) Representative (*n* = 5) micrographs of HeLa cells expressing empty vector (pcDNA) or mt-PKA-BioID2-HA and labeled with HA (green) and TOM20 (red) as a mitochondrial marker. Scale bars = 10 μm. (**B**) Schematic representation of the biotinylation-based proximity labeling combined with mass spectrometry analysis showing that mt-PKA-BioID2-HA add biotin to proteins located within a radius of 10 nm. (**C**) Representative immunoblots (*n* = 5) of HA, Streptavidin coupled to horseradish peroxidase (HRP) and α-Tubulin in total cell lysates of HEK cells expressing empty vector (pcDNA) or mt-PKA-BioID2-HA and treated as indicated. (**D**) Representative immunoblots (*n* = 4) of V5 and Myc after immunoprecipitation (IP) of V5 from HEK cells expressing pcDNA, TIM44-V5 and/or mt-PKA-Myc. (TPL: total protein load).

Table 1. Potential mitochondrial targets of PKA located inside mitochondria. Twenty-one potential interactors were identified by BioID2 coupled to LC-MS/MS.

Protein Name	Accession Number	Predicted Phosphoresidue	Molecular Function
AKAP1	Q92667	S107, S191, S235, S577	Kinase anchoring
DRP1	G8JLD5	T79, S637	Mitochondrial shape
TOM70	O94826	S81, S375	Protein import
TIM44	O43615	S80	
Hsp60	P10809	S159	Protein folding and stress response
TRAP1	Q12931	S180	
TCP-1-eta	Q99832	T526	
CLPX	O76031	S220	Protein degradation
IleRS	Q9NSE4	S818	tRNA maturation and mRNA metabolism
TRMT10C	Q7L0Y3	T382	
LRP130	P42704	S1022, S1302, S1393	
RNS4I	P61221	S560	
Citrin	Q9UJS0	S469	Metabolites exchange
P5CR1	P32322	S109	Response to oxidative stress
SHMT2	P34897	S184	

Table 1. *Cont.*

Protein Name	Accession Number	Predicted Phosphoresidue	Molecular Function
GDH1	P00367	T206	TCA cycle regulation
ACL	P53396	S455	
ATAD3A	Q9NVI7	T166, T221	Mitochondrial protein synthesis and maintenance of cristae junctions
MIC19	C9JRZ6	T11, S29, S50	
HAX-1	O00165	S53	Cell survival
MAVS	Q7Z434	S100, S238, S373	

7.3. PKA Interacts with TIM44

In silico analyses suggest that PKA could phosphorylate TIM44-S80 (Table 1). To confirm the interaction of TIM44 with mt-PKA, we generated V5 tagged TIM44 constructs. Interestingly, immunoprecipitation of V5 from whole cell lysates of HEK cells expressing TIM44-WT-V5 and mt-PKA-Myc showed that mt-PKA-Myc coimmunoprecipitated with TIM44-V5 (Figure 3D), confirming the physical interaction between the two proteins.

7.4. Impact of TIM44-S80 Phosphomutants on Mitochondria

The role of TIM44-S80 phosphorylation in mitochondrial physiology was then examined using phosphomimetic TIM44-S80D-V5 and phosphodeficient TIM44-S80A-V5 mutants. Immunoblotting revealed similar levels of TIM44-V5 variants, suggesting that TIM44-S80 phosphomutations do not impact on TIM44 level (Figure 4A). Cycloheximide treatment followed by Western blotting showed similar degradation rates of TIM44-V5 and phosphomutants (Figure 4B), indicating that the stability of TIM44-V5 is not affected by TIM44-S80 phosphomutations. Similarly, TIM44-S80 variants were all correctly imported inside mitochondria since they colocalized with the mitochondrial protein ATP5B (Figure 4C). Altogether, these findings suggest that the potential phosphorylation of TIM44-S80 would not alter TIM44 homeostasis.

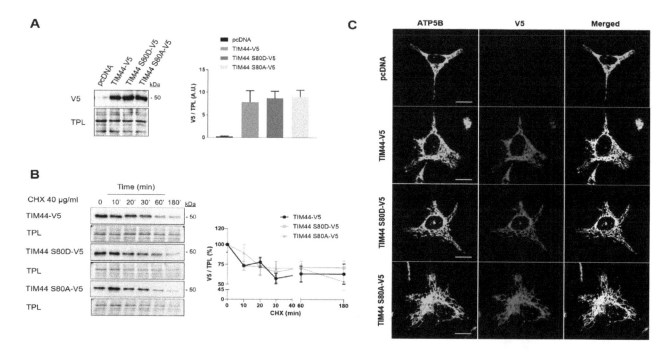

Figure 4. Impact of TIM44-S80 phosphomutations on the stability and localization of TIM44-V5. (**A**) Left, representative immunoblots (*n* = 5) of V5 in total cell lysates of HeLa cells expressing empty

vector (pcDNA), TIM44-V5, TIM44 S80D-V5 or TIM44 S80A-V5. Right, quantification of V5 levels normalized by total protein load (TPL) shown on left (A.U.: arbitrary unit). (**B**) Left, representative immunoblots ($n = 5$) and right, corresponding quantification of normalized V5 levels to total protein load (TPL) in total cell lysates of HeLa cells expressing empty vector (pcDNA), TIM44-V5, TIM44 S80D-V5 or TIM44 S80A-V5 and treated with cycloheximide (CHX) as indicated. (**C**) Representative micrographs ($n = 5$) of HeLa cells expressing empty vector (pcDNA), TIM44-V5, TIM44 S80D-V5 or TIM44 S80A-V5 and labeled with V5 (red) and ATP5B (green) as a mitochondrial marker. Scale bars = 10 µm.

In order to assess the impact of TIM44-S80 phosphomutations on mitochondria, several proteins located in different mitochondrial compartments and involved in different mitochondrial functions were examined in HeLa cells expressing pcDNA or TIM44 variants. Immunoblottings showed that the level of these proteins remained unchanged among TIM44 mutants, suggesting that the phosphomutations of TIM44-S80 do not impact on mitochondrial content (Figure 5A and Figure S1A). Next, we examined the impact of these phosphomutations on cellular respiration and viability. We observed that the expression of the phosphodeficient TIM44-S80A-V5 mutant significantly decreased basal and uncoupled respiration rates (Figure 5B). These results suggest that phosphorylation of TIM44-S80 is needed to maintain optimal respiration. Trypan blue exclusion assay revealed that phosphomutants of TIM44-S80 did not affect cellular viability (Figure 5C). Overall, our findings suggest that the phosphorylation status of TIM44-S80 could impact on OXPHOS.

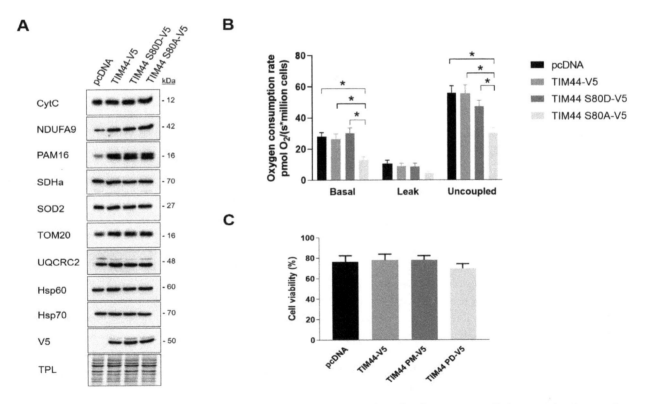

Figure 5. Impact of TIM44-S80 phosphomutants on mitochondrial content, cellular respiration and viability. (**A**) Representative immunoblots ($n = 4$) of mitochondrial proteins CytC, NDUFA9, PAM16, SDHa, SOD2, TOM20, UQCRC2, Hsp60 and Hsp70 in total cell lysates of HeLa cells expressing empty vector (pcDNA), TIM44-V5, TIM44 S80D-V5 or TIM44 S80A-V5. See Figure S1A for quantifications. (**B**) Basal, leak and uncoupled cellular respiration ($n = 6$) of HeLa cells expressing empty vector (pcDNA), TIM44-V5, TIM44 S80D-V5 or TIM44 S80A-V5. Data were analyzed by two-way ANOVA followed by Tukey's post hoc test (* $p < 0.05$). (**C**) Cell viability ($n = 6$) determined by trypan blue dye exclusion assay of HeLa cells expressing empty vector (pcDNA), TIM44-V5, TIM44 S80D-V5 or TIM44 S80A-V5.

We then examined the impact of TIM44-S80 phosphomutations on the functional relationship linking TIM44 to PKA. Immunoblottings from BN-PAGE first revealed similar levels of native complexes containing TIM44-V5 and the catalytic subunit α of PKA (Figure 6A and Figure S1B). Similarly, immunoblottings from SDS-PAGE showed no change in levels of the catalytic subunit α of PKA (Figure 6B) and of PKA activity (Figure S1C) among cells expressing the different TIM44-V5 constructs. However, TIM44-V5 levels globally decreased in presence of mt-PKA-Myc (Figure 6C and Figure S1D). Mt-PKA-Myc coimmunoprecipitated at higher levels with the phosphodeficient TIM44-S80A-V5 (Figure 6C). These results demonstrate that serine to alanine substitution at position 80 strengthens the interaction of TIM44-V5 to mt-PKA-Myc, suggesting that TIM44-S80 is involved in the physical interaction between the protein and PKA.

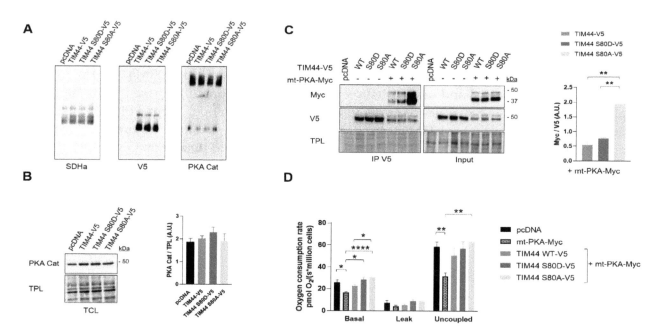

Figure 6. Functional interaction between TIM44-V5 phosphomutants and mt-PKA. (**A**) Representative BN-PAGE immunoblots ($n = 4$) of SDHa, V5 and PKA catalytic (Cat) subunit in TCLs of HeLa cells expressing empty vector (pcDNA), TIM44-V5, TIM44 S80D-V5 or TIM44 S80A-V5. See Figure S1B for quantifications. (**B**) Left, representative immunoblots ($n = 4$) of PKA catalytic (Cat) subunit levels and (right) corresponding quantifications on total protein loads (TPL) in total cell lysates of HeLa cells expressing empty vector (pcDNA), TIM44-V5, TIM44 S80D-V5 or TIM44 S80A-V5. (**C**) Left, representative immunoblots ($n = 5$) and (right) quantification of coimmunoprecipitated V5 with Myc from total cell lysates of HEK cells expressing pcDNA, TIM44 phosphomutants and/or mt-PKA-Myc. TPL: total protein load. Data were analyzed by one-way ANOVA followed Tukey's post hoc test (** $p < 0.01$). (**D**) Basal, leak and uncoupled cellular respiration ($n = 5$) of HeLa cells expressing empty vector (pcDNA), mt-PKA-Myc, or co-expressing TIM44-V5, TIM44 S80D-V5 and TIM44 S80A-V5 with mt-PKA-Myc. Data were analyzed by two-way ANOVA test followed by Tukey's post hoc test (* $p < 0.05$, ** $p < 0.01$, **** $p < 0.0001$).

Interestingly, we observed that expression of TIM44-WT only partly and non-significantly rescued the mtPKA-dependent decrease of cellular respiration (Figure 6D). Also, both phosphomutants rescued, at least partly, the decrease of basal and uncoupled respiration induced by the mt-PKA overexpression (Figure 6D) without any effect on cell viability (Figure S1E). These findings suggest that mt-PKA could affect respiration via TIM44. It is however difficult to link these effects to the potential phosphorylation of TIM44-S80 considering that both phosphomutants reversed the alteration of respiration induced by the overexpression of mt-PKA.

8. Discussion

Although it is still debated, evidence suggest that PKA is localized within mitochondrial matrix where it can directly target proteins and modulate the activity of the organelle [38,39]. The aim of the present work was to characterize the interactome of PKA specifically targeted to the mitochondrial matrix using BioID. Our findings indicate that the overexpression of mt-PKA decreases cellular respiration and increases superoxide levels. Twenty-one mitochondrial proteins were then identified as potential substrates of mt-PKA using BioID, LC-MS/MS and in silico phospho-site prediction. Using coimmunoprecipitation, we confirmed the interaction of mt-PKA-Myc with TIM44-V5. The phosphodeficient mutant of TIM44-S80 negatively impacted cellular respiration and strengthened the interaction between mt-PKA and TIM44 constructs. Future studies will be needed to understand whether potential phosphorylation of TIM44-S80 affects the functions of the protein and its capacity to import proteins across the IMM.

PKA-C-α and PKA RI-α/β subunits were detected in mitochondria-enriched fractions of HeLa cells, although only the PKA-C-α subunit seems localized inside the organelle. These results are consistent with previous findings demonstrating the localization of AKAP121, PKA-RII and PKA-C at the IMM/matrix in both isolated mitochondria and cardiomyocytes [16]. Thus, we can speculate that the PKA-C-α subunits are coupled with PKA-RII and not PKA-RI subunits in mitochondria of Hela cells. Future studies should provide insights into the specific composition and distribution of PKA holoenzymes among the mitochondrial subcompartments of different cell types.

The present work shows that the overexpression of mt-PKA decreased OXPHOS and increased mitochondrial ROS levels (Figure 2A,B). Several studies previously showed that PKA impacts on mitochondrial activity via direct phosphorylation of OXPHOS components. Notably, PKA regulates OXPHOS via phosphorylation of NDUFS4, which is important to allow import of this protein inside the organelle and assembly of complex I [37]. PKA-mediated phosphorylation of NDUFA1-S55 is also necessary for appropriate complex I assembly [60,61]. In the presence of a high ATP/ADP ratio, PKA phosphorylates subunits of complex IV subunits and promotes its inhibition by ATP, consequently allowing a decrease in membrane potential and an increase in ROS production [34,62]. Mitochondrial PKA modulates complex V activity through phosphorylation of its inhibitor AIF1 [35]. These findings illustrate that PKA can modulate metabolism via direct phosphorylation of multiple OXPHOS components, although it is not clear whether these proteins are phosphorylated directly inside the organelle. The present work did not however identify proteins involved in OXPHOS as potential PKA substrates, suggesting that PKA can also modulate mitochondrial activity upstream of OXPHOS. For instance, GDH1 and ACL which are involved in the regulation of TCA cycle were targeted by mt-BioID2-HA. Deletion of the yeast PKA catalytic subunit Tpk3p reduces respiratory activity [63], whereas increased Tpk3p activity is sufficient to induce formation of dysfunctional mitochondria with striking morphological abnormalities that produce high levels of ROS [49]. These mitochondrial dysfunctions are induced by transcriptional changes that inhibit mitochondrial biogenesis, alter the electron transport system and inhibit stress response mechanisms [49]. Interestingly, excess ROS induce the sequestration of the PKA-C-α subunit into the matrix, leads to the hyper-phosphorylation of complex IV subunits I, IVi1, and Vb and reduces the activity of complex IV under hypoxia/ischemia [64,65]. Since these proteins were not identified here, the alterations of ROS and cellular respiration induced by the overexpression of mt-PKA observed in the present work were likely linked to different mechanisms. Overall, our findings suggest that the overexpression of mt-PKA impacted metabolism not via phosphorylation of OXPHOS components.

Using BioID, LC-MS/MS and in silico phospho-site prediction, we identified 21 potential substrates of mt-PKA. These proteins are involved in various processes such as import machinery, stress response, mitochondrial dynamics, TCA cycle, protein synthesis and degradation as well cell survival (Table 1). Several identified potential targets are involved in mitochondrial protein turnover (such as CLPX, IleRS, TRMT10C, LRP130, RNS4I, ATAD3A) and stress response (such as Hsp60, TRAP1, TCP-1-eta, P5CR1, SHMT2). For instance, P5CR1 has been reported to antagonize oxidative insults and promote cell survival [66]. SHMT2 is required for the assembly of complex I and the maintenance of mitochondrial respiration [67]. Expression of SHMT2 is altered by OXPHOS dysfunction [68] and is involved in protective response to mitochondrial toxicity [69]. Proteins involved in the mitochondrial unfolded stress response (UPR^{mt}) were also identified as potential targets of mt-PKA, including the chaperones Hsp60 and TRAP1 [70] as well as the protease CLPX [71]. This stress response signals from perturbed mitochondria to increase the expression of genes encoding mitochondrial proteins involved in quality control [72,73]. UPR^{mt} protects cells from diverse mitochondrial stresses, including OXPHOS dysfunction and mitochondrial protein misfolding [74]. Overall, the potential substrates of mt-PKA identified here are involved in multiple steps of the mitochondrial proteins stress-responses and could be linked to the alterations of mitochondrial metabolism induced by the overexpression of mt-PKA.

Two-third of the potential targets of mt-PKA identified in this study (AKAP1, TOM70, TIM44, TRAP1, CLPX, TRMT10C, LRP130, SHMT2, ACL, ATAD3A, MIC19, HAX-1, MAVS) overlap with interactors of the mitochondrial prohibitin (PHB) complex. PHBs are IMM proteins that form ringlike structures composed of multiple PHB1 and PHB2 subunits. The PHB complex functions as a membrane scaffold important for the recruitment and stability of proteins within mitochondria [75]. For instance, PHBs control OPA-1 processing to regulate mitochondrial fusion [76]. They have also been involved in the organization and stability of mitochondrial nucleoids via their interaction with TFAM, mtSSB [77,78], and mtDNA-binding proteins such as ATAD3 [79]. Interestingly, deficiency in PHBs alters the assembly of (super)complexes of the electron transport system [80–82] and increases ROS production [83,84], suggesting that the metabolic alterations induced by the overexpression of mt-PKA observed here could be linked to PHB complexes. Given the similar impact of both PKA and PHB complex on mitochondrial dynamics, cristae structure and OXPHOS, it would be interesting to examine whether mt-PKA affects PHBs phosphorylation and functions. Such works would clearly improve our understanding of how PHB interactors are involved in mt-PKA signaling and mitochondrial physiology.

Our in silico analyses identified TIM44-S80 as a potential target of PKA. Interestingly, overexpression of TIM44 alone partly rescued the alteration of respiration by mt-PKA, which could indicate that mt-PKA affects OXPHOS via TIM44. However, both the phosphomimetic and the phosphodeficient of TIM44-S80 reversed the mt-PKA-dependent decrease of respiration, suggesting that phosphorylation of TIM-S80 is not involved in the effect of mt-PKA. It is nevertheless possible that the substitution of serine for aspartic acid was not optimal to mimic the phosphorylation of S80 in terms of charge and protein conformation. A better characterization of the TIM44-S80 phosphomimetic should be important to address the role of PKA-TIM44 signaling in future studies.

TIM44 is a peripheral IMM protein interacting with the matrix face of the TIM17/23 translocon [85]. TIM44 recruits different motor proteins, including the mitochondrial Hsp70 (mtHsp70) and Pam16 to the translocon to allow proper import of proteins in the mitochondrial matrix [86]. TIM44-S80 is located at the N-terminal domain which is known to bind TIM17/23, mtHsp70 and Pam16 [86,87] and serves as a dynamic arm to drive efficient translocation of matrix proteins [88], suggesting that phosphorylation of TIM44-S80 by PKA could induce conformational changes of the scaffold TIM44 and impact on import of matrix proteins. We did not observe alterations in the levels of the matrix protein SOD2 or the TIM44 interactors Hsp70 and Pam16 upon expression of TIM44-S80 phosphomutants. Although these results suggest that phosphorylation of TIM44-S80 do not impact on the stability of matrix proteins, it will be important in future studies to examine the impact of (de)phosphorylation of TIM44-S80 on its interaction with the motor proteins mtHsp70 and Pam16 and in the regulation of mitochondrial protein translocation across the IMM.

Among the targets of PKA previously identified, only AKAP1 and MIC19 were confirmed by our work. AKAP1 anchors PKA to mitochondrial membranes [16,23], whereas PKA-mediated phosphorylation of MIC19 negatively regulates mitophagy by impairing Parkin recruitment to damaged mitochondria [89]. Future studies should examine whether such mechanisms could have been involved in the PKA-mediated alterations of ROS levels and cellular respiration observed here. Numerous other known PKA substrates, as described above, were however not identified by our approach. The criteria that we used to filter MS data could have led to elimination of PKA substrates previously identified. For instance, AIF was detected by LC-MS/MS but was discarded since there was not a difference of five detected peptides between cells expressing pcDNA and mt-PKA-BioID2-HA. Several other components of OXPHOS were detected (Tables S1–S5) but were also discarded according to our filters. It is thus possible that these proteins could be involved in the mt-PKA-mediated effects on mitochondrial activity.

Our BioID approach does not discard the possibility that the mt-PKA-BioID construct interacted with the identified interactors, such as TIM44-S80, outside of the mitochondrial matrix. For instance, the OMM proteins Drp1 and TOM70 were identified as potential targets of PKA. PKA phosphorylates Drp1-S637 and induce detachment of Drp1 from OMM and inhibition of mitochondrial fission [90]. TOM70 is part of the translocase of the outer membrane complex and recognizes preproteins with hydrophobic domains destined to the IMM [91,92]. Moreover, several non-mitochondrial proteins were detected by our BioID assay (Table S4). In fact, the mt-PKA-BioID2 constructs targeted 208 proteins, among which 33 are mitochondrial proteins (Tables S3 and S4). These findings suggest that our BioID2 construct likely interacted with several extra-mitochondrial proteins before its complete translocation within the organelle. The interactomes of the different mitochondrial subcompartments were recently characterized using BioID2. Similar to our findings, this work identified a total of 1465 proteins, among which only 528 were mitochondrial proteins [93]. Considering that BioID2 is constitutively active and that cells are treated with biotin during 24 h, it is thus possible that BioID2 constructs identify interactors from the site of synthesis to their final destination, as previously discussed [94,95]. Recent development in proximity biotinylation techniques could help to circumvent these biases. For instance, the TurboID enables shorter biotin treatment [96] whereas 2c-BioID allows to keep the protein of interest separated from the biotin ligase until the biotin treatment [94].

9. Conclusions

Taken together, our results demonstrate that the overexpression of mt-PKA decreases mitochondrial respiration and enhances ROS levels. We also identified various mitochondrial proteins involved in metabolism and stress response as potential targets of mt-PKA. It will be important in future studies to examine whether the 21 targets of the mt-PKA-BioID2 construct are phosphorylated by endogenous PKA. It will also be important to verify whether intramitochondrial phosphatases, such as Pptc7 [97], PGAM5 [98,99], PP2B [33,100] and PPM1K [101], are able to counteract PKA-mediated phosphorylation of these proteins. These works will provide insights about potential new signaling pathways triggered and regulated inside mitochondria.

Supplementary Materials:
Figure S1: Supplementary quantification and data, Table S1: Total proteins identified by BioID and LC-MS/MS, Table S2: Proteins identified with a minimum of 5 peptides, a protein false discovery rate (FDR) threshold of 1% and a peptide FDR threshold of 95%, Table S3: Proteins identified with a minimum difference of five peptides between cells expressing pcDNA and mt-PKA-BioID2-HA, Table S4: Mitochondrial proteins identified with a minimum difference of five peptides between cells expressing pcDNA and mt-PKA-BioID2-HA, Table S5: In silico phosphoprediction analysis on mitochondrial proteins interacting with mt-PKA-BioID2-HA.

Author Contributions: Formal analysis, Y.O.A.; funding acquisition, E.H.-C.; investigation, Y.O.A.; methodology, Y.O.A.; project administration, E.H.-C.; supervision, E.H.-C.; writing—original draft, Y.O.A.; writing—review and editing, Y.O.A. and E.H.-C. All authors have read and agreed to the published version of the manuscript.

Acknowledgments: LC-MS/MS was performed by the Proteomics Platform of the CHU de Quebec research center, Quebec City, Quebec, Canada. The pcDNA3-TIM44-V5 construct was kindly provided by Elena Bonora (University of Bologna, Bologna, Italy).

References

1. Mitchell, P.J. Coupling of Phosphorylation to Electron and Hydrogen Transfer by a Chemi-Osmotic type of Mechanism. *Nat. Cell Biol.* **1961**, *191*, 144–148. [CrossRef]
2. Mitchell, P.; Moyle, J. Chemiosmotic Hypothesis of Oxidative Phosphorylation. *Nat. Cell Biol.* **1967**, *213*, 137–139. [CrossRef]
3. Duchen, M.R. Mitochondria and calcium: From cell signalling to cell death. *J. Physiol.* **2000**, *529*, 57–68. [CrossRef]
4. Padrão, A.I.; Vitorino, R.; Duarte, J.A.; Ferreira, R.; Lemos-Amado, F. Unraveling the Phosphoproteome Dynamics in Mammal Mitochondria from a Network Perspective. *J. Proteome Res.* **2013**, *12*, 4257–4267. [CrossRef]
5. Kruse, R.; Højlund, K. Mitochondrial phosphoproteomics of mammalian tissues. *Mitochondrion* **2017**, *33*, 45–57. [CrossRef]
6. Lucero, M.; Suarez, A.E.; Chambers, J.W. Phosphoregulation on mitochondria: Integration of cell and organelle responses. *CNS Neurosci. Ther.* **2019**, *25*, 837–858. [CrossRef] [PubMed]
7. Fischer, F.; Hamann, A.; Osiewacz, H.D. Mitochondrial quality control: An integrated network of pathways. *Trends Biochem. Sci.* **2012**, *37*, 284–292. [CrossRef]
8. Held, N.M.; Houtkooper, R.H. Mitochondrial quality control pathways as determinants of metabolic health. *BioEssays* **2015**, *37*, 867–876. [CrossRef]
9. Lim, S.; Smith, K.R.; Lim, S.-T.S.; Tian, R.; Lu, J.; Tan, M. Regulation of mitochondrial functions by protein phosphorylation and dephosphorylation. *Cell Biosci.* **2016**, *6*, 1–15. [CrossRef]
10. Mahapatra, G.; Varughese, A.; Ji, Q.; Lee, I.; Liu, J.; Vaishnav, A.; Sinkler, C.; Kapralov, A.; Moraes, C.T.; Sanderson, T.H.; et al. Phosphorylation of Cytochrome c Threonine 28 Regulates Electron Transport Chain Activity in Kidney: Implications For AMP Kinase. *J. Biol. Chem.* **2017**, *292*, 64–79. [CrossRef]
11. Yeaman, S.J.; Hutcheson, E.T.; Roche, T.E.; Pettit, F.H.; Brown, J.R.; Reed, L.J.; Watson, D.C.; Dixon, G.H. Sites of phosphorylation on pyruvate dehydrogenase from bovine kidney and heart. *Biochemistry* **1978**, *17*, 2364–2370. [CrossRef]
12. Ledee, L.; Kang, M.A.; Kajimoto, M.; Purvine, S.O.; Brewer, H.; Pasa-Tolic, L.; Portman, M.A. Quantitative cardiac phosphoproteomics profiling during ischemia-reperfusion in an immature swine model. *Am. J. Physiol. Circ. Physiol.* **2017**, *313*, H125–H137. [CrossRef]
13. Bjorn, S.S. Specificity in the cAMP/PKA signaling pathway. differential expression, regulation, and subcellular localization of subunits of PKA. *Front. Biosci.* **2000**, *5*, D678–D693. [CrossRef]
14. Zhang, P.; Smith-Nguyen, E.V.; Keshwani, M.M.; Deal, M.S.; Kornev, A.P.; Taylor, S.S. Structure and Allostery of the PKA RII Tetrameric Holoenzyme. *Science* **2012**, *335*, 712–716. [CrossRef]
15. E Lehnart, S.; Marks, A.R. Phosphodiesterase 4D and heart failure: A cautionary tale. *Expert Opin. Ther. Targets* **2006**, *10*, 677–688. [CrossRef]
16. Sardanelli, A.M.; Signorile, A.; Nuzzi, R.; De Rasmo, D.; Technikova-Dobrova, Z.; Drahota, Z.; Occhiello, A.; Pica, A.; Papa, S. Occurrence of A-kinase anchor protein and associated cAMP-dependent protein kinase in the inner compartment of mammalian mitochondria. *FEBS Lett.* **2006**, *580*, 5690–5696. [CrossRef] [PubMed]
17. Papa, S.; Sardanelli, A.M.; Cocco, T.; Speranza, F.; Scacco, S.C.; Technikova-Dobrova, Z. The nuclear-encoded 18 kDa (IP) AQDQ subunit of bovine heart complex I is phosphorylated by the mitochondrial cAMP-dependent protein kinase. *FEBS Lett.* **1996**, *379*, 299–301. [CrossRef]
18. Sardanelli, A.M.; Technikova-Dobrova, Z.; Scacco, S.; Speranza, F.; Papa, S. Characterization of proteins phosphorylated by the cAMP-dependent protein kinase of bovine heart mitochondria. *FEBS Lett.* **1995**, *377*, 470–474. [CrossRef]
19. Technikova-Dobrova, Z.; Sardanelli, A.M.; Stanca, M.R.; Papa, S. cAMP-dependent protein phosphorylation in mitochondria of bovine heart. *FEBS Lett.* **1994**, *350*, 187–191. [CrossRef]
20. Sardanelli, A.M.; Technikova-Dobrova, Z.; Speranza, F.; Mazzocca, A.; Scacco, S.; Papa, S. Topology of the mitochondrial cAMP-dependent protein kinase and its substrates. *FEBS Lett.* **1996**, *396*, 276–278. [CrossRef]
21. Wong, W.; Scott, J.D. AKAP signalling complexes: Focal points in space and time. *Nat. Rev. Mol. Cell Biol.* **2004**, *5*, 959–970. [CrossRef] [PubMed]

22. Scott, J.D.; Dessauer, C.W.; Taskén, K. Creating order from chaos: Cellular regulation by kinase anchoring. *Annu. Rev. Pharmacol. Toxicol.* **2012**, *53*, 187–210. [CrossRef]

23. Livigni, A.; Scorziello, A.; Agnese, S.; Adornetto, A.; Carlucci, A.; Garbi, C.; Castaldo, I.; Annunziato, L.; Avvedimento, E.V.; Feliciello, A. Mitochondrial AKAP121 Links cAMP and src Signaling to Oxidative Metabolism. *Mol. Biol. Cell* **2006**, *17*, 263–271. [CrossRef] [PubMed]

24. Robin, M.-A.; Prabu, S.K.; Raza, H.; Anandatheerthavarada, H.K.; Avadhani, N.G. Phosphorylation Enhances Mitochondrial Targeting of GSTA4-4 through Increased Affinity for Binding to Cytoplasmic Hsp. *J. Biol. Chem.* **2003**, *278*, 18960–18970. [CrossRef]

25. Gerbeth, C.; Schmidt, O.; Rao, S.; Harbauer, A.B.; Mikropoulou, D.; Opalińska, M.; Guiard, B.; Pfanner, N.; Meisinger, C. Glucose-Induced Regulation of Protein Import Receptor Tom22 by Cytosolic and Mitochondria-Bound Kinases. *Cell Metab.* **2013**, *18*, 578–587. [CrossRef]

26. Rao, S.; Schmidt, O.; Harbauer, A.B.; Schönfisch, B.; Guiard, B.; Pfanner, N.; Meisinger, C. Biogenesis of the preprotein translocase of the outer mitochondrial membrane: Protein kinase A phosphorylates the precursor of Tom40 and impairs its import. *Mol. Biol. Cell* **2012**, *23*, 1618–1627. [CrossRef]

27. Young, J.C.; Hoogenraad, N.J.; Hartl, F. Molecular Chaperones Hsp90 and Hsp70 Deliver Preproteins to the Mitochondrial Import Receptor Tom. *Cell* **2003**, *112*, 41–50. [CrossRef]

28. Appukuttan, A.; Kasseckert, S.A.; Micoogullari, M.; Flacke, J.-P.; Kumar, S.; Woste, A.; Abdallah, Y.; Pott, L.; Reusch, H.P.; Ladilov, Y. Type 10 adenylyl cyclase mediates mitochondrial Bax translocation and apoptosis of adult rat cardiomyocytes under simulated ischaemia/reperfusion. *Cardiovasc. Res.* **2011**, *93*, 340–349. [CrossRef]

29. Harada, H.; Becknell, B.; Wilm, M.; Mann, M.; Huang, L.J.-S.; Taylor, S.S.; Scott, J.D.; Korsmeyer, S.J. Phosphorylation and Inactivation of BAD by Mitochondria-Anchored Protein Kinase A. *Mol. Cell* **1999**, *3*, 413–422. [CrossRef]

30. Moujalled, D.; Weston, R.; Anderton, H.; Ninnis, R.; Goel, P.; Coley, A.; Huang, D.C.S.; Wu, L.; Strasser, A.; Puthalakath, H. Cyclic-AMP-dependent protein kinase A regulates apoptosis by stabilizing the BH3-only protein Bim. *EMBO Rep.* **2010**, *12*, 77–83. [CrossRef] [PubMed]

31. Tan, Y.; Demeter, M.R.; Ruan, H.; Comb, M.J. BAD Ser-155 Phosphorylation Regulates BAD/Bcl-XL Interaction and Cell Survival. *J. Biol. Chem.* **2000**, *275*, 25865–25869. [CrossRef] [PubMed]

32. Chang, C.-R.; Blackstone, C. Cyclic AMP-dependent Protein Kinase Phosphorylation of Drp1 Regulates Its GTPase Activity and Mitochondrial Morphology. *J. Biol. Chem.* **2007**, *282*, 21583–21587. [CrossRef]

33. Cribbs, J.T.; Strack, S. Reversible phosphorylation of Drp1 by cyclic AMP-dependent protein kinase and calcineurin regulates mitochondrial fission and cell death. *EMBO Rep.* **2007**, *8*, 939–944. [CrossRef] [PubMed]

34. Acin-Perez, R.; Gatti, D.L.; Bai, Y.; Manfredi, G. Protein Phosphorylation and Prevention of Cytochrome Oxidase Inhibition by ATP: Coupled Mechanisms of Energy Metabolism Regulation. *Cell Metab.* **2011**, *13*, 712–719. [CrossRef]

35. García-Bermúdez, J.; Sánchez-Aragó, M.; Soldevilla, B.; Del Arco, A.; Nuevo-Tapioles, C.; Cuezva, J.M. PKA Phosphorylates the ATPase Inhibitory Factor 1 and Inactivates Its Capacity to Bind and Inhibit the Mitochondrial H+-ATP Synthase. *Cell Rep.* **2015**, *12*, 2143–2155. [CrossRef]

36. Lu, B.; Lee, J.; Nie, X.; Li, M.; Morozov, Y.I.; Venkatesh, S.; Bogenhagen, D.F.; Temiakov, D.; Suzuki, C.K. Phosphorylation of Human TFAM in Mitochondria Impairs DNA Binding and Promotes Degradation by the AAA+ Lon Protease. *Mol. Cell* **2013**, *49*, 121–132. [CrossRef]

37. De Rasmo, D.; Panelli, D.; Sardanelli, A.M.; Papa, S. cAMP-dependent protein kinase regulates the mitochondrial import of the nuclear encoded NDUFS4 subunit of complex I. *Cell. Signal.* **2008**, *20*, 989–997. [CrossRef] [PubMed]

38. Means, C.K.; Lygren, B.; Langeberg, L.K.; Jain, A.; Dixon, R.E.; Vega, A.L.; Gold, M.G.; Petrosyan, S.; Taylor, S.S.; Murphy, A.N.; et al. An entirely specific type I A-kinase anchoring protein that can sequester two molecules of protein kinase A at mitochondria. *Proc. Natl. Acad. Sci. USA* **2011**, *108*, E1227–E1235. [CrossRef]

39. Acin-Perez, R.; Salazar, E.; Kamenetsky, M.; Buck, J.; Levin, L.R.; Manfredi, G. Cyclic AMP Produced inside Mitochondria Regulates Oxidative Phosphorylation. *Cell Metab.* **2009**, *9*, 265–276. [CrossRef]

40. Acín-Pérez, R.; Salazar, E.; Brosel, S.; Yang, H.; Schon, E.A.; Manfredi, G. Modulation of mitochondrial protein phosphorylation by soluble adenylyl cyclase ameliorates cytochrome oxidase defects. *EMBO Mol. Med.* **2009**, *1*, 392–406. [CrossRef]

41. Di Benedetto, G.; Scalzotto, E.; Mongillo, M.; Pozzan, T. Mitochondrial Ca2+ Uptake Induces Cyclic AMP Generation in the Matrix and Modulates Organelle ATP Levels. *Cell Metab.* **2013**, *17*, 965–975. [CrossRef]

42. Hebert-Chatelain, E.; Desprez, T.; Serrat, R.; Bellocchio, L.; Soria-Gomez, E.; Busquets-Garcia, A.; Zottola, A.C.P.; Delamarre, A.; Cannich, A.; Vincent, P.; et al. A cannabinoid link between mitochondria and memory. *Nat. Cell Biol.* **2016**, *539*, 555–559. [CrossRef]

43. Kim, D.I.; Jensen, S.C.; Noble, K.A.; Kc, B.; Roux, K.H.; Motamedchaboki, K.; Roux, K.J. An improved smaller biotin ligase for BioID proximity labeling. *Mol. Biol. Cell* **2016**, *27*, 1188–1196. [CrossRef] [PubMed]

44. Roux, K.J.; Kim, D.I.; Raida, M.; Burke, B. A promiscuous biotin ligase fusion protein identifies proximal and interacting proteins in mammalian cells. *J. Cell Biol.* **2012**, *196*, 801–810. [CrossRef]

45. Guedouari, H.; Daigle, T.; Scorrano, L.; Hebert-Chatelain, E. Sirtuin 5 protects mitochondria from fragmentation and degradation during starvation. *Biochim. Biophys. Acta (BBA)-Bioenerg.* **2017**, *1864*, 169–176. [CrossRef]

46. Bradford, M.M. A rapid and sensitive method for the quantitation of microgram quantities of protein utilizing the principle of protein-dye binding. *Anal. Biochem.* **1976**, *72*, 248–254. [CrossRef]

47. Choo, Y.S.; Johnson, G.V.; Macdonald, M.; Detloff, P.J.; Lesort, M. Mutant huntingtin directly increases susceptibility of mitochondria to the calcium-induced permeability transition and cytochrome c release. *Hum. Mol. Genet.* **2004**, *13*, 1407–1420. [CrossRef]

48. Ladner, C.L.; Yang, J.; Turner, R.J.; A Edwards, R. Visible fluorescent detection of proteins in polyacrylamide gels without staining. *Anal. Biochem.* **2004**, *326*, 13–20. [CrossRef]

49. Leadsham, J.E.; Gourlay, C.W. cAMP/PKA signaling balances respiratory activity with mitochondria dependent apoptosis via transcriptional regulation. *BMC Cell Biol.* **2010**, *11*, 92. [CrossRef]

50. Lark, D.; Reese, L.R.; Ryan, T.E.; Torres, M.J.; Smith, C.D.; Lin, C.-T.; Neufer, P.D. Protein Kinase A Governs Oxidative Phosphorylation Kinetics and Oxidant Emitting Potential at Complex I. *Front. Physiol.* **2015**, *6*. [CrossRef]

51. Zhang, F.; Zhang, L.; Qi, Y.; Xu, H. Mitochondrial cAMP signaling. *Cell. Mol. Life Sci.* **2016**, *73*, 4577–4590. [CrossRef]

52. Lefkimmiatis, K.; Leronni, D.; Hofer, A.M. The inner and outer compartments of mitochondria are sites of distinct cAMP/PKA signaling dynamics. *J. Cell Biol.* **2013**, *202*, 453–462. [CrossRef]

53. Cooper, D.M.F. Compartmentalization of adenylate cyclase and cAMP signalling. *Biochem. Soc. Trans.* **2005**, *33*, 1319–1322. [CrossRef]

54. Burdyga, A.; Surdo, N.C.; Monterisi, S.; Di Benedetto, G.; Grisan, F.; Penna, E.; Pellegrini, L.; Zaccolo, M.; Bortolozzi, M.; Swietach, P.; et al. Phosphatases control PKA-dependent functional microdomains at the outer mitochondrial membrane. *Proc. Natl. Acad. Sci. USA* **2018**, *115*, E6497–E6506. [CrossRef]

55. Zong, S.; Wu, M.; Gu, J.; Liu, T.; Guo, R.; Yang, M. Structure of the intact 14-subunit human cytochrome c oxidase. *Cell Res.* **2018**, *28*, 1026–1034. [CrossRef]

56. Jakobs, S.; Wurm, C.A. Super-resolution microscopy of mitochondria. *Curr. Opin. Chem. Biol.* **2014**, *20*, 9–15. [CrossRef]

57. Neuberger, G.; Schneider, G.; Eisenhaber, F. pkaPS: Prediction of protein kinase A phosphorylation sites with the simplified kinase-substrate binding model. *Biol. Direct* **2007**, *2*, 1. [CrossRef]

58. Blom, N.; Gammeltoft, S.; Brunak, S. Sequence and structure-based prediction of eukaryotic protein phosphorylation sites. *J. Mol. Biol.* **1999**, *294*, 1351–1362. [CrossRef]

59. Xue, Y.; Liu, Z.; Cao, J.; Ma, Q.; Gao, X.; Wang, Q.; Jin, C.; Zhou, Y.; Wen, L.; Ren, J. GPS 2.1: Enhanced prediction of kinase-specific phosphorylation sites with an algorithm of motif length selection. *Protein Eng. Des. Sel.* **2010**, *24*, 255–260. [CrossRef]

60. Chen, R.; Fearnley, I.M.; Peak-Chew, S.Y.; Walker, J.E. The Phosphorylation of Subunits of Complex I from Bovine Heart Mitochondria. *J. Biol. Chem.* **2004**, *279*, 26036–26045. [CrossRef]

61. Yadava, N.; Potluri, P.; Scheffler, I.E. Investigations of the potential effects of phosphorylation of the MWFE and ESSS subunits on complex I activity and assembly. *Int. J. Biochem. Cell Biol.* **2008**, *40*, 447–460. [CrossRef] [PubMed]

62. Helling, S.; Vogt, S.; Rhiel, A.; Ramzan, R.; Wen, L.; Marcus, K.; Kadenbach, B. Phosphorylation and Kinetics of Mammalian CytochromecOxidase. *Mol. Cell. Proteom.* **2008**, *7*, 1714–1724. [CrossRef] [PubMed]

63. Chevtzoff, C.; Vallortigara, J.; Avéret, N.; Rigoulet, M.; Devin, A. The yeast cAMP protein kinase Tpk3p is involved in the regulation of mitochondrial enzymatic content during growth. *Biochim. Biophys. Acta (BBA)-Gen. Subj.* **2005**, *1706*, 117–125. [CrossRef]

64. Prabu, S.K.; Anandatheerthavarada, H.K.; Raza, H.; Srinivasan, S.; Spear, J.F.; Avadhani, N.G. Protein Kinase A-mediated Phosphorylation Modulates CytochromecOxidase Function and Augments Hypoxia and Myocardial Ischemia-related Injury. *J. Biol. Chem.* **2006**, *281*, 2061–2070. [CrossRef]

65. Srinivasan, S.; Spear, J.; Chandran, K.; Joseph, J.; Kalyanaraman, B.; Avadhani, N.G. Oxidative Stress Induced Mitochondrial Protein Kinase A Mediates Cytochrome C Oxidase Dysfunction. *PLoS ONE* **2013**, *8*, e77129. [CrossRef]

66. Kuo, M.-L.; Lee, M.B.-E.; Tang, M.; Besten, W.D.; Hu, S.; Sweredoski, M.J.; Hess, S.; Chou, C.-M.; Changou, C.A.; Su, M.; et al. PYCR1 and PYCR2 Interact and Collaborate with RRM2B to Protect Cells from Overt Oxidative Stress. *Sci. Rep.* **2016**, *6*, 18846. [CrossRef]

67. Lucas, S.; Chen, G.; Aras, S.; Wang, J. Serine catabolism is essential to maintain mitochondrial respiration in mammalian cells. *Life Sci. Alliance* **2018**, *1*, e201800036. [CrossRef]

68. Bao, X.R.; Ong, S.-E.; Goldberger, O.; Peng, J.; Sharma, R.; A Thompson, D.; Vafai, S.B.; Cox, A.G.; Marutani, E.; Ichinose, F.; et al. Mitochondrial dysfunction remodels one-carbon metabolism in human cells. *eLife* **2016**, *5*, e10575. [CrossRef]

69. Celardo, I.; Lehmann, S.; Costa, A.C.; Loh, S.H.Y.; Martins, L.M. dATF4 regulation of mitochondrial folate-mediated one-carbon metabolism is neuroprotective. *Cell Death Differ.* **2017**, *24*, 638–648. [CrossRef]

70. Baqri, R.M.; Pietron, A.V.; Gokhale, R.H.; Turner, B.A.; Kaguni, L.S.; Shingleton, A.W.; Kunes, S.; Miller, K.E. Mitochondrial chaperone TRAP1 activates the mitochondrial UPR and extends healthspan in Drosophila. *Mech. Ageing Dev.* **2014**, *141*, 35–45. [CrossRef]

71. Al-Furoukh, N.; Ianni, A.; Nolte, H.; Hölper, S.; Krüger, M.; Wanrooij, S.; Braun, T. ClpX stimulates the mitochondrial unfolded protein response (UPRmt) in mammalian cells. *Biochim. Biophys. Acta Bioenerg.* **2015**, *1853*, 2580–2591. [CrossRef]

72. Zhao, Q.; Wang, J.; Levichkin, I.V.; Stasinopoulos, S.; Ryan, M.T.; Hoogenraad, N.J. A mitochondrial specific stress response in mammalian cells. *EMBO J.* **2002**, *21*, 4411–4419. [CrossRef]

73. Hunt, R.J.; Bateman, J.M. Mitochondrial retrograde signaling in the nervous system. *FEBS Lett.* **2018**, *592*, 663–678. [CrossRef]

74. Nargund, A.M.; Pellegrino, M.W.; Fiorese, C.J.; Baker, B.M.; Haynes, C.M. Mitochondrial Import Efficiency of ATFS-1 Regulates Mitochondrial UPR Activation. *Science* **2012**, *337*, 587–590. [CrossRef] [PubMed]

75. Tatsuta, T.; Model, K.; Langer, T. Formation of Membrane-bound Ring Complexes by Prohibitins in Mitochondria. *Mol. Biol. Cell* **2005**, *16*, 248–259. [CrossRef]

76. Merkwirth, C.; Dargazanli, S.; Tatsuta, T.; Geimer, S.; Löwer, B.; Wunderlich, F.T.; Von Kleist-Retzow, J.-C.; Waisman, A.; Westermann, B.; Langer, T. Prohibitins control cell proliferation and apoptosis by regulating OPA1-dependent cristae morphogenesis in mitochondria. *Genes Dev.* **2008**, *22*, 476–488. [CrossRef] [PubMed]

77. Bogenhagen, D.F.; Wang, Y.; Shen, E.L.; Kobayashi, R. Protein Components of Mitochondrial DNA Nucleoids in Higher Eukaryotes. *Mol. Cell. Proteom.* **2003**, *2*, 1205–1216. [CrossRef]

78. Kasashima, K.; Sumitani, M.; Satoh, M.; Endo, H. Human prohibitin 1 maintains the organization and stability of the mitochondrial nucleoids. *Exp. Cell Res.* **2008**, *314*, 988–996. [CrossRef]

79. He, J.; Cooper, H.M.; Reyes, A.; Di Re, M.; Sembongi, H.; Litwin, T.R.; Gao, J.; Neuman, K.C.; Fearnley, I.M.; Spinazzola, A.; et al. Mitochondrial nucleoid interacting proteins support mitochondrial protein synthesis. *Nucleic Acids Res.* **2012**, *40*, 6109–6121. [CrossRef] [PubMed]

80. Miwa, S.; Jow, H.; Baty, K.; Johnson, A.; Czapiewski, R.; Saretzki, G.; Treumann, A.; Von Zglinicki, T. Low abundance of the matrix arm of complex I in mitochondria predicts longevity in mice. *Nat. Commun.* **2014**, *5*, 3837. [CrossRef]

81. Nijtmans, L.G.; De Jong, L.; Sanz, M.A.; Coates, P.J.; Berden, J.A.; Back, J.W.; Muijsers, A.O.; Van Der Spek, H.; Grivell, L.A. Prohibitins act as a membrane-bound chaperone for the stabilization of mitochondrial proteins. *EMBO J.* **2000**, *19*, 2444–2451. [CrossRef]

82. Jian, C.; Xu, F.; Hou, T.; Sun, T.; Li, J.; Cheng, H.; Wang, X.; Jinghang, L. Deficiency of PHB complex impairs respiratory supercomplex formation and activates mitochondrial flashes. *J. Cell Sci.* **2017**, *130*, 2620–2630. [CrossRef]

83. Strub, G.M.; Paillard, M.; Liang, J.; Gomez, L.; Allegood, J.C.; Hait, N.C.; Maceyka, M.; Price, M.M.; Chen, Q.; Simpson, D.C.; et al. Sphingosine-1-phosphate produced by sphingosine kinase 2 in mitochondria interacts with prohibitin 2 to regulate complex IV assembly and respiration. *FASEB J.* **2010**, *25*, 600–612. [CrossRef]

84. Bourges, I.; Ramus, C.; De Camaret, B.M.; Beugnot, R.; Remacle, C.; Cardol, P.; Hofhaus, G.; Issartel, J.-P. Structural organization of mitochondrial human complex I: Role of the ND4 and ND5 mitochondria-encoded subunits and interaction with prohibitin. *Biochem. J.* **2004**, *383*, 491–499. [CrossRef]

85. Schmidt, O.; Pfanner, N.; Meisinger, C. Mitochondrial protein import: From proteomics to functional mechanisms. *Nat. Rev. Mol. Cell Biol.* **2010**, *11*, 655–667. [CrossRef]

86. Schiller, D.; Cheng, Y.C.; Liu, Q.; Walter, W.; Craig, E.A. Residues of Tim44 Involved in both Association with the Translocon of the Inner Mitochondrial Membrane and Regulation of Mitochondrial Hsp70 Tethering. *Mol. Cell. Biol.* **2008**, *28*, 4424–4433. [CrossRef]

87. Schilke, B.A.; Hayashi, M.; Craig, E.A. Genetic Analysis of Complex Interactions Among Components of the Mitochondrial Import Motor and Translocon in Saccharomyces cerevisiae. *Genetics* **2012**, *190*, 1341–1353. [CrossRef]

88. Ting, S.-Y.; Yan, N.L.; A Schilke, B.; Craig, E.A. Dual interaction of scaffold protein Tim44 of mitochondrial import motor with channel-forming translocase subunit Tim23. *eLife* **2017**, *6*, 23609. [CrossRef]

89. Akabane, S.; Uno, M.; Tani, N.; Shimazaki, S.; Ebara, N.; Kato, H.; Kosako, H.; Oka, T. PKA Regulates PINK1 Stability and Parkin Recruitment to Damaged Mitochondria through Phosphorylation of MIC. *Mol. Cell* **2016**, *62*, 371–384. [CrossRef]

90. Flippo, K.H.; Gnanasekaran, A.; Perkins, G.; Ajmal, A.; Merrill, R.A.; Dickey, A.S.; Taylor, S.S.; McKnight, G.S.; Chauhan, A.K.; Usachev, Y.M.; et al. AKAP1 Protects from Cerebral Ischemic Stroke by Inhibiting Drp1-Dependent Mitochondrial Fission. *J. Neurosci.* **2018**, *38*, 8233–8242. [CrossRef]

91. Yamamoto, H.; Fukui, K.; Takahashi, H.; Kitamura, S.; Shiota, T.; Terao, K.; Uchida, M.; Esaki, M.; Nishikawa, S.-I.; Yoshihisa, T.; et al. Roles of Tom70 in Import of Presequence-containing Mitochondrial Proteins. *J. Biol. Chem.* **2009**, *284*, 31635–31646. [CrossRef]

92. Fan, A.C.Y.; Gava, L.M.; Ramos, C.H.I.; Young, J.C. Human mitochondrial import receptor Tom70 functions as a monomer. *Biochem. J.* **2010**, *429*, 553–563. [CrossRef]

93. Antonicka, H.; Lin, Z.-Y.; Janer, A.; Weraarpachai, W.; Gingras, A.-C.; Shoubridge, E.A. A high-density human mitochondrial proximity interaction network. *bioRxiv* **2020**. [CrossRef]

94. Chojnowski, A.; Sobota, R.M.; Ong, P.F.; Xie, W.; Wong, X.; Dreesen, O.; Burke, B.; Stewart, C.L. 2C-BioID: An Advanced Two Component BioID System for Precision Mapping of Protein Interactomes. *iScience* **2018**, *10*, 40–52. [CrossRef]

95. Kim, D.I.; Roux, K.J. Filling the Void: Proximity-Based Labeling of Proteins in Living Cells. *Trends Cell Biol.* **2016**, *26*, 804–817. [CrossRef]

96. Branon, T.C.; Bosch, J.A.; Sanchez, A.D.; Udeshi, N.D.; Svinkina, T.; Carr, S.A.; Feldman, J.L.; Perrimon, N.; Ting, A.Y. Efficient proximity labeling in living cells and organisms with TurboID. *Nat. Biotechnol.* **2018**, *36*, 880–887. [CrossRef]

97. Niemi, N.M.; Wilson, G.M.; Overmyer, K.A.; Vögtle, F.-N.; Myketin, L.; Lohman, D.C.; Schueler, K.L.; Attie, A.D.; Meisinger, C.; Coon, J.J.; et al. Pptc7 is an essential phosphatase for promoting mammalian mitochondrial metabolism and biogenesis. *Nat. Commun.* **2019**, *10*, 3197. [CrossRef]

98. Wang, Z.; Jiang, H.; Chen, S.; Du, F.; Wang, X. The Mitochondrial Phosphatase PGAM5 Functions at the Convergence Point of Multiple Necrotic Death Pathways. *Cell* **2012**, *148*, 228–243. [CrossRef] [PubMed]

99. Sekine, S.; Kanamaru, Y.; Koike, M.; Nishihara, A.; Okada, M.; Kinoshita, H.; Kamiyama, M.; Maruyama, J.; Uchiyama, Y.; Takeda, K.; et al. Rhomboid Protease PARL Mediates the Mitochondrial Membrane Potential Loss-induced Cleavage of PGAM. *J. Biol. Chem.* **2012**, *287*, 34635–34645. [CrossRef]

100. Cereghetti, G.M.; Stangherlin, A.; De Brito, O.M.; Chang, C.R.; Blackstone, C.; Bernardi, P.; Scorrano, L. Dephosphorylation by calcineurin regulates translocation of Drp1 to mitochondria. *Proc. Natl. Acad. Sci. USA* **2008**, *105*, 15803–15808. [CrossRef]

3

AMP-Activated Protein Kinase as a Key Trigger for the Disuse-Induced Skeletal Muscle Remodeling

Natalia A. Vilchinskaya [1], Igor I. Krivoi [2] and Boris S. Shenkman [1,*]

[1] Myology Laboratory, Institute of Biomedical Problems RAS, Moscow 123007, Russia; vilchinskayanatalia@gmail.com

[2] Department of General Physiology, St. Petersburg State University, St. Petersburg 199034, Russia; iikrivoi@gmail.com

* Correspondence: bshenkman@mail.ru

Abstract: Molecular mechanisms that trigger disuse-induced postural muscle atrophy as well as myosin phenotype transformations are poorly studied. This review will summarize the impact of 5' adenosine monophosphate -activated protein kinase (AMPK) activity on mammalian target of rapamycin complex 1 (mTORC1)-signaling, nuclear-cytoplasmic traffic of class IIa histone deacetylases (HDAC), and myosin heavy chain gene expression in mammalian postural muscles (mainly, soleus muscle) under disuse conditions, i.e., withdrawal of weight-bearing from ankle extensors. Based on the current literature and the authors' own experimental data, the present review points out that AMPK plays a key role in the regulation of signaling pathways that determine metabolic, structural, and functional alternations in skeletal muscle fibers under disuse.

Keywords: AMPK; HDAC4/5; p70S6K; MyHC I(β), motor endplate remodeling; soleus muscle; mechanical unloading; hindlimb suspension

---

1. Introduction

Skeletal muscle is a highly plastic organ, which is able to change its structure and metabolism depending on the mode of contractile activity. Such conditions as hypokinesia, immobilization, paralysis, and weightlessness can lead to a complex of atrophic changes (most pronounced in postural muscles), resulting from a significant reduction in muscle mass and contractile function [1,2]. Skeletal muscle disuse also leads to a reduction in muscle stiffness and slow-to-fast myosin phenotype transformations [2–7]. Muscle atrophy observed during muscle inactivation under conditions of real and simulated microgravity, joint immobilization, or spinal isolation is associated with an increase in proteolytic processes and a decrease in protein synthesis [4,8–11]. Myosin phenotype shift occurs as a result of a decrease in the gene expression of the slow isoform of myosin heavy chain (MyHC) and an increase in the expression of the fast MyHC isoforms [12–14].

To study the mechanisms of muscle disuse atrophy, a variety of experimental models with the different rate of reduction in muscle electrical and contractile activity are used. In this sense, one of the most suitable models is a rodent hindlimb suspension (HS) technique, which prevents the hindlimbs from touching any supporting surface, resulting in a cessation of rat soleus neuromuscular activity [15–17]. Similar effects are observed during dry immersion in human skeletal muscle [1,18,19]. These models not only provide an almost complete cessation of the soleus muscle contractile activity, simulating the effects of weightlessness, but also allow the experimentalist to avoid invasive procedures (denervation, spinal isolation, administration of toxins, etc.). Hence, this review will mainly summarize the data obtained in HS and dry immersion models.

Despite a large number of studies aimed at the analysis of disuse muscle atrophy, the triggering mechanisms of its development within a few hours/days after withdrawal of weight-bearing from postural muscles are still poorly studied [1,15,16]. The earliest effects of unloading (the first 24 h) on postural muscle include: (1) Depolarization of the sarcolemma due to an inactivation of the $\alpha2$ subunit of the Na,K-ATPase [20,21], (2) disintegration of cholesterol rafts [22], and (3) translocation of the neuronal NO-synthase from the subsarcolemmal compartment to the cytoplasm [23]. However, the mechanisms of development of these changes and, most importantly, the dependence of these mechanisms on molecular triggers determined by the level of muscle contractile activity/inactivity remain unknown.

It seems natural that the reduction/cessation of electrical and, accordingly, contractile activity of the muscle can lead to changes in the basic physiological mechanisms that directly depend on the activity of muscle fibers.

1. Changes in electrogenic signaling mechanisms due to termination of electrical activity (hypothetical decrease in Na^+ concentration inside the muscle fibers, the expected temporary cessation of Ca^{2+} flow through voltage-dependent L-type calcium channels).

2. Mechanosensory molecular changes due to termination of the mechanical action of extracellular matrix structures on mechanosensory molecules (changes in the state of integrins, etc. [24], termination of the active state of actin stress-fibers, inactivation of mechanosensitive channels, inactivation of mechanosensory myofibrillar proteins).

3. Changes in energy metabolism as a result of termination of ATP expenditure (changes in the ratio of ATP/ADP/AMP and PCr/Cr, accumulation of glycogen and reactive oxygen species).

It is not yet possible to trace the entire complex of processes that link the cessation of electrical activity and the elimination of mechanical loading of muscle fibers with the development of early molecular events during mechanical unloading. As for the consequences of the cessation of metabolic energy expenditure during unloading, the situation is somewhat different. In the 1990s, it was hypothesized that mechanical unloading leads to a change in the balance of high-energy phosphate compounds towards the accumulation of fully phosphorylated compounds [25]. Wakatsuki and co-authors have shown the accumulation of phosphocreatine in the soleus muscle of rats after 10 days of HS [26]. In 1987, Henriksen and Tischler reported a 25% increase in glycogen content in rat soleus during the first three days of HS [27].

If the described changes in energy metabolism are the potential triggers for signaling processes leading to the development of postural muscle atrophy, reduced intrinsic muscle stiffness, and myosin phenotype shift, there should be a specific sensor of the state of energy metabolism. Such a sensor has long been known. It is 5′ adenosine monophosphate -activated protein kinase (AMPK), the cell's main energy sensor reacting to the changes in the ratio of high-energy phosphates (Figure 1). Therefore, the termination of electrical activity in soleus muscle at the initial stage of gravitational unloading should affect AMPK activity.

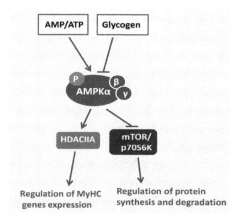

Figure 1. Key physiological regulators of 5′ adenosine monophosphate -activated protein kinase (AMPK) activity in skeletal muscle and physiologically-relevant AMPK targets. (original scheme)

2. AMPK Is a Key Energy Sensor and Metabolic Regulator of Signaling Pathways in Skeletal Muscle Fibers

AMPK is involved in transmission of extracellular and intracellular signals by phosphorylation of various substrates in many metabolic reactions in skeletal muscle. AMRK is a heterotrimeric complex consisting of three proteins: α-subunit, which has its own kinase activity, and two regulatory subunits, β and γ [28].

AMPK activity is regulated both allosterically and by post-translational modifications (phosphorylation). Allosteric activation of AMPK is carried out with the help of AMP and its analogues. Phosphorylation AMPK at Thr172 of the α-subunit leads to its activation. This phosphorylation is regulated by calcium-/calmodulin-dependent kinase kinase 2 (CaMKK2) and liver kinase B1 (LKB1). Further activation of AMPK occurs due to conformational changes occurring upon binding of AMP or ADP to the γ-subunit of AMPK, promoting Thr172 phosphorylation at the α-subunit. The combined effect of Thr172 phosphorylation of the α-subunit of AMPK and allosteric regulation leads to more than a 1000-fold increase in AMPK activity, which makes AMPK highly sensitive to alternations in the energy status of the cell [29].

Activation of AMPK may also occur under the action of extracellular regulatory factors, such as interleukin-6 (IL-6) and brain-derived neurotrophic factor (BDNF) [30]. Interestingly, AMPK is involved in the regulation of IL-6 expression in skeletal muscles [31]. One of the factors of AMPK activation is nitric oxide (NO), the production of which is determined by the activity of neuronal and endothelial NO synthases [32,33]. There is evidence that AMPK activation can result from mechanical stretch via components of the dystrophin-glycoprotein complex (at least in cardiomyocytes) [34].

AMPK can also be phosphorylated on Ser485/491 sites by protein kinase D and some isoforms of protein kinase C [35], which leads to inhibition of AMPK activity. A decrease in AMPK activity is associated with increased glycogen content, as well as accumulation of ATP and creatine phosphate [36,37]. AMPK has a number of molecular targets in skeletal muscle. It is known that AMPK can activate Na,K-ATPase [38,39] and phosphorylate neuronal NO-synthase [40]. AMPK is also involved in the regulation of protein synthesis and degradation [36]. AMPK is a negative regulator of protein synthesis in skeletal muscle. This kinase can inhibit the key regulator of protein synthesis, the mammalian target of rapamycin complex 1 (mTORC1), through phosphorylation of TSC2 [41] and raptor [42]. AMPK can also be involved in the degradation of myofibrillar proteins [43]. Nakashima and co-authors have shown that AMPK participates in the degradation of myofibrillar proteins through the activation of forkhead box proteins (FOXO) transcription factors and subsequent up-regulation of muscle-specific E3 ubiquitin-ligases atrogin-1/MAFbx and MuRF-1 [44].

In addition, AMPK, as a key energy sensor of the cell regulating energy metabolism, participates in the initiation of autophagy [45]. AMPK can directly phosphorylate Unc-51-like kinase (ULK-1) across several sites as well as activate autophagy by inhibiting mTOR activity [46,47].

In recent years, it has been shown that AMPK can influence the expression of a number of genes by phosphorylation of class IIA histone deacetylases (HDAC4, HDAC5, HDAC7), leading to their exclusion from the nucleus and activation of gene expression [48–50].

Thus, according to modern concepts, AMPK activity is mainly determined by the state of energy metabolism: AMPK activity increases with increased ATP consumption, AMP accumulation, and glycogen depletion, and decreases with the accumulation of ATP and glycogen in muscle fibers. Activated AMPK phosphorylates and retains class IIA HDACs outside the myonuclei (thereby contributing to the expression of a number of genes) and inhibits the activity of mTORC1 and its primary targets (Figure 1). Dephosphorylated AMPK, on the contrary, promotes HDACs nuclear import and transcriptional suppression of gene expression, while reducing the degree of mTORC1 suppression.

3. AMPK Activity under Conditions of Mechanical Unloading

Until recently, AMPK activity/phosphorylation in skeletal muscle under unloading conditions has been poorly studied. Moreover, the literature reveals contradictory results concerning AMPK Thr172 phosphorylation at various time-points in different models of disuse. It has been shown that four- and seven-day denervation resulted in a significant increase in AMPK phosphorylation in rodent skeletal muscles [51,52], while deletion of AMPKα2 significantly attenuated denervation-induced skeletal muscle wasting and protein degradation [51]. In human skeletal muscle with recent complete cervical spinal cord injury, AMPKα2 protein abundance decreased by 25% during the first year after injury, without significant change in AMPKα1 content. Furthermore, AMPK phosphorylation on Thr172 was significantly decreased during the first year post-spinal cord injury in human vastus lateralis muscle [53]. Thirty-day space flight and subsequent recovery did not affect AMPK Thr172 phosphorylation in murine longissimus dorsi muscle [54]. In terms of fiber-type composition, murine longissimus dorsi is similar to soleus muscle. However, it should be noted that it is not quite correct to compare the data obtained from rat and mouse soleus muscle. It is known that rat soleus muscle, as well as human soleus, comprises about 85% of slow-twitch fibers expressing the slow isoform of MyHC, while mouse soleus consists of approximately 40% slow-twitch fibers [55]. Obviously, this fact determines the essential features of metabolism in the mouse soleus muscle and its response to gravitational unloading. Therefore, unloading-induced changes in mouse soleus can significantly differ from that of rats and humans.

Vilchinskaya et al. (2015) have shown for the first time that short-term (three days) gravitational unloading via dry immersion leads to a significant decrease in the level of AMPK Thr172 phosphorylation in human soleus muscle [56]. The literature data on the AMPK activity in rat soleus following HS are inconsistent. AMPK activity, which is usually assessed by the level of Thr172 phosphorylation [57–59], was reported to be reduced in rat soleus after two weeks of HS [60], whereas Hilder and co-authors showed that 14-day HS results in a significant increase in AMPK Thr172 phosphorylation in rat soleus [61]. At the same time, Egawa and others did not find any changes in AMPKα1 и AMPKα2 activity in mouse soleus after 14-day HS, however, the level of ACC phosphorylation was upregulated [62,63].

A significant reduction in AMPK Thr172 phosphorylation was previously observed in rat soleus at the early stage (24 h) of hindlimb unloading [11]. A recent study has also demonstrated a significant decrease in AMPK Thr172 phosphorylation in rat soleus muscle already after 6- and 12-h mechanical unloading [64]. In an inactive skeletal muscle, a rapid accumulation of completely phosphorylated high-energy phosphates can occur, resulting in reduced AMPK activity. Thus, AMPK dephosphorylation at the early stage of gravitational unloading can be caused by a decrease in postural muscle energy consumption and a corresponding change in the ratio of phosphorylated and dephosphorylated adenine nucleotides (ATP, ADP, AMP).

It is known that binding of AMPK to glycogen results in reduced AMPK activity [37]. Therefore, it is possible that a decrease in the activity of AMPK at the initial stages of mechanical unloading may be associated with glycogen accumulation. Indeed, glycogen concentration in rat soleus muscle during the first three days of HS is significantly increased [27].

After seven-day HS, AMPK phosphorylation does not differ from that of control [11], which correlates well with the restoration of electromyographic activity of rat soleus following six to seven days of HS [15]. Moreover, 14-day HS resulted in a significant increase in AMPK Thr172 phosphorylation in rat soleus [61,65]. It is notable that after 14-day HS, the increase of AMPK activity (judging by ACC phosphorylation) [62,63] is less pronounced in murine vs rat soleus.

Now, it is difficult to establish the precise mechanisms that cause an increase in AMPK activity by 14 days of HS. However, some assumptions can be made concerning potential signaling mechanisms leading to this phenomenon. There is evidence that the concentration of interleukin-6 (IL-6) is significantly increased in rodent skeletal muscle after 5- and 14-day HS [66,67] as well as in human skeletal muscle following 60-day bed rest [68]. It is known that IL-6 can increase AMPK activity in rodent skeletal muscle [69], and it is likely that increased concentration of IL-6 during long-term gravitational unloading promotes AMPK hyperphosphorylation. BDNF was also shown to increase AMPK phosphorylation in isolated rat extensor digitorum muscle [30]. Therefore, an increase in BDNF mRNA expression in the spinal cord and soleus muscle of rats after 14 days of HS [70] could be the cause of the increased AMPK Thr172 phosphorylation during this period.

Thus, HS experiments show complex time-course changes in AMPK activity in the rodent soleus muscle under mechanical unloading. It is important to note that the level of AMPK phosphorylation is significantly reduced during the first day of HS, a period that precedes atrophy development. Additionally, the most likely cause of such a decrease is a shift in the ratio of phosphorylated and dephosphorylated adenine nucleotides (AMP/ATP and ADP/ATP ratios).

4. The Role of AMPK in the Regulation of mTOR/p70S6K and Akt/FOXO3/MuRF-1/MAFbx/Atrogin-1 Signaling Pathways under Gravitational Unloading

Anabolic processes in skeletal muscle fibers are regulated by a number of signaling pathways, the most important of which is the mammalian target of rapamycin complex 1 (mTORC1) signaling pathway. mTORC1, through its downstream targets (p70S6K, 4E-BP1), stimulates mRNA translation initiation on a ribosome. Activation of protein synthesis following resistance exercise is associated with the increased level of p70S6K (Thr389) phosphorylation, leading to subsequent phosphorylation of ribosomal protein S6 and initiation of protein synthesis. Some authors reported a decrease in p70S6K (Thr389) phosphorylation in rat soleus after four to five days of HS [71,72]. Other studies showed that even 7–10 days of HS did not affect p70S6K phosphorylation, the level of which decreased only following 14 days of unloading [73–76].

It is well known that activated AMPK has an inhibitory effect on the anabolic processes via suppression of mTORC1 and its key substrate, p70S6K [41–43,77–80]. Summing up the available data, we can assume that changes in p70S6K and AMPK phosphorylation in rat soleus muscle during the first two weeks of HS show reciprocal relations and complex dynamics. It is important to note that during the early stage of HS (one to three days), an increase in p70S6K Thr389 phosphorylation is accompanied by a decrease in the level of AMPK Thr172 phosphorylation. However, after seven days of HS, there is no difference between these parameters and control values [11]. Two-week HS results in an opposite effect: A decrease in p70S6K phosphorylation is accompanied by an increase in AMPK phosphorylation. These data are consistent with the report by Sugiura and co-authors (2005) that showed no changes in p70S6K phosphorylation after 10 days of HS [74], whereas 14-day HS led to a significant decrease in p70S6K phosphorylation as compared to control levels [71,72,75,76,81]. Interestingly, according to Hilder et al. (2005) and Zhang et al. (2018), the level of AMPK phosphorylation following 14-day unloading is significantly increased [61,65]. The high level of AMPK phosphorylation is accompanied by a decrease in phosphorylation of not only p70S6K, but also Akt and FOXO3, which can lead to an upregulation of muscle-specific E3-ubiquitin ligases, MuRF-1 and MAFbx/atrogin-1 [65]. These results are consistent with earlier studies showing the ability of AMPK to stimulate FOXO3 dephosphorylation and the expression of E3-ubiquitin ligases [44,82]. Time-course changes in the level of p70S6K and AMPK phosphorylation during mechanical unloading suggest that an increase in p70S6K phosphorylation at the first day of HS may be due to the low AMPK activity. This hypothesis has been recently tested by Vilchinskaya and co-authors [83]. Pretreatment of rats with 5-aminoimidazole-4-carboxamide ribonucleotide (AICAR) several days before and during 24-h HS prevented both a decrease in AMPK Th172 phosphorylation as well as an increase in p70S6K Thr389 phosphorylation [83]. This result fully confirmed the hypothesis and once again demonstrated the role of AMPK as a negative regulator of mTORC1-signaling [41,42]. The results obtained by Vilchinskaya et al. (2017) suggest that AMPK dephosphorylation in postural soleus muscle at the initial stage of mechanical unloading is one of the reasons for a paradoxical increase in the level of p70S6K phosphorylation. Increased p70S6K phosphorylation is usually considered to be a consequence of (1) inactivation of endogenous mTORC1 inhibitor tuberous sclerosis complex (TSC1/2) due to

AMPK dephosphorylation [41] and (2) accumulation of sphingolipid ceramide [84]. Interestingly, it was previously shown that activation of AMPK with AICAR prevents ceramide accumulation in skeletal muscle fibers [85]. Therefore, it can be assumed that AMPK dephosphorylation can contribute to the accumulation of ceramide in rat soleus under gravitational unloading [86–88]. Hsieh and co-authors (2014) observed an interesting effect of p70S6K hyperphosphorylation. The authors showed that activated p70S6K promotes phosphorylation of insulin receptor substrate (IRS1) on Ser636-639, which leads to a reduction in IRS-1 activity and, accordingly, dephosphorylation of downstream protein kinase Akt (Figure 2).

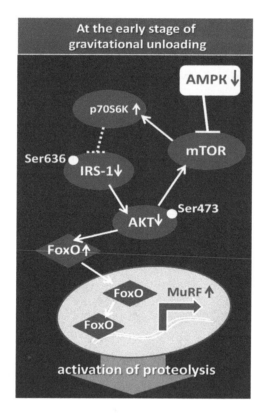

Figure 2. Hypothetical role of AMPK in the activation of signaling pathways regulating the expression of E3-ubiquitin ligases during gravitational unloading.

Dephosphorylation of Akt on Ser473, as a rule, causes an increased expression of E3-ubiquitin ligases (MuRF-1 and MAFbx/atrogin-1) [88]. It is well known that even short-term (one to three days) gravitational unloading leads to Akt dephosphorylation and upregulation of E3-ubiquitin ligases [10,66]. According to a number of authors, gravitational unloading results in a significant decrease in IRS-1 content [54,89]. Based on these literature data, it can be assumed that AMPK dephosphorylation during the first day of unloading leads to an increase in p70S6K phosphorylation resulting in the increased E3-ubiquitin ligases' expression and enhanced proteolysis. Recent experiments with inhibition of p70S6K phosphorylation during the first day of gravitational unloading do not contradict this hypothesis: The use of rapamycin (mTORC1 inhibitor) resulted in a significant decrease in MuRF-1 and MAFbx/atrogin-1 expression [88].

Thus, AMPK dephosphorylation at the initial period of HS leads to an increase in the level of p70S6K phosphorylation, which can contribute to the subsequent upregulation of proteolytic enzymes.

AMPK is known to activate autophagy by ULK phosphorylation (see above). Therefore, one would expect to see a reduction in autophagy markers at the initial period of unloading when AMPK activity is downregulated. However, it has been recently shown that most of autophagy markers, except for ULK, were upregulated in rat soleus following 6- and 12-h HS [64]. The authors of the cited report emphasize that this state of autophagy markers is not consistent with a reduced level of AMPK phosphorylation. They explain this phenomenon by the possible unequal phosphorylation of different AMPK isoforms. It is clear that this issue needs further investigation.

At a later stage of unloading, an increase in AMPK phosphorylation is accompanied by a decrease in p70S6K phosphorylation and activation of the proteolytic signaling system [65].

At the same time, no significant changes in phosphorylation of Akt and p70S6K and the rate of protein synthesis were found in the soleus of transgenic mice overexpressing the dominant-negative mutant of AMPK following 14-day HS [62]. Thus, the lack of AMPK activity in the transgenic mice did not affect anabolic signaling following a relatively long period of unloading (14 days). However, 14-day unloading of these transgenic mice induced a significant increase in the expression of both markers of the ubiquitin-proteasome system and ULK, a marker of autophagy. The differences between dominant-negative AMPK mutants and wild-type mice were significant [62]. There were also significant differences in the severity of atrophic changes. The weight of soleus muscle in dominant-negative AMPK mutants after 14-day HS was significantly reduced, but to a lesser extent than that in wild-type mice (difference of 10–15%) [62,63]. Thus, AMPK can contribute to the development of muscle atrophy during unloading. This contribution to muscle atrophy seems to be carried out via the activation of catabolic processes rather than through the suppression of anabolic regulation.

This fact allows us to suggest that, at later stages of gravitational unloading, AMPK activity can act as a key signaling node of protein homeostasis in skeletal muscle.

5. AMPK Is Involved in the Regulation of Myosin Phenotype under Mechanical Unloading

Unloading-induced changes in skeletal muscle myosin phenotype have been observed by a number of researchers. It is well established that there is an increase in the expression of fast-type myosin isoforms and a decrease in the expression of slow-type myosin isoform in skeletal muscles of space-flown rodents as well as astronauts [3,4,90,91]. HS results in a significant increase in the content of fast-twitch (type II) fibers and a significant decrease in the proportion of slow-twitch (type I) fibers in rat soleus muscle [7,92–95]. In samples of soleus muscle, taken from astronauts after a six-month spaceflight, there was a decrease in the proportion of fibers expressing slow MyHC isoform and an increase in the proportion of fibers expressing fast MyHC isoforms [96]. Mechanisms of molecular regulation of myosin phenotype transformation remain largely unexplored. It is known that AMPK can affect the expression of a number of genes by phosphorylation of class IIA histone deacetylases (HDAC4, HDAC5, HDAC7), which leads to HDACs dissociation from gene promoters and removal from the nucleus, thereby allowing for increased gene expression [48–50]. It was earlier shown that at the first day of mechanical unloading, mRNA expression of the slow MyHC isoform in rat soleus is reduced [13]. One of the ways of MyHC expression regulation in muscle fibers is associated with phosphorylation of histone deacetylase 4 (HDAC4) [7,97]. However, it remains unclear what mechanisms are implicated in such a rapid decrease in gene expression. Until recently, there have been no studies on the role of AMPK in the regulation of myosin phenotype in muscle fibers under conditions of gravitational unloading. Can AMPK be involved in the regulation of MyHC gene expression in skeletal muscle fibers? To answer this question, Vilchinskaya et al. (2017) carried out an experiment with AICAR pretreatment of Wistar male rats [98]. There was a significant decrease in MyHC I (β) pre-mRNA expression and a pronounced tendency to a decrease in the mature MyHC I(β) mRNA expression in rat soleus muscle after 24 h of HS. These results are in good agreement with the data previously obtained in skeletal muscles of Sprague-Dawley rats under similar conditions [13]. However, when exposed to AICAR, there were not any significant decreases in the MyHC I(β) pre-mRNA and mature mRNA expression. Since one of the possible mechanisms of AMPK gene expression regulation is linked to histone deacetylase 4 and 5 (HDAC4/HDAC5) phosphorylation status [49,50], it was hypothesized that a significant decrease in AMPK Thr172 phosphorylation after one-day HS would result in HDAC4 and HDAC5 nuclear accumulation [83]. This hypothesis is supported by a previous report that showed a significant deacetylation of histone H3 at the MyHC I (β) gene in rat soleus following seven-day HS [99]. In addition, it has been previously shown that HDAC4 is predominantly localized to the nuclei in fast-twitch fibers in contrast to the sarcoplasm in slow-twitch fibers [100]. Indeed, in our experiment with 24-h HS, HDAC4 accumulation in the nuclear fraction was found; however, in the AICAR-pretreated group, the accumulation of HDAC4 in the nuclei did not occur, which correlates well with data on AMPK phosphorylation and confirms the hypothesis of AMPK-dependent control of nucleocytoplasmic trafficking of HDAC4 [98,101]. Recently, Yoshihara and co-authors have found nuclear accumulation of HDAC4 in rat gastrocnemius muscle following 10 days of ankle joint immobilization. This accumulation (as in the experiment of Vilchinskaya and co-authors) was accompanied by a decrease in the level of AMPK phosphorylation [102]. As for HDAC5, even a slight increase in AMPK activity in the control AICAR-pretreated rat was accompanied by a decrease in HDAC5 content in the nuclear fraction. This phenomenon could be associated with HDAC5 phosphorylation via AMPK. However, 24-h HS resulted in a significant decrease in the nuclear HDAC5 content.

Such a reduction in HDAC5 content in the nuclear fraction could be associated with HDAC5 degradation or HDAC5 phosphorylation and subsequent nuclear export. HDAC5 can also be a target for protein kinase D (PKD) [103,104]. Since the AMPK activity is significantly reduced within the first day of unloading (see above), it appears that HDAC5 nuclear export during unloading is not directly linked to AMPK. It has been shown that a decrease in AMPK activity can result in an upregulation of protein kinase D (PKD) phosphorylation [103]. Indeed, one-day HS resulted in a significant increase in PKD Ser916 phosphorylation, however, in the HS+AICAR group, PKD phosphorylation did not differ from the control levels.

Interestingly, such a decrease in PKD phosphorylation in the HS+AICAR group vs. the HS group can significantly attenuate the loss of nuclear HDAC5. It is noteworthy that an increase in PKD phosphorylation in unloaded animals did not affect the content of nuclear HDAC4. Obviously, under unloading conditions, HDAC4 does not appear to be a target of PKD. This study was the first to observe the reciprocal relationship between AMPK and PKD in an inactivated skeletal muscle. In addition, gravitational unloading led to an increase in the level of negative AMPK phosphorylation on Ser485/491, which is known to be associated with PKD [35]. However, allosteric activation of AMPK by AICAR reduced the intensity of negative phosphorylation, possibly due to a reduction in PKD phosphorylation.

AMPK is known to stimulate the expression of peroxisome proliferator-activated receptor gamma coactivator 1-alpha (PGC1α), the most important regulator of the expression and activity of signaling proteins. In particular, PGC1α is involved in the control of the expression of the slow isoform of MyHC [105]. A significant decrease in PGC1α expression was found in mouse soleus after four days of hindlimb unloading [106]. However, Vilchinskaya et al. (2017) found no changes in PGC1α mRNA expression after 24-h HS [98]. AMPK activation by AICAR pretreatment also did not induce any changes in PGC1α expression. Obviously, the impact of AMPK on the expression of the slow isoform of MyHC during the first day of unloading is carried out primarily through the trafficking of HDAC4, without the involvement of PGC1α.

Thus, the results obtained by Vilchinskaya and the co-authors (2017) clearly show that AMPK Thr172 dephosphorylation within the first day of gravitational unloading has a significant effect on the regulation of myosin phenotype in rat postural muscle. In particular, AMPK Thr172 dephosphorylation led to a decrease in MyHC I(β) pre-mRNA and mRNA expression. The study of Vilchinskaya et al. (2017) indicates that HDAC4 is not a target of PKD at the early stages of unloading, and, probably, HDAC4 nuclear import results from a decrease in the AMPK activity (Figure 3). It is possible that an increase in PKD activity can lead to HDAC5 nuclear export.

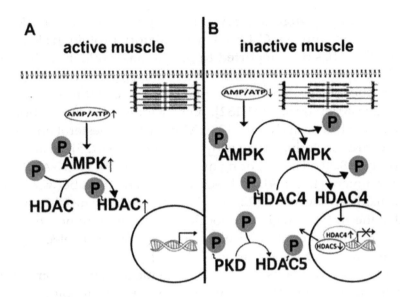

Figure 3. The role of AMPK in class IIa histone deacetylases traffic in rat soleus muscle at the initial stages of gravitational unloading (modified from [98]). (**A**): active muscle; (**B**): inactive muscle. Time-course studies on the MyHCI (β) expression demonstrate that a reduction in MyHCI (β) mRNA expression begins on the first day of unloading and then steadily decreases during, at least, two weeks of hindlimb unloading [13,14]. As shown in dominant-negative AMPK mutants, AMPK has no effect on the expression of slow MyHC after 14-day unloading [63]. It is clear that an experiment with transgenic animals does not allow for tracing the effect of AMPK on the expression of MyHC at the different time-points of unloading. However, the results of this experiment indicate that the decrease in the expression of slow isoform of MyHC after the completion of the initial stage of hindlimb unloading is determined not by AMPK, but by some other mechanisms, for example, via inhibition of the calcineurin/ Nuclear factor of activated T-cells (NFAT) signaling pathway [14,107].

6. AMPK and Disuse-Induced Motor Endplate Remodeling

Reliable neuromuscular transmission is essential for normal bodily functions. Motor endplate is a highly specialized sarcolemma region in which the transmission of activity from motor neurons to striated muscle fiber is realized. Endplate structure, including features of nicotinic acetylcholine receptors' (nAChRs) localization and distribution, are among the factors crucial for maintenance of highly efficient neuromuscular transmission [108,109].

Ultrastructure of the neuromuscular junction strongly depends on the motor activity, and exhibits high morphological and functional plasticity. Different modes of increased activity are manifested in the morphological remodeling, such as the expansion of the neuromuscular junction size. Reduced patterns of neuromuscular activity also trigger endplate remodeling. Alterations in neuromuscular junction stability and integrity progressively increase with age [110–114] and myodystrophy [115–117], in animal models of muscle injures [118,119], after denervation [120] and prolonged (four weeks) HS [121]. Although the molecular mechanisms underlying the structural and functional endplate plasticity are intensively studied, they are not completely clear [113,115,116,122–125].

Currently, many facts point to the involvement of AMPK in neuromuscular junction remodeling. AMPK is linked to a variety of cellular processes and is also considered a crucial activator of autophagy and its downstream target, ULK1 [126]. Autophagy is involved in neuromuscular junction preservation during aging, and impairments in autophagy exacerbate synaptic structure degeneration [111]. AMPK and AMPK-activated autophagy are among the most important factors that maintain neuromuscular junction stability [111,127,128]. Pharmacological activation of AMPK by AICAR administration improves integrity of neuromuscular junctions and prevents skeletal muscle pathology in a mouse model of severe spinal muscular atrophy [128]. Additionally, AICAR treatment has been shown to stimulate autophagy and ameliorate muscular dystrophy in the mdx mice, a model of Duchenne muscular dystrophy, suggesting AMPK as a powerful therapeutic target [129,130]. Also, AMPK may play a role in activating PGC-1α, a key regulatory protein in skeletal muscle adaptation to physical activity. PGC-1α has been reported to play a major role in maintaining neuromuscular junction integrity [122,123].

AMPK also plays a beneficial role in the regulation of skeletal muscle cholesterol synthesis as well as sarcolemma cholesterol levels [131,132]. Direct molecular interaction between membrane cholesterol and the nAChRs has been shown [133]. Moreover, cholesterol and lipid rafts contribute to the orchestration of nAChRs clustering at the endplate region [134,135]. In addition, AMPK can affect the Na,K-ATPase activity [38,39]. The targeting and activity of the Na,K-ATPase are also influenced by the cholesterol environment [136–139], and reciprocal interactions between cholesterol and the α2 Na,K-ATPase isozyme has been suggested [22,140]. Notably, both the nAChRs and the α2 Na,K-ATPase isozyme are enriched at the endplate region, co-localized, and co-immunoprecipitated, suggesting that these proteins exist as a functional multimolecular complex to regulate electrogenesis and to maintain the effectiveness of neuromuscular transmission [141–143].

The loss of the α2 Na,K-ATPase isozyme electrogenic activity accompanied by disturbances in lipid rafts and endplate structure stability were observed in rat soleus muscle even after 6–12 h of HS [21,22,64]. Such acute disuse decreased the endplate area and increased the density of the nAChRs distribution. These changes were accompanied by decreased phosphorylation of AMPK and its substrate, ACC, and increased autophagy [64]. Autophagy is known to be involved in the nAChRs turnover regulation [127] and has a major impact on neuromuscular synaptic function [111]. So, an increase in autophagy after acute HS can reflect an adaptive response to compensate for the endplate area loss through increasing the density of the nAChRs distribution [64].

Notably, pretreatment of the rats with AMPK activator, AICAR, followed by HS, stabilized the resting membrane potential, endplates area, and the nAChRs distribution density, indicating that AMPK activation can prevent disuse-induced endplate structural and functional reorganization (unpublished observation).

In summary, these novel findings indicate that endplate functional and structural characteristics rapidly (within hours) respond to skeletal muscle disuse. Decreased phosphorylation of AMPK accompanied by increased autophagy is the earliest disuse-induced remodeling event preceding the overt skeletal muscle atrophy.

7. Conclusions

AMPK demonstrates multidirectional changes in a mammalian soleus muscle during gravitational unloading: A deep decrease in the level of phosphorylation and kinase activity at the early stages of the process and a significant increase in activity at the later stages. Time-course changes in AMPK activity under unloading are nonlinear and require a more detailed analysis at each stage of the process. The experiments discussed in this review show that changes in AMPK activity under unloading conditions can have a significant impact on the key signaling pathways and molecular structures in skeletal muscle fiber. As a result of reduction in AMPK activity at the initial stage of unloading, paradoxical hyperphosphorylation of p70S6K occurs, which, according to in vitro experiments, can lead to the activation of proteolytic processes in muscle fiber [83,87]. A decrease in the level of phosphorylation and kinase activity of AMPK at the early stages of unloading also affects the change in nucleocytoplasmic traffic of class IIA HDACs, resulting in a decrease in the expression of the myh7 gene (slow isoform of MyHC) and, possibly, a number of other genes controlling energy metabolism. At the later time-points of unloading, an increase in the level of AMPK phosphorylation/activity is accompanied by a decrease in the activity of p70S6K [76], FOXO3 dephosphorylation, and an increase in the expression of the key enzymes of the ubiquitin-proteasome pathway [65], which, obviously, should lead to decreased muscle protein synthesis and enhanced proteolysis. Unfortunately, to date, the precise physiological mechanisms, both intramuscular and systemic, underlying such a complex and nonlinear nature of AMPK activity during gravitational unloading are not known. It is possible that glycogen accumulation [27] or a change in the ratio of dephosphorylated and phosphorylated high-energy phosphates could lead to a deep AMPK dephosphorylation. The cause of a gradual increase in AMPK phosphorylation in rat soleus during the first week of HS remains unclear. It is possible that this increase is due to a gradual increase in the electromyographic activity of the postural muscle [15]. It is difficult to explain a significant increase in AMPK phosphorylation in rat soleus by the end of the second week of HS. Possibly, such systemic factors as BDNF and/or interleukin-6 could play a role in these processes. All these questions have yet to be answered.

Thus, recent studies have revealed the key role of AMPK in the processes of deep remodeling of signaling pathways, leading to changes in metabolism, structure, and function of postural muscle fibers under unloading conditions. Further studies are needed to elucidate new signaling mechanisms that trigger, determine, and limit atrophy development and intrinsic muscle stiffness, as well as myosin phenotype changes, in a mammalian postural muscle under mechanical unloading.

Acknowledgments: The authors express their deep gratitude to Timur M. Mirzoev for the participation in the manuscript preparation and creative criticism of the paper.

Abbreviations

AMPK	5′ adenosine monophosphate -activated protein kinase
AICAR	5-aminoimidazole-4-carboxamide ribonucleotide
BGPA	β-Guanidinopropionic acid
HS	hindlimb suspension
mTOR	mammalian/mechanistic target of rapamycin
p70S6K	ribosomal protein S6 kinase beta-1
CaMKKβ	Ca(2+)/calmodulin-dependent protein kinase kinase β
LKB1	liver kinase B1
BDNF	brain-derived neurotrophic factor
MuRF-1	Muscle RING-finger protein-1
FOXO	forkhead box proteins
IRS-1	insulin receptor substrate 1
MyHC	myosin heavy chain isoforms
nAChRs	nicotinic acetylcholine receptors
ULK-1	Unc-51-like kinase
PGC1α	peroxisome proliferator-activated receptor gamma coactivator 1-alpha

References

1. Grigor'ev, A.I.; Kozlovskaia, I.B.; Shenkman, B.S. The role of support afferents in organisation of the tonic muscle system. *Rossiiskii fiziologicheskii zhurnal imeni IM Sechenova* **2004**, *90*, 508–521.
2. Kozlovskaia, I.B.; Grigor'eva, L.S.; Gevlich, G.I. Comparative analysis of the effect of weightlessness and its model on the velocity-strength properties and tonus of human skeletal muscles. *Kosm. Biol. Aviakosm. Med.* **1984**, *18*, 22–26. [PubMed]
3. Oganov, V.S.; Skuratova, S.A.; Murashko, L.M.; Guba, F.; Takach, O. Effect of short-term space flights on physiological properties and composition of myofibrillar proteins of the skeletal muscles of rats. *Kosm. Biol. Aviakosm. Med.* **1988**, *22*, 50–54. [PubMed]
4. Baldwin, K.M.; Haddad, F.; Pandorf, C.E.; Roy, R.R.; Edgerton, V.R. Alterations in muscle mass and contractile phenotype in response to unloading models: Role of transcriptional/pretranslational mechanisms. *Front. Physiol.* **2013**, *4*, 284. [CrossRef] [PubMed]
5. Kandarian, S.C.; Stevenson, E.J. Molecular events in skeletal muscle during disuse atrophy. *Exerc. Sport Sci. Rev.* **2002**, *30*, 111–116. [CrossRef] [PubMed]
6. Adams, G.R.; Baldwin, K.M. Age dependence of myosin heavy chain transitions induced by creatine depletion in rat skeletal muscle. *J. Appl. Physiol.* **1995**, *78*, 368–371. [CrossRef] [PubMed]

7.   Shenkman, B.S. From Slow to Fast: Hypogravity-Induced Remodeling of Muscle Fiber Myosin Phenotype. *Acta Nat.* **2016**, *8*, 47–59.

8.   Chopard, A.; Hillock, S.; Jasmin, B.J. Molecular events and signalling pathways involved in skeletal muscle disuse-induced atrophy and the impact of countermeasures. *J. Cell Mol. Med.* **2009**, *13*, 3032–3050. [CrossRef] [PubMed]

9.   Bodine, S.C. Disuse-induced muscle wasting. *Int. J. Biochem. Cell Biol.* **2013**, *45*, 2200–2208. [CrossRef] [PubMed]

10. Kachaeva, E.V.; Shenkman, B.S. Various jobs of proteolytic enzymes in skeletal muscle during unloading: Facts and speculations. *J. Biomed. Biotechnol.* **2011**, *2012*, 493618. [CrossRef] [PubMed]

11. Mirzoev, T.; Tyganov, S.; Vilchinskaya, N.; Lomonosova, Y.; Shenkman, B. Key Markers of mTORC1-Dependent and mTORC1-Independent Signaling Pathways Regulating Protein Synthesis in Rat Soleus Muscle during Early Stages of Hindlimb Unloading. *Cell. Physiol. Biochem.* **2016**, *39*, 1011–1020. [CrossRef] [PubMed]

12. Stevens, L.; Sultan, K.R.; Peuker, H.; Gohlsch, B.; Mounier, Y.; Pette, D. Time dependent changes in myosin heavy chain mRNA and protein isoforms 94 in unloaded soleus muscle of rat. *Am. J. Physiol. Cell. Physiol.* **1999**, *277*, 1044–1049. [CrossRef] [PubMed]

13. Giger, J.M.; Bodell, P.W.; Zeng, M.; Baldwin, K.M.; Haddad, F. Rapid muscle atrophy response to unloading: Pretranslational processes involving MHC and actin. *J. Appl. Physiol.* **2009**, *107*, 1204–1212. [CrossRef] [PubMed]

14. Lomonosova, Y.N.; Turtikova, O.V.; Shenkman, B.S. Reduced expression of MyHC slow isoform in rat soleus during unloading is accompanied by alterations of endogenous inhibitors of calcineurin/NFAT signaling pathway. *J. Muscle Res. Cell Motil.* **2016**, *37*, 7–16. [CrossRef] [PubMed]

15. Alford, E.K.; Roy, R.R.; Hodgson, J.A.; Edgerton, V.R. Electromyography of rat soleus, medial gastrocnemius, and tibialis anterior during hind limb suspension. *Exp. Neurol.* **1987**, *96*, 635–649. [CrossRef]

16. Kawano, F.; Matsuoka, Y.; Oke, Y.; Higo, Y.; Terada, M.; Wang, X.D.; Nakai, N.; Fukuda, H.; Imajoh-Ohmi, S.; Ohira, Y. Role(s) of nucleoli and phosphorylation of ribosomal protein S6 and/or HSP27 in the regulation of muscle mass. *Am. J. Physiol. Cell Physiol.* **2007**, *293*, 35–44. [CrossRef] [PubMed]

17. De-Doncker, L.; Kasri, M.; Picquet, F.; Falempin, M. Physiologically adaptive changes of the L5afferent neurogram and of the rat soleus EMG activity during 14 days of hindlimb unloading and recovery. *J. Exp. Biol.* **2005**, *208*, 4585–4592. [CrossRef] [PubMed]

18. Kozlovskaya, I.; Dmitrieva, I.; Grigorieva, L.; Kirenskaya, A.; Kreidich, Y. Gravitational mechanisms in the motor system studies in real and simulated weightlessness. In *Stance and Motion Facts and Concepts*; Gurfinkel, V.S., Loffe, M.E., Massion, J., Eds.; Springer: New York, NY, USA, 1988; ISBN 978-1-4899-0823-0.

19. Navasiolava, N.M.; Custaud, M.A.; Tomilovskaya, E.S.; Larina, I.M.; Mano, T.; Gauquelin-Koch, G.; Gharib, C.; Kozlovskaya, I.B. Long-term dry immersion: Review and prospects. *Eur. J. Appl. Physiol.* **2011**, *111*, 1235–1260. [CrossRef] [PubMed]

20. Kravtsova, V.V.; Matchkov, V.V.; Bouzinova, E.V.; Vasiliev, A.N.; Razgovorova, I.A.; Heiny, J.A.; Krivoi, I.I. Isoform-specific Na,K-ATPase alterations precede disuse-induced atrophy of rat soleus muscle. *Biomed. Res. Int.* **2015**, 720172. [CrossRef] [PubMed]

21. Kravtsova, V.V.; Petrov, A.M.; Matchkov, V.V.; Bouzinova, E.V.; Vasiliev, A.N.; Benziane, B.; Zefirov, A.L.; Chibalin, A.V.; Heiny, J.A.; Krivoi, I.I. Distinct α2 Na,K-ATPase membrane pools are differently involved in early skeletal muscle remodeling during disuse. *J. Gen. Physiol.* **2016**, *147*, 175–188. [CrossRef] [PubMed]

22. Petrov, A.M.; Kravtsova, V.V.; Matchkov, V.V.; Vasiliev, A.N.; Zefirov, A.L.; Chibalin, A.V.; Heiny, J.A.; Krivoi, I.I. Membrane lipid rafts are disturbed in the response of rat skeletal muscle to short-term disuse. *Am. J. Physiol. Cell Physiol.* **2017**, *312*, 627–C637. [CrossRef] [PubMed]

23. Lechado, I.; Terradas, A.; Vitadello, M.; Traini, L.; Namuduri, A.V.; Gastaldello, S.; Gorza, L. Sarcolemmal loss of active nNOS (Nos1) is an oxidative stress-dependent, early event driving disuse atrophy. *J. Pathol.* **2018**. [CrossRef]

24. Gordon, S.E.; Flück, M.; Booth, F.W. Selected Contribution: Skeletal muscle focal adhesion kinase, paxillin, and serum response factor are loading dependent. *J. Appl. Physiol.* **2001**, *90*, 1174–1183. [CrossRef] [PubMed]

25. Ohira, Y.; Yasui, W.; Kariya, F.; Wakatsuki, T.; Nakamura, K.; Asakura, T.; Edgerton, V.R. Metabolic adaptation of skeletal muscles to gravitational unloading. *Acta Astronaut.* **1994**, *33*, 113–117. [CrossRef]

26. Wakatsuki, T.; Ohira, Y.; Yasui, W.; Nakamura, K.; Asakura, T.; Ohno, H.; Yamamoto, M. Responses of contractile properties in rat soleus to high-energy phosphates and/or unloading. *Jpn. J. Physiol.* **1994**, *44*, 193–204. [CrossRef] [PubMed]

27. Henriksen, E.J.; Tischler, M.E. Time Course of the Response of Carbohydrate Metabolism to Unloading of the Soleus. *Metabolism* **1988**, *37*, 201–208. [CrossRef]

28. Witczak, C.A.; Sharoff, C.G.; Goodyear, L.J. AMP-activated protein kinase in skeletal muscle: From structure and localization to its role as a master regulator of cellular metabolism. *Cell Mol. Life Sci.* **2008**, *65*, 3737–3755. [CrossRef] [PubMed]

29. Mounier, R.; Théret, M.; Lantier, L.; Foretz, M.; Viollet, B. Expanding roles for AMPK in skeletal muscle plasticity. *Trends Endocrinol. Metab.* **2015**, *26*, 275–286. [CrossRef] [PubMed]

30. Matthews, V.B.; Aström, M.B.; Chan, M.H.; Bruce, C.R.; Krabbe, K.S.; Prelovsek, O.; Akerström, T.; Yfanti, C.; Broholm, C.; Mortensen, O.H.; et al. Brain-derived neurotrophic factor is produced by skeletal muscle cells in response to contraction and enhances fat oxidation via activation of AMP-activated protein kinase. *Diabetologia* **2009**, *52*, 1409–1418. [CrossRef] [PubMed]

31. Glund, S.; Treebak, J.T.; Long, Y.C.; Barres, R.; Viollet, B.; Wojtaszewski, J.F.; Zierath, J.R. Role of adenosine 5′-monophosphate-activated protein kinase in interleukin-6 release from isolated mouse skeletal muscle. *Endocrinology* **2009**, *150*, 600–606. [CrossRef] [PubMed]

32. Lira, V.A.; Soltow, Q.A.; Long, J.H.; Betters, J.L.; Sellman, J.E.; Criswell, D.S. Nitric oxide increases GLUT4 expression and regulates AMPK signaling in skeletal muscle. *Am. J. Physiol. Endocrinol. Metab.* **2007**, *293*, 1062–1068. [CrossRef] [PubMed]

33. Lira, V.A.; Brown, D.L.; Lira, A.K.; Kavazis, A.N.; Soltow, Q.A.; Zeanah, E.H.; Criswell, D.S. Nitric oxide and AMPK cooperatively regulate PGC-1α in skeletal muscle. *J. Physiol.* **2010**, *588*, 3551–3566. [CrossRef] [PubMed]

34. Garbincius, J.F.; Michele, D.E. Dystrophin-glycoprotein complex regulates muscle nitric oxide production through mechanoregulation of AMPK signaling. *Proc. Natl. Acad. Sci. USA* **2015**, *112*, 13663–13668. [CrossRef] [PubMed]

35. Coughlan, K.A.; Valentine, R.J.; Sudit, B.S.; Allen, K.; Dagon, Y.; Kahn, B.B.; Ruderman, N.B.; Saha, A.K. PKD1 Inhibits AMPKα2 through Phosphorylation of Serine 491 and Impairs Insulin Signaling in Skeletal Muscle Cells. *Biochem. J.* **2016**, *473*, 4681–4697. [CrossRef] [PubMed]

36. Hardie, D.G.; Ross, F.A.; Hawley, S.A. AMP-activated protein kinase: A target for drugs both ancient and modern. *Chem. Biol.* **2012**, *19*, 1222–1236. [CrossRef] [PubMed]

37. Koay, A.; Woodcroft, B.; Petrie, E.J.; Yue, H.; Emanuelle, S.; Bieri, M.; Bailey, M.F.; Hargreaves, M.; Park, J.T.; Park, K.H.; et al. AMPK beta subunits display isoform specific affinities for carbohydrates. *FEBS Lett.* **2010**, *584*, 3499–3503. [CrossRef] [PubMed]

38. Ingwersen, M.S.; Kristensen, M.; Pilegaard, H.; Wojtaszewski, J.F.; Richter, E.A.; Juel, C. Na, K-ATPase activity in mouse muscle is regulated by AMPK and PGC-1α. *J. Membr. Biol.* **2011**, *242*, 1–10. [CrossRef] [PubMed]

39. Benziane, B.; Bjornholm, M.; Pirkmajer, S.; Austin, R.L.; Kotova, O.; Viollet, B.; Zierath, J.R.; Chibalin, A.V. Activation of AMP-activated protein kinase stimulates Na⁺,K⁺-ATPase activity in skeletal muscle cells. *J. Biol. Chem.* **2012**, *287*, 23451–23463. [CrossRef] [PubMed]

40. Chen, Z.P.; McConell, G.K.; Belinda, J.; Snow, R.J.; Canny, B.J.; Kemp, B.E. AMPK signaling in contracting human skeletal muscle: Acetyl-CoA carboxylase and NO synthase phosphorylation. *Am. J. Physiol. Endocrinol. Metab.* **2000**, *279*, E1202–E1206. [CrossRef] [PubMed]

41. Inoki, K.; Zhu, T.; Guan, K.L. TSC2 mediates cellular energy response to control cell growth and survival. *Cell* **2003**, *115*, 577–590. [CrossRef]

42. Gwinn, D.M.; Shackelford, D.B.; Egan, D.F.; Mihaylova, M.M.; Mery, A.; Vasquez, D.S.; Turk, B.E.; Shaw, R.J. AMPK phosphorylation of raptor mediates a metabolic checkpoint. *Mol. Cell* **2008**, *30*, 214–226. [CrossRef] [PubMed]

43. Nystrom, G.J.; Lang, C.H. Sepsis and AMPK Activation by AICAR Differentially Regulate FoxO-1, -3 and -4 mRNA in Striated Muscle. *Int. J. Clin. Exp. Med.* **2008**, *1*, 50–63. [PubMed]

44. Nakashima, K.; Yakabe, Y.; Ishida, A.; Katsumata, M. Effects of orally administered glycine on myofibrillar proteolysis and expression of proteolytic-related genes of skeletal muscle in chicks. *Amino Acids* **2008**, *35*, 451–456. [CrossRef] [PubMed]

45. Kim, J.; Kundu, M.; Viollet, B.; Guan, K. AMPK and mTOR regulate autophagy via direct phosphorylation of ULK1. *Nat. Cell Biol.* **2011**, *13*, 132–141. [CrossRef] [PubMed]
46. Egan, D.F.; Shackelford, D.B.; Mihaylova, M.M.; Gelino, S.; Kohnz, R.A.; Mair, W.; Vasquez, D.S.; Joshi, A.; Gwinn, D.M.; Taylor, R.; et al. Phosphorylation of ULK1 (hATG1) by AMP-activated protein kinase connects energy sensing to mitophagy. *Science* **2011**, *331*, 456–461. [CrossRef] [PubMed]
47. Dunlop, E.A.; Tee, A.R. The kinase triad, AMPK, mTORC1 and ULK1, maintains energy and nutrient homoeostasis. *Biochem. Soc. Trans.* **2013**, *41*, 939–943. [CrossRef] [PubMed]
48. Mihaylova, M.M.; Shaw, R.J. The AMPK signalling pathway coordinates cell growth, autophagy and metabolism. *Nat. Cell Biol.* **2011**, *13*, 1016–1023. [CrossRef] [PubMed]
49. Röckl, K.S.; Hirshman, M.F.; Brandauer, J.; Fujii, N.; Witters, L.A.; Goodyear, L.J. Skeletal muscle adaptation to exercise training: AMP-activated protein kinase mediates muscle fiber type shift. *Diabetes* **2007**, *56*, 2062–2069. [CrossRef] [PubMed]
50. McGee, S.L.; Hargreaves, M. AMPK-mediated regulation of transcription in skeletal muscle. *Clin. Sci. (London)* **2010**, *118*, 507–518. [CrossRef] [PubMed]
51. Guo, Y.; Meng, J.; Tang, Y.; Wang, T.; Wei, B.; Feng, R.; Gong, B.; Wang, H.; Ji, G.; Lu, Z. AMP-activated kinase α2 deficiency protects mice from denervation-induced skeletal muscle atrophy. *Arch. Biochem. Biophys.* **2016**, *600*, 56–60. [CrossRef] [PubMed]
52. Gao, H.; Li, Y.F. Distinct signal transductions in fast- and slow- twitch muscles upon denervation. *Physiol. Rep.* **2018**, *6*. [CrossRef] [PubMed]
53. Kostovski, E.; Boon, H.; Hjeltnes, N.; Lundell, L.S.; Ahlsén, M.; Chibalin, A.V.; Krook, A.; Iversen, P.O.; Widegren, U. Altered content of AMP-activated protein kinase isoforms in skeletal muscle from spinal cord injured subjects. *Am. J. Physiol. Endocrinol. Metab.* **2013**, *305*, E1071–E1080. [CrossRef] [PubMed]
54. Mirzoev, T.M.; Vil'chinskaia, N.A.; Lomonosova, I.N.; Nemirovskaia, T.L.; Shenkman, B.S. Effect of 30-day space flight and subsequent readaptation on the signaling processes in m. longissimus dorsi of mice. *Aviakosm. Ekolog. Med.* **2014**, *48*, 12–15. [PubMed]
55. Augusto, V.; Padovani, R.C.; Campos, G.E.R. Skeletal muscle fiber types in C57BL6J mice. *Braz. J. Morphol. Sci.* **2004**, *21*, 89–94.
56. Vilchinskaya, N.A.; Mirzoev, T.M.; Lomonosova, Y.N.; Kozlovskaya, I.B.; Shenkman, B.S. Human muscle signaling responses to 3-day head-out dry immersion. *J. Musculoskelet. Neuronal Interact.* **2015**, *15*, 286–293. [PubMed]
57. Hawley, S.A.; Boudeau, J.; Reid, J.L.; Mustard, K.J.; Udd, L.; Mäkelä, T.P.; Alessi, D.R.; Hardie, D.G. Complexes between the LKB1 tumor suppressor, STRAD alpha/beta and MO25 alpha/beta are upstream kinases in the AMP-activated protein kinase cascade. *J. Biol.* **2003**, *2*, 28. [CrossRef] [PubMed]
58. Hawley, S.A.; Pan, D.A.; Mustard, K.J.; Ross, L.; Bain, J.; Edelman, A.M.; Frenguelli, B.G.; Hardie, D.G. Calmodulin-dependent protein kinase kinase-beta is an alternative upstream kinase for AMP-activated protein kinase. *Cell Metab.* **2005**, *2*, 9–19. [CrossRef] [PubMed]
59. Woods, A.; Vertommen, D.; Neumann, D.; Turk, R.; Bayliss, J.; Schlattner, U.; Wallimann, T.; Carling, D.; Rider, M.H. Identification of phosphorylation sites in AMP-activated protein kinase (AMPK) for upstream AMPK kinases and study of their roles by site-directed mutagenesis. *J Biol. Chem.* **2003**, *278*, 28434–28442. [CrossRef] [PubMed]
60. Han, B.; Zhu, M.J.; Ma, C.; Du, M. Rat hindlimb unloading down-regulates insulin like growth factor-1 signaling and AMP-activated protein kinase, and leads to severe atrophy of the soleus muscle. *Appl. Physiol. Nutr. Metab.* **2007**, *32*, 1115–1123. [CrossRef] [PubMed]
61. Hilder, T.L.; Baer, L.A.; Fuller, P.M.; Fuller, C.A.; Grindeland, R.E.; Wade, C.E.; Graves, L.M. Insulinindependent pathways mediating glucose uptake in hindlimbsuspended skeletal muscle. *J. Appl. Physiol.* **2005**, *99*, 2181–2188. [CrossRef] [PubMed]
62. Egawa, T.; Goto, A.; Ohno, Y.; Yokoyama, S.; Ikuta, A.; Suzuki, M.; Sugiura, T.; Ohira, Y.; Yoshioka, T.; Hayashi, T.; et al. Involvement of AMPK in regulating slow-twitch muscle atrophy during hindlimb unloading in mice. *Am. J. Physiol. Endocrinol. Metab.* **2015**, *309*, 651–662. [CrossRef] [PubMed]
63. Egawa, T.; Ohno, Y.; Goto, A.; Yokoyama, S.; Hayashi, T.; Goto, K. AMPK Mediates Muscle Mass Change But Not the Transition of Myosin Heavy Chain Isoforms during Unloading and Reloading of Skeletal Muscles in Mice. *Int. J. Mol. Sci.* **2018**, *19*, 2954. [CrossRef] [PubMed]

64.　Chibalin, A.V.; Benziane, B.; Zakyrjanova, G.F.; Kravtsova, V.V.; Krivoi, I.I. Early endplate remodeling and skeletal muscle signaling events following rat hindlimb suspension. *J. Cell. Physiol.* **2018**, *233*, 6329–6336. [CrossRef] [PubMed]

65.　Zhang, S.F.; Zhang, Y.; Li, B.; Chen, N. Physical inactivity induces the atrophy of skeletal muscle of rats through activating AMPK/FoxO3 signal pathway. *Eur. Rev. Med. Pharmacol. Sci.* **2018**, *22*, 199–209. [PubMed]

66.　Grano, M.; Mori, G.; Minielli, V.; Barou, O.; Colucci, S.; Giannelli, G.; Alexandre, C.; Zallone, A.Z.; Vico, L. Rat hindlimb unloading by tail suspension reduces osteoblast differentiation, induces IL-6 secretion, and increases bone resorption in ex vivo cultures. *Calcif. Tissue Int.* **2002**, *70*, 176–185. [CrossRef] [PubMed]

67.　Yakabe, M.; Ogawa, S.; Ota, H.; Iijima, K.; Eto, M.; Ouchi, Y.; Akishita, M. Inhibition of interleukin-6 decreases atrogene expression and ameliorates tail suspension-induced skeletal muscle atrophy. *PLoS ONE* **2018**, *13*, e0191318. [CrossRef] [PubMed]

68.　Mutin-Carnino, M.; Carnino, A.; Roffino, S.; Chopard, A. Effect of muscle unloading, reloading and exercise on inflammation during a head-down bed rest. *Int. J. Sports Med.* **2014**, *35*, 28–34. [CrossRef] [PubMed]

69.　Ruderman, N.B.; Keller, C.; Richard, A.M.; Saha, A.K.; Luo, Z.; Xiang, X.; Giralt, M.; Ritov, V.B.; Menshikova, E.V.; Kelley, D.E.; et al. Interleukin-6 regulation of AMP-activated protein kinase. Potential role in the systemic response to exercise and prevention of the metabolic syndrome. *Diabetes* **2006**, *55*, S48–54. [CrossRef] [PubMed]

70.　Yang, W.; Zhang, H. Effects of hindlimb unloading on neurotrophins in the rat spinal cord and soleus muscle. *Brain Res.* **2016**, *1630*, 1–9. [CrossRef] [PubMed]

71.　Dupont, E.; Cieniewski-Bernard, C.; Bastide, B.; Stevens, L. Electrostimulation during hindlimb unloading modulates PI3K-AKT downstream targets without preventing soleus atrophy and restores slow phenotype through ERK. *Am. J. Physiol. Regul. Integr. Comp. Physiol.* **2011**, *300*, 408–417. [CrossRef] [PubMed]

72.　Bajotto, G.; Sato, Y.; Kitaura, Y.; Shimomura, Y. Effect of branched-chain amino acid supplementation during unloading on regulatory components of protein synthesis in atrophied soleus muscles. *Eur. J. Appl. Physiol.* **2011**, *111*, 1815–1828. [CrossRef] [PubMed]

73.　Childs, T.E.; Spangenburg, E.E.; Vyas, D.R.; Booth, F.W. Temporal alterations in protein signaling cascades during recovery from muscle atrophy. *Am. J. Physiol. Cell. Physiol.* **2003**, *285*, 391–398. [CrossRef] [PubMed]

74.　Sugiura, T.; Abe, N.; Nagano, M.; Goto, K.; Sakuma, K.; Naito, H.; Yoshioka, T.; Powers, S.K. Changes in PKB/Akt and calcineurin signaling during recovery in atrophied soleus muscle induced by unloading. *Am. J. Physiol. Regul. Integr. Comp. Physiol.* **2005**, *288*, 1273–1278. [CrossRef] [PubMed]

75.　Gwag, T.; Lee, K.; Ju, H.; Shin, H.; Lee, J.W.; Choi, I. Stress and signaling responses of rat skeletal muscle to brief endurance exercise during hindlimb unloading: A catch-up process for atrophied muscle. *Cell Physiol. Biochem.* **2009**, *24*, 537–546. [CrossRef] [PubMed]

76.　Lysenko, E.A.; Turtikova, O.V.; Kachaeva, E.V.; Ushakov, I.B.; Shenkman, B.S. Time course of ribosomal kinase activity during hindlimb unloading. *Dokl. Biochem. Biophys.* **2010**, *434*, 223–226. [CrossRef] [PubMed]

77.　Hamilton, D.L.; Philp, A.; MacKenzie, M.G.; Patton, A.; Towler, M.C.; Gallagher, I.J.; Bodine, S.C.; Baar, K. Molecular brakes regulating mTORC1 activation in skeletal muscle following synergist ablation. *Am. J. Physiol. Endocrinol. Metab.* **2014**, *307*, E365–E373. [CrossRef] [PubMed]

78.　Xu, J.; Ji, J.; Yan, X.H. Cross-talk between AMPK and mTOR in regulating energy balance. *Crit. Rev. Food. Sci. Nutr.* **2012**, *52*, 373–381. [CrossRef] [PubMed]

79.　McGee, S.L.; Mustard, K.J.; Hardie, D.G.; Baar, K. Normal hypertrophy accompanied by phosphoryation and activation of AMP-activated protein kinase alpha1 following overload in LKB1 knockout mice. *J. Physiol.* **2008**, *586*, 1731–1741. [CrossRef] [PubMed]

80.　Mounier, R.; Lantier, L.; Leclerc, J.; Sotiropoulos, A.; Pende, M.; Daegelen, D.; Sakamoto, K.; Foretz, M.; Viollet, B. Important role for AMPKalpha1 in limiting skeletal muscle cell hypertrophy. *FASEB J.* **2009**, *23*, 2264–2273. [CrossRef] [PubMed]

81.　Hornberger, T.A.; Hunter, R.B.; Kandarian, S.C.; Esser, K.A. Regulation of translation factors during hindlimb unloading and denervation of skeletal muscle in rats. *Am. J. Physiol. Cell Physiol.* **2001**, *281*, 179–187. [CrossRef] [PubMed]

82.　Krawiec, B.J.; Nystrom, G.J.; Frost, R.A.; Jefferson, L.S.; Lang, C.H. AMP-activated protein kinase agonists increase mRNA content of the muscle-specific ubiquitin ligases MAFbx and MuRF1 in C2C12 cells. *Am. J. Physiol. Endocrinol. Metab.* **2007**, *292*, E1555–E1567. [CrossRef] [PubMed]

83. Vilchinskaya, N.A.; Mochalova, E.P.; Belova, S.P.; Shenkman, B.S. Dephosphorylation of AMP-activated protein kinase in a postural muscle: A key signaling event on the first day of functional unloading. *Biophysics* **2016**, *61*, 1019–1025. [CrossRef]

84. Hsieh, C.T.; Chuang, J.H.; Yang, W.C.; Yin, Y.; Lin, Y. Ceramide inhibits insulin-stimulated Akt phosphorylation through activation of Rheb/mTORC1/S6K signaling in skeletal muscle. *Cell. Signal.* **2014**, *26*, 1400–1408. [CrossRef] [PubMed]

85. Erickson, K.A.; Smith, M.E.; Anthonymuthu, T.S.; Evanson, M.J.; Brassfield, E.S.; Hodson, A.E.; Bressler, M.A.; Tucker, B.J.; Thatcher, M.O.; Prince, J.T.; et al. AICAR inhibits ceramide biosynthesis in skeletal muscle. *Diabetol. Metab. Syndr.* **2012**, *4*, 45. [CrossRef] [PubMed]

86. Bryndina, I.G.; Shalagina, M.N.; Ovechkin, S.V.; Ovchinina, N.G. Sphingolipids in skeletal muscles of C57B1/6 mice after short-term simulated microgravity. *Rossiiskii fiziologicheskii zhurnal imeni IM Sechenova* **2014**, *100*, 1280–1286.

87. Salaun, E.; Lefeuvre-Orfila, L.; Cavey, T.; Martin, B.; Turlin, B.; Ropert, M.; Loreal, O.; Derbré, F. Myriocin prevents muscle ceramide accumulation but not muscle fiber atrophy during short-term mechanical unloading. *J. Appl. Physiol.* **2016**, *120*, 178–187. [CrossRef] [PubMed]

88. Shenkman, B.; Vilchinskaya, N.; Mochalova, E.; Belova, S.; Nemirovskaya, T. Signaling events at the early stage of muscle disuse. *FEBS J.* **2017**, *284*, 367. [CrossRef]

89. Nakao, R.; Hirasaka, K.; Goto, J.; Ishidoh, K.; Yamada, C.; Ohno, A.; Okumura, Y.; Nonaka, I.; Yasutomo, K.; Baldwin, K.M.; et al. Ubiquitin ligase Cbl-b is a negative regulator for insulin-like growth factor 1 signaling during muscle atrophy caused by unloading. *Mol. Cell. Biol.* **2009**, *29*, 4798–4811. [CrossRef] [PubMed]

90. Shenkman, B.S.; Nemirovskaya, T.L. Calcium-dependent signaling mechanisms and soleus fiber remodeling under gravitational unloading. *J. Muscle Res. Cell Motil.* **2008**, *29*, 221–230. [CrossRef] [PubMed]

91. Fitts, R.H.; Riley, D.R.; Widrick, J.J. Physiology of a microgravity environment invited review: Microgravity and skeletal muscle. *J. Appl. Physiol.* **2000**, *89*, 823–839. [CrossRef] [PubMed]

92. Templeton, G.H.; Sweeney, H.L.; Timson, B.F.; Padalino, M.; Dudenhoeffer, G.A. Changes in fiber composition of soleus muscle during rat hindlimb suspension. *J. Appl. Physiol.* **1988**, *65*, 1191–1195. [CrossRef] [PubMed]

93. Desplanches, D.; Mayet, M.H.; Sempore, B.; Flandrois, R. Structural and functional responses to prolonged hindlimb suspension in rat muscle. *J. Appl. Physiol.* **1987**, *63*, 558–563. [CrossRef] [PubMed]

94. Desplanches, D.; Kayar, S.R.; Sempore, B.; Flandrois, R.; Hoppeler, H. Rat soleus muscle ultrastructure after hindlimb suspension. *J. Appl. Physiol.* **1990**, *69*, 504–508. [CrossRef] [PubMed]

95. Riley, D.A.; Slocum, G.R.; Bain, J.L.; Sedlak, F.R.; Sedlak, F.R.; Sowa, T.E.; Mellender, J.W. Rat hindlimb unloading: Soleus histochemistry, ultrastructure, and electromyography. *J. Appl. Physiol.* **1990**, *69*, 58–66. [CrossRef] [PubMed]

96. Trappe, S.; Costill, D.; Gallagher, P.; Creer, A.; Peters, J.R.; Evans, H.; Riley, D.A.; Fitts, R.H. Exercise in space: Human skeletal muscle after 6 months aboard the International Space Station. *J. Appl. Physiol.* **2009**, *106*, 1159–1168. [CrossRef] [PubMed]

97. Liu, Y.; Randall, W.R.; Martin, F.; Schneider, M.F. Activity-dependent and -independent nuclear fluxes of HDAC4 mediated by different kinases in adult skeletal muscle. *J. Cell Biol.* **2005**, *168*, 887–897. [CrossRef] [PubMed]

98. Vilchinskaya, N.A.; Mochalova, E.P.; Nemirovskaya, T.L.; Mirzoev, T.M.; Turtikova, O.V.; Shenkman, B.S. Rapid decline in MyHC I(β) mRNA expression in rat soleus during hindlimb unloading is associated with AMPK dephosphorylation. *J. Physiol.* **2017**, *595*, 7123–7134. [CrossRef] [PubMed]

99. Pandorf, C.E.; Haddad, F.; Wright, C.; Bodell, P.W.; Baldwin, K.M. Differential epigenetic modifications of histones at the myosin heavy chain genes in fast and slow skeletal muscle fibers and in response to muscle unloading. *Am. J. Physiol. Cell Physiol.* **2009**, *297*, 6–16. [CrossRef] [PubMed]

100. Cohen, T.J.; Choi, M.C.; Kapur, M.; Lira, V.A.; Lira, V.A.; Yan, Z.; Yao, T.P. HDAC4 regulates muscle fiber type-specific gene expression programs. *Mol. Cells* **2015**, *38*, 343–348. [CrossRef] [PubMed]

101. Vilchinskaya, N.A.; Turtikova, O.V.; Shenkman, B.S. Regulation of the Nuclear–Cytoplasmic Traffic of Class IIa Histone Deacetylases in Rat Soleus Muscle at the Early Stage of Gravitational Unloading. *Biol. Membr.* **2017**, *34*, 109–115. [CrossRef]

102. Yoshihara, T.; Machida, S.; Kurosaka, Y.; Kakigi, R.; Sugiura, T.; Naito, H. Immobilization induces nuclear accumulation of HDAC4 in rat skeletal muscle. *J. Physiol. Sci.* **2016**, *66*, 337–343. [CrossRef] [PubMed]

103. McGee, S.L.; Swinton, C.; Morrison, S.; Gaur, V.; Gaur, V.; Campbell, D.E.; Jorgensen, S.B.; Kemp, B.E.; Baar, K.; Steinberg, G.R.; Hargreaves, M. Compensatory regulation of HDAC5 in muscle maintains metabolic adaptive responses and metabolism in response to energetic stress. *Faseb, J.* **2014**, *28*, 3384–3395. [CrossRef] [PubMed]

104. Ya, F.; Rubin, C.S. Protein kinase D: Coupling extracellular stimuli to the regulation of cell physiology. *EMBO Rep.* **2011**, *12*, 785–796.

105. Gan, Z.; Rumsey, J.; Hazen, B.C.; Lai, L.; Leone, T.C.; Vega, R.B.; Xie, H.; Conley, KE.; Auwerx, J.; Smith, S.R.; et al. Nuclear receptor/microRNA circuitry links muscle fiber type to energy metabolism. *J. Clin. Investig.* **2013**, *123*, 2564–2575. [CrossRef] [PubMed]

106. Cannavino, J.; Brocca, L.; Sandri, M.; Bottinelli, R.; Pellegrino, M.A. PGC1-α over-expression prevents metabolic alterations and soleus muscle atrophy in hindlimb unloaded mice. *J. Physiol.* **2014**, *592*, 4575–4589. [CrossRef] [PubMed]

107. Sharlo, C. A.; Lomonosova, Y. N.; Turtikova, O.V.; Mitrofanova, O.V.; Kalamkarov, G.R.; Bugrova, A.E.; Shevchenko, T.F. The Role of GSK-3β Phosphorylation in the Regulation of Slow Myosin Expression in Soleus Muscle during Functional Unloading. *Biochem. (Mosc.) Suppl. Ser. A Membr. Cell Biol.* **2018**, *1*, 85–91. [CrossRef]

108. Wood, S.J.; Slater, C.R. Safety factor at the neuromuscular junction. *Prog. Neurobiol.* **2001**, *64*, 393–429. [CrossRef]

109. Ruff, R.L. Endplate contributions to the safety factor for neuromuscular transmission. *Muscle Nerve* **2011**, *44*, 854–861. [CrossRef] [PubMed]

110. Cheng, A.; Morsch, M.; Murata, Y.; Ghazanfari, N.; Reddel, S.W.; Phillips, W.D. Sequence of age-associated changes to the mouse neuromuscular junction and the protective effects of voluntary exercise. *PLoS ONE* **2013**, *8*, e67970. [CrossRef] [PubMed]

111. Carnio, S.; LoVerso, F.; Baraibar, M.A.; Longa, E.; Khan, M.M.; Maffei, M.; Reischl, M.; Canepari, M.; Loefler, S.; Kern, H.; et al. Autophagy impairment in muscle induces neuromuscular junction degeneration and precocious aging. *Cell Rep.* **2014**, *8*, 1509–1521. [CrossRef] [PubMed]

112. Willadt, S.; Nash, M.; Slater, C.R. Age-related fragmentation of the motor endplate is not associated with impaired neuromuscular transmission in the mouse diaphragm. *Sci. Rep.* **2016**, *6*, 24849. [CrossRef] [PubMed]

113. Lee, K.M.; Chand, K.K.; Hammond, L.A.; Lavidis, N.A.; Noakes, P.G. Functional decline at the aging neuromuscular junction is associated with altered laminin-α4 expression. *Aging* **2017**, *9*, 880–899. [CrossRef] [PubMed]

114. Hughes, D.C.; Marcotte, G.R.; Marshall, A.G.; West, D.W.D.; Baehr, L.M.; Wallace, M.A.; Saleh, P.M.; Bodine, S.C.; Baar, K. Age-related differences in dystrophin: Impact on force transfer proteins, membrane integrity, and neuromuscular junction stability. *J. Gerontol. A Biol. Sci. Med. Sci.* **2017**, *72*, 640–648. [CrossRef] [PubMed]

115. Rudolf, R.; Khan, M.M.; Labeit, S.; Deschenes, M.R. Degeneration of neuromuscular junction in age and dystrophy. *Front. Aging Neurosci.* **2014**, *6*, 99. [CrossRef] [PubMed]

116. Pratt, S.J.; Valencia, A.P.; Le, G.K.; Shah, S.B.; Lovering, R.M. Pre- and postsynaptic changes in the neuromuscular junction in dystrophic mice. *Front. Physiol.* **2015**, *6*, 252. [CrossRef] [PubMed]

117. van der Pijl, E.M.; van Putten, M.; Niks, E.H.; Verschuuren, J.J.; Aartsma-Rus, A.; Plomp, J.J. Characterization of neuromuscular synapse function abnormalities in multiple Duchenne muscular dystrophy mouse models. *Eur. J. Neurosci.* **2016**, *43*, 1623–1635. [CrossRef] [PubMed]

118. Falk, D.J.; Todd, A.G.; Lee, S.; Soustek, M.S.; ElMallah, M.K.; Fuller, D.D.; Notterpek, L.; Byrne, B.J. Peripheral nerve and neuromuscular junction pathology in Pompe disease. *Hum. Mol. Genet.* **2015**, *24*, 625–636. [CrossRef] [PubMed]

119. Tu, H.; Zhang, D.; Corrick, R.M.; Muelleman, R.L.; Wadman, M.C.; Li, Y.L. Morphological regeneration and functional recovery of neuromuscular junctions after tourniquet-induced injuries in mouse hindlimb. *Front. Physiol.* **2017**, *8*, 207. [CrossRef] [PubMed]

120. Yampolsky, P.; Pacifici, P.G.; Witzemann, V. Differential muscle-driven synaptic remodeling in the neuromuscular junction after denervation. *Eur. J. Neurosci.* **2010**, *31*, 646–658. [CrossRef] [PubMed]

121. Deschenes, M.R.; Wilson, M.H. Age-related differences in synaptic plasticity following muscle unloading. *J. Neurobiol.* **2003**, *57*, 246–256. [CrossRef] [PubMed]

122. Nishimune, H.; Stanford, J.A.; Mori, Y. Role of exercise in maintaining the integrity of the neuromuscular junction. *Muscle Nerve* **2014**, *49*, 315–324. [CrossRef] [PubMed]

123. Gonzalez-Freire, M.; de Cabo, R.; Studenski, S.A.; Ferrucci, L. The neuromuscular junction: Aging at the crossroad between nerves and muscle. *Front. Aging Neurosci.* **2014**, *6*, 208. [CrossRef] [PubMed]

124. Tintignac, L.A.; Brenner, H.R.; Rüegg, M.A. Mechanisms regulating neuromuscular junction development and function and causes of muscle wasting. *Physiol. Rev.* **2015**, *95*, 809–852. [CrossRef] [PubMed]

125. Wang, J.; Wang, F.; Zhang, P.; Liu, H.; He, J.; Zhang, C.; Fan, M.; Chen, X. PGC-1α over-expression suppresses the skeletal muscle atrophy and myofiber-type composition during hindlimb unloading. *Biosci. Biotechnol. Biochem.* **2017**, *81*, 500–513. [CrossRef] [PubMed]

126. Martin-Rincon, M.; Morales-Alamo, D.; Calbet, J.A.L. Exercise-mediated modulation of autophagy in skeletal muscle. *Scand. J. Med. Sci. Sports* **2018**, *28*, 772–781. [CrossRef] [PubMed]

127. Khan, M.M.; Strack, S.; Wild, F.; Hanashima, A.; Gasch, A.; Brohm, K.; Reischl, M.; Carnio, S.; Labeit, D.; Sandri, M.; et al. Role of autophagy, SQSTM1, SH3GLB1, and TRIM63 in the turnover of nicotinic acetylcholine receptors. *Autophagy* **2014**, *10*, 123–136. [CrossRef] [PubMed]

128. Cerveró, C.; Montull, N.; Tarabal, O.; Piedrafita, L.; Esquerda, J.E.; Calderó, J. Chronic treatment with the AMPK agonist AICAR prevents skeletal muscle pathology but fails to improve clinical outcome in a mouse model of severe spinal muscular atrophy. *Neurotherapeutics* **2016**, *13*, 198–216. [CrossRef] [PubMed]

129. Pauly, M.; Daussin, F.; Burelle, Y.; Li, T.; Godin, R.; Fauconnier, J.; Koechlin-Ramonatxo, C.; Hugon, G.; Lacampagne, A.; Coisy-Quivy, M.; et al. AMPK activation stimulates autophagy and ameliorates muscular dystrophy in the mdx mouse diaphragm. *Am. J. Pathol.* **2012**, *181*, 583–592. [CrossRef] [PubMed]

130. Ljubicic, V.; Jasmin, B.J. AMP-activated protein kinase at the nexus of therapeutic skeletal muscle plasticity in Duchenne muscular dystrophy. *Trends Mol. Med.* **2013**, *19*, 614–624. [CrossRef] [PubMed]

131. Habegger, K.M.; Hoffman, N.J.; Ridenour, C.M.; Brozinick, J.T.; Elmendorf, J.S. AMPK enhances insulin-stimulated GLUT4 regulation via lowering membrane cholesterol. *Endocrinology* **2012**, *153*, 2130–2141. [CrossRef] [PubMed]

132. Ambery, A.G.; Tackett, L.; Penque, B.A.; Brozinick, J.T.; Elmendorf, J.S. Exercise training prevents skeletal muscle plasma membrane cholesterol accumulation, cortical actin filament loss, and insulin resistance in C57BL/6J mice fed a western-style high-fat diet. *Physiol. Rep.* **2017**, *5*, e13363. [CrossRef] [PubMed]

133. Brannigan, G.; Hénin, J.; Law, R.; Eckenhoff, R.; Klein, M.L. Embedded cholesterol in the nicotinic acetylcholine receptor. *Proc. Natl. Acad. Sci. USA* **2008**, *105*, 14418–14423. [CrossRef] [PubMed]

134. Zhu, D.; Xiong, W.C.; Mei, L. Lipid rafts serve as a signaling platform for nicotinic acetylcholine receptor clustering. *J. Neurosci.* **2006**, *26*, 4841–4851. [CrossRef] [PubMed]

135. Willmann, R.; Pun, S.; Stallmach, L.; Sadasivam, G.; Santos, A.F.; Caroni, P.; Fuhrer, C. Cholesterol and lipid microdomains stabilize the postsynapse at the neuromuscular junction. *EMBO J.* **2006**, *25*, 4050–4060. [CrossRef] [PubMed]

136. Chen, Y.; Li, X.; Ye, Q.; Tian, J.; Jing, R.; Xie, Z. Regulation of α1 Na/K-ATPase expression by cholesterol. *J. Biol. Chem.* **2011**, *286*, 15517–15524. [CrossRef] [PubMed]

137. Kapri-Pardes, E.; Katz, A.; Haviv, H.; Mahmmoud, Y.; Ilan, M.; Khalfin-Penigel, I.; Carmeli, S.; Yarden, O.; Karlish, S.J.D. Stabilization of the α2 isoform of Na,K-ATPase by mutations in a phospholipid binding pocket. *J. Biol. Chem.* **2011**, *286*, 42888–42899. [CrossRef] [PubMed]

138. Haviv, H.; Habeck, M.; Kanai, R.; Toyoshima, C.; Karlish, S.J. Neutral phospholipids stimulate Na,K-ATPase activity: A specific lipid-protein interaction. *J. Biol. Chem.* **2013**, *288*, 10073–10081. [CrossRef] [PubMed]

139. Cornelius, F.; Habeck, M.; Kanai, R.; Toyoshima, C.; Karlish, S.J. General and specific lipid-protein interactions in Na,K-ATPase. *Biochim. Biophys. Acta* **2015**, *1848*, 1729–1743. [CrossRef] [PubMed]

140. Kravtsova, V.V.; Petrov, A.M.; Vasiliev, A.N.; Zefirov, A.L.; Krivoi, I.I. Role of cholesterol in the maintenance of endplate electrogenesis in rat diaphragm. *Bull. Exp. Biol. Med.* **2015**, *158*, 298–300. [CrossRef] [PubMed]

141. Heiny, J.A.; Kravtsova, V.V.; Mandel, F.; Radzyukevich, T.L.; Benziane, B.; Prokofiev, A.V.; Pedersen, S.E.; Chibalin, A.V.; Krivoi, I.I. The nicotinic acetylcholine receptor and the Na,K-ATPase α2 isoform interact to regulate membrane electrogenesis in skeletal muscle. *J. Biol. Chem.* **2010**, *285*, 28614–28626. [CrossRef] [PubMed]

142. Chibalin, A.V.; Heiny, J.A.; Benziane, B.; Prokofiev, A.V.; Vasiliev, A.N.; Kravtsova, V.V.; Krivoi, I.I. Chronic nicotine exposure modifies skeletal muscle Na,K-ATPase activity through its interaction with the nicotinic acetylcholine receptor and phospholemman. *PLoS ONE* **2012**, *7*, e33719. [CrossRef] [PubMed]
143. Matchkov, V.V.; Krivoi, I.I. Specialized functional diversity and interactions of the Na,K-ATPase. *Front. Physiol.* **2016**, *7*, 179. [CrossRef] [PubMed]

4

Endothelial AMP-Activated Kinase α1 Phosphorylates eNOS on Thr495 and Decreases Endothelial NO Formation

Nina Zippel [1], Annemarieke E. Loot [1], Heike Stingl [1,2], Voahanginirina Randriamboavonjy [1,2], Ingrid Fleming [1,2] and Beate Fisslthaler [1,2,*]

[1] Institute for Vascular Signalling, Centre for Molecular Medicine, Johann Wolfgang Goethe University, 60590 Frankfurt, Germany; NZippel@gmx.de (N.Z.); a.loot@certe.nl (A.E.L.); Stingl@vrc.uni-frankfurt.de (H.S.); Voahangy@vrc.uni-frankfurt.de (V.R.); fleming@em.uni-frankfurt.de (I.F.)
[2] DZHK (German Centre for Cardiovascular Research) partner site RhineMain, Theodor Stern Kai 7, 60590 Frankfurt, Germany
* Correspondence: fisslthaler@vrc.uni-frankfurt.de

Abstract: AMP-activated protein kinase (AMPK) is frequently reported to phosphorylate Ser1177 of the endothelial nitric-oxide synthase (eNOS), and therefore, is linked with a relaxing effect. However, previous studies failed to consistently demonstrate a major role for AMPK on eNOS-dependent relaxation. As AMPK also phosphorylates eNOS on the inhibitory Thr495 site, this study aimed to determine the role of AMPKα1 and α2 subunits in the regulation of NO-mediated vascular relaxation. Vascular reactivity to phenylephrine and acetylcholine was assessed in aortic and carotid artery segments from mice with global (AMPKα$^{-/-}$) or endothelial-specific deletion (AMPKα$^{\Delta EC}$) of the AMPKα subunits. In control and AMPKα1-depleted human umbilical vein endothelial cells, eNOS phosphorylation on Ser1177 and Thr495 was assessed after AMPK activation with thiopental or ionomycin. Global deletion of the AMPKα1 or α2 subunit in mice did not affect vascular reactivity. The endothelial-specific deletion of the AMPKα1 subunit attenuated phenylephrine-mediated contraction in an eNOS- and endothelium-dependent manner. In in vitro studies, activation of AMPK did not alter the phosphorylation of eNOS on Ser1177, but increased its phosphorylation on Thr495. Depletion of AMPKα1 in cultured human endothelial cells decreased Thr495 phosphorylation without affecting Ser1177 phosphorylation. The results of this study indicate that AMPKα1 targets the inhibitory phosphorylation Thr495 site in the calmodulin-binding domain of eNOS to attenuate basal NO production and phenylephrine-induced vasoconstriction.

Keywords: endothelial nitric-oxide synthase; vasodilation; phenylephrine; vasoconstriction; endothelial cells; ionomycin

1. Introduction

AMP-activated protein kinase (AMPK) is activated in response to intracellular energy depletion, e.g., during insulin resistance when cellular glucose uptake is limited—especially in contracting skeletal muscle [1] or in cultured cells in the absence of extracellular glucose or hypoxia [2]. Once activated, AMPK acts to conserve energy by stimulating glucose uptake and mitochondrial biosynthesis, as well as by stimulating autophagy to provide substrates for metabolism. At the same time, AMPK inhibits anabolic pathways, such as cholesterol biosynthesis and fatty-acid synthesis, which are not essential for survival (for a recent review, see Reference [3]). In addition to activation by energy-sensitive stimuli, AMPK can also be stimulated following cell exposure to cytokines, growth factors, and mechanical stimuli [4]. In endothelial cells, AMPK was implicated in the inhibition of cell activation [5,6], as well as in angiogenesis in vitro [7] and in vivo [8]. These effects are, at least partially, attributed to the phosphorylation and stimulation of the endothelial nitric-oxide (NO) synthase (eNOS) by AMPK. This claim was backed up by reports of AMPK-dependent phosphorylation of eNOS (on Ser1177) following the exposure of cultured endothelial cells to agonists such as the vascular endothelial growth factor (VEGF) [9] and adiponectin [10], or pharmacological agents including peroxisome proliferator-activated receptor (PPAR) agonists [11] and statins [12]. Similar reports were also published using AMPK activators such as 5-aminoimidazole-4-carboxamide ribonucleotide (AICAR) [13] and metformin [14,15], or natural polyphenols like amurensin G [16] or resveratrol [17]. However, the effects are generally weak and much less impressive than the stimulation seen in response to hypoxia [7], shear stress [18–20], and thrombin [21] which result in robust AMPK activation.

Evidence for a link between AMPK- and NO-dependent alterations in vascular reactivity is also not consistent and depends on the model studied. For example, in resistance arteries in rat hindlimb and cremaster muscles, AICAR induces an NO- and endothelium-independent relaxation [22]. In mice, small-molecule AMPK activators, PT-1 or A769662, elicit the vasodilation of mesenteric arteries by decreasing intracellular Ca^{2+} levels and inducing depolymerization of the actin cytoskeleton [23,24]. In other studies, AICAR was reported to impair the relaxation elicited by sodium nitroprusside (SNP), indicating a general effect on smooth-muscle contractility [25]. In genetic models, the situation is not any clearer as the deletion of the AMPKα1 subunit did not affect acetylcholine (ACh)-induced NO production and relaxation unless mice were treated with angiotensin II over seven days [26]. Also, in isolated phenylephrine-contracted rings of murine aorta, AICAR elicited a profound dose-dependent relaxation that was independent of either the endothelium or the inhibition of eNOS, and mediated by the AMPKα1 subunit in smooth-muscle cells [27]. The most thorough study investigating the role of endothelial AMPKα subunits on vascular function and blood pressure reported hypertension in endothelial-specific AMPKα1 knockout mice; however, in the mesentery artery, the effect was attributed to the opening of charybdotoxin-sensitive potassium channels and smooth-muscle hyperpolarization [28]. The global deletion of the AMPKα2 subunit was also reported to attenuate the ACh-induced relaxation of murine aorta. This effect was attributed to eNOS uncoupling via an AMPKα2-mediated proteasomal degradation of the GTP cyclohydrolase [29], which generates the eNOS cofactor, tetrahydrobiopterin. Also, other researchers failed to detect any evidence for the AMPK-dependent activation of eNOS [30,31]. In this study, we set out to make a more thorough analysis of the effects of AMPKα1 and AMPKα2 deletion on NO-mediated vascular function. We also carefully studied changes in eNOS phosphorylation in cultured and native endothelial cells.

2. Results

2.1. Consequences of Global AMPKα Deletion on Vascular Responsiveness

The maximal KCl- and phenylephrine-induced contractions of isolated aortic rings were indistinguishable between wild-type mice and their corresponding AMPKα1$^{-/-}$ littermates (Figure 1A,B). However, there was a tendency toward an attenuated contraction in the aortic rings from the AMPKα1$^{-/-}$ mice and −log half maximal effective concentration (pEC$_{50}$) values were -6.899 ± 0.082 and -6.711 ± 0.099 ($n = 7$, not significant (n.s.)) in rings from wild-type and AMPKα1$^{-/-}$ mice, respectively. The endothelium- and NO-dependent relaxation elicited by ACh (Figure 1C), as well as the endothelium-independent relaxation elicited by SNP (Figure 1D), was identical in vessels from both strains. When experiments were repeated using carotid arteries, samples from AMPKα1$^{-/-}$ mice demonstrated a slightly weaker contractile response to KCl than the wild-type mice, as well as a slightly attenuated response to phenylephrine (pEC$_{50}$ values were -6.393 ± 0.065 and -5.895 ± 0.093, respectively, ($n = 7$, n.s.) in rings from wild-type and AMPKα1$^{-/-}$ mice (Figure S1). Again, there was no apparent difference in relaxant responsiveness to ACh or SNP.

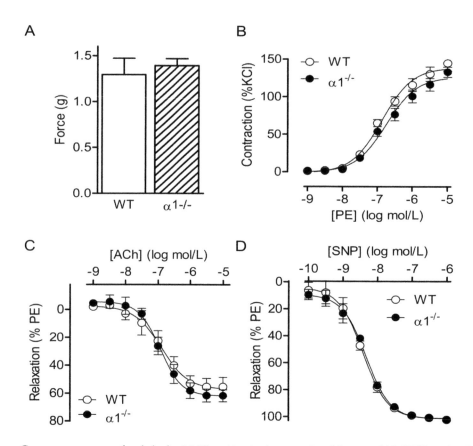

Figure 1. Consequences of global AMP-activated protein kinase (AMPK) $\alpha1$ deletion on vascular reactivity. Vascular reactivity in aortic rings from wild-type (WT) and AMPK$\alpha1^{-/-}$ ($\alpha1^{-/-}$) mice. (**A**) Responsiveness of endothelium-intact aortic rings to KCl (80 mmol/L). (**B–D**) Concentration–response curves to (**B**) phenylephrine (PE), (**C**) acetylcholine (ACh), and (**D**) sodium nitroprusside (SNP). The graphs summarize data obtained from seven animals in each group.

Similar experiments using arteries from AMPK$\alpha2^{-/-}$ mice gave essentially the same results, i.e., no significant difference in either the agonist-induced contraction or relaxation of the aorta in either the presence or absence of a functional endothelium (Figure 2).

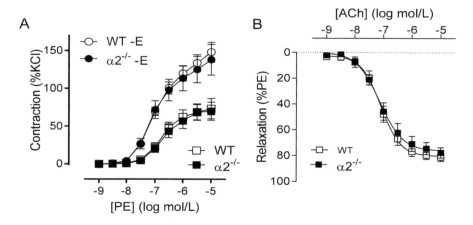

Figure 2. Consequences of global AMPK$\alpha2$ deletion on vascular reactivity. Vascular reactivity in aortic rings from wild-type (WT) and AMPK$\alpha2^{-/-}$ ($\alpha2^{-/-}$) mice: (**A**) contractile response to phenylephrine (PE) in the presence and absence (−E) of endothelium. (**B**) Concentration-dependent relaxation due to acetylcholine. The graphs summarize data obtained from seven animals in each group.

2.2. Consequence of Endothelial-Specific AMPKα Deletion on Vascular Responsiveness

As the global deletion of AMPK seemed to affect vascular smooth-muscle contraction rather than NO-mediated relaxation, animals lacking the AMPKα1 or AMPKα2 subunits specifically in endothelial cells (i.e., AMPKα1$^{\Delta EC}$ and AMPKα2$^{\Delta EC}$ mice) were generated, and the specificity of the deletion verified in isolated cluster of differentiation 144 (CD144)-positive pulmonary endothelial cells (Figure S2A). The deletion of endothelial AMPKα1 did not influence the expression level of AMPKα1 in whole aortic lysates (Figure S2B).

Endothelial-specific deletion of AMPKα1 did not affect the N$^{\omega}$-nitro-L-arginine methyl ester (L-NAME)-induced contraction of aortic rings (Figure 3A), which is an index of basal Ca^{2+}-independent NO production under isometric stretch conditions [32], or that induced by KCl (not shown). However, the ACh-induced relaxation was slightly improved by endothelial-specific AMPKα1 deletion (Figure 3B) with pEC$_{50}$ values for ACh of -7.217 ± 0.095 and -7.360 ± 0.076 ($n = 14$; n.s.) in rings from wild-type and AMPKα1$^{-/-}$ mice, respectively. An increased production of NO was evident as a markedly impaired contraction of aortic rings from AMPKα1$^{\Delta EC}$ mice compared to rings from wild-type mice to phenylephrine that was abolished by L-NAME (Figure 3C, Table 1). Similarly, removal of the endothelium with 3-[(3-cholamidopropyl)dimethylammonio]-1-propanesulfonate (CHAPS) also abrogated the improved relaxation that was dependent on AMPKα1 deletion (Figure 3D, Table 1).

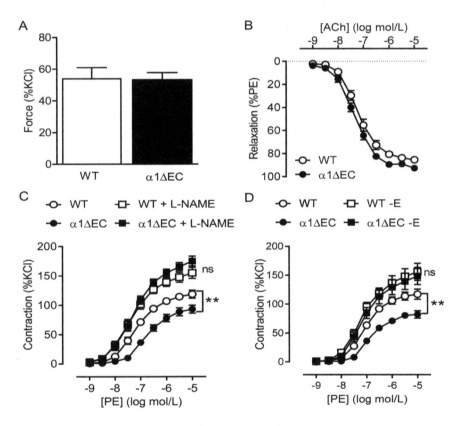

Figure 3. Consequences of endothelial-specific deletion of AMPKα1 on vascular reactivity. (**A**) Effect of N$^{\omega}$-nitro-L-arginine methyl ester (L-NAME; 300 µmol/L) on the tone of aortic rings from wild-type (WT) and AMPKα1$^{\Delta EC}$ (α1ΔEC) mice pre-contracted to 30% of the maximal KCl-induced contraction by phenylephrine. (**B**) Concentration-dependent relaxation due to acetylcholine in aortic rings pre-constricted with phenylephrine from wild-type (WT) and AMPKα1$^{\Delta EC}$ (α1ΔEC) mice. (**C,D**) Concentration-dependent contraction of aortic rings from wild-type (WT) and AMPKα1$^{\Delta EC}$ (α1ΔEC) mice due to phenylephrine. Experiments were performed in the absence and presence of L-NAME (300 µmol/L, (**C**) and in the presence and absence (−E) of functional endothelium (**D**); $n = 10$ to 16, ** $p < 0.01$ AMPKα1$^{\Delta EC}$ versus wild type.

Table 1. The $-\log$ half maximal effective concentration (pEC_{50}) values relating to the consequences of endothelial-specific deletion of AMP activated protein kinase (AMPK) $\alpha1$ on vascular response to phenylephrine. Experiments were performed in endothelium intact rings the presence of solvent or N^{ω}-nitro-L-arginine methyl ester (L-NAME; 300 μmol/L), as well as in endothelium-denuded ($-$E) samples from the same animals; $n = 10$–16.

pEC_{50} Values	Solvent		L-NAME		$-$E	
Wild type	-7.04 ± 0.13		-7.43 ± 0.10		-7.21 ± 0.05	
AMPK$^{\Delta EC}$	-6.77 ± 0.05	*	-7.31 ± 0.09	§§	-7.13 ± 0.05	§§

* $p < 0.05$ versus wild type; §§ $p < 0.001$ versus solvent.

Endothelial-specific deletion of the AMPK$\alpha2$ subunit had no consequence on the phenylephrine-induced contraction, ACh-induced relaxation, or the SNP-induced relaxation of isolated aortic rings from wild type versus the respective AMPK$\alpha2^{\Delta EC}$ littermates (Figure S3).

2.3. Vascular Responses to AMPK Activators

One reason for the lack of consequence of AMPK$\alpha1$ deletion on agonist-induced relaxation may be related to the fact that the ACh-induced phosphorylation and activation of eNOS is, like that of other agonists, largely regulated by the activity of Ca^{2+}/calmodulin-dependent kinase II [33]. Therefore, responses to two potential AMPK activators, i.e., resveratrol [34] and amurensin G [16], as well as two reportedly specific small-molecule AMPK activators, 991 and PT-1, were studied.

Resveratrol elicited the almost complete relaxation of aortic rings from wild-type and AMPK$\alpha1^{\Delta EC}$ mice (Figure S4A), but these responses were insensitive to NOS inhibition, and therefore, unrelated to its activation. Amurensin G is reported to activate AMPK in endothelial cells and increase eNOS phosphorylation [16]. While it was able to elicit the NOS inhibitor-sensitive relaxation of aortic rings from wild-type mice, it was equally effective and equally sensitive to NOS inhibition in aortic rings from corresponding AMPK$\alpha1^{\Delta EC}$ mice (Figure S4B). Thus, amurensin G exerted its relaxation in an eNOS-dependent manner and the activity of both AMPK activators was AMPK$\alpha1$-independent. The situation was similar when PT-1 and 991 were studied. The compounds elicited phosphorylation of AMPK in endothelial cells of murine aortic rings from wild-type cells (Figure S4E). Although these compounds elicited vascular relaxation, the responses were slow, and although they were sensitive to NOS inhibition, the effects were comparable in aortic rings from wild-type and AMPK$\alpha1^{\Delta EC}$ mice (Figure S4C,D).

2.4. AMPK$\alpha1$ and eNOS Phosphorylation

The activity of eNOS is reciprocally regulated by its phosphorylation on the activator site, Ser1177 [35,36], and the inhibitory site, Thr495 [37–39]. The next step was, therefore, to analyze the ability of AMPK to phosphorylate eNOS on these two residues in vitro. Wild-type (Myc-tagged) eNOS or eNOS mutants in which Thr495 was substituted with alanine or aspartate (Thr495A, Thr495D), or Ser1177 was substituted with alanine or aspartate (Ser1177A, Ser1177D) were overexpressed in HEK293 cells and used as the substrate for in vitro kinase assays for recombinant AMPK$\alpha1$. While the phosphorylation of wild-type eNOS and Ser1177 mutants was clearly detectable, there was only minimal phosphorylation of the Thr495 mutants (Figure 4A). These findings could be confirmed using phospho-specific antibodies to assess eNOS phosphorylation on Ser1177 and Thr495 on immunoprecipitated eNOS from human endothelial cells (Figure 4B).

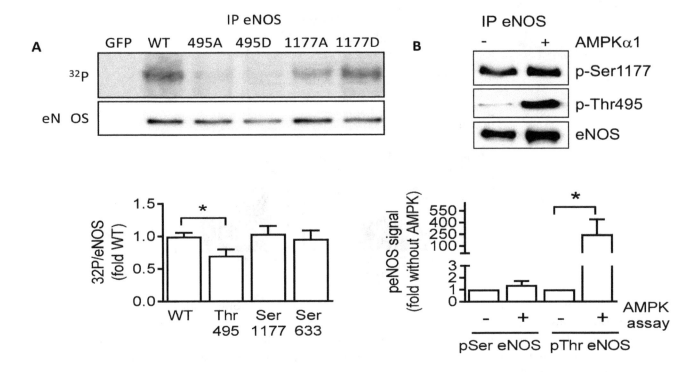

Figure 4. Endothelial nitric-oxide synthase (eNOS) is a substrate of AMPK in vitro. (**A**) Wild-type eNOS, as well as Thr495 and Ser1177 mutants, was overexpressed in HEK293 cells, then immunoprecipitated and used as substrate for AMPKα1 in in vitro kinase assays. The upper panel shows the autoradiograph of eNOS proteins. The lower panel shows the Western blot of the immunoprecipitated (IP) eNOS protein used as input. The graph summarizes the data from four independent experiments. (**B**) Wild-type eNOS (Flag-tagged) overexpressed in human umbilical vein endothelial cells was immunoprecipitated and used as a substrate for AMPKα1. Phosphorylation was assessed using specific antibodies for phosphorylated Ser1177 (p-Ser1177) and Thr945 eNOS (p-Thr495). The graph summarizes the data from five independent kinase reactions. * $p < 0.05$.

To transfer the observations to a more physiological system, changes in eNOS phosphorylation were studied in response to thiopental in cultured endothelial cells. Thiopental elicited the rapid and pronounced phosphorylation of AMPK (Figure 5A) in primary cultures of human endothelial cells. At the same time, thiopental decreased the phosphorylation of eNOS on Ser1177 and increased eNOS phosphorylation on Thr495 (Figure 5B). In contrast, the Ca^{2+}-elevating agonist and NO-dependent vasodilator, bradykinin, elicited a significant increase in the phosphorylation of Ser1177, and had no significant effect on Thr495 phosphorylation.

Next, AMPKα1 was deleted in human endothelial cells using a "clustered regularly interspaced short palindromic repeats" (CRISPR)/CRISPR-associated protein 9 (Cas9)-based approach. After each passaging, protein expression levels of AMPK α1 were analyzed with Western blotting. After passages 4–5, the protein was no longer detectable ($n = 6$, Figure 6), and agonist-induced changes in eNOS phosphorylation were assessed. Given that bradykinin and ACh receptors are rapidly lost during culture, cells were stimulated with the Ca^{2+} ionophore, ionomycin. In AMPKα1-expressing cells, ionomycin elicited the phosphorylation of eNOS on Ser1177 and the dephosphorylation of Thr495, followed by a rapid re-phosphorylation on Thr495, similar to the effects of other Ca^{2+}-elevating agonists [39]. In AMPKα1-depleted cells, basal Thr495 phosphorylation of eNOS was significantly impaired, and the re-phosphorylation after 2 min was also less pronounced than in control cells. The ionomycin-induced phosphorylation of eNOS Ser1177 was not affected by the depletion of AMPKα1 (Figure 6).

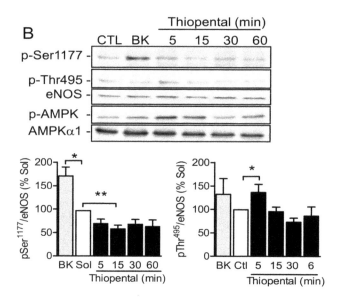

Figure 5. Effect of AMPK activation on eNOS activity and phosphorylation. Cultured human endothelial cells were incubated with solvent (Sol), thiopental (1 mmol/L, 5–60 min), or bradykinin (BK; 1 μmol/L, 2 min). Thereafter, the phosphorylation of (**A**) AMPK and (**B**) phosphorylation of eNOS at Ser1177 and Thr495 were assessed. Bar graphs summarize the data obtained in four to five different cell batches; * $p < 0.05$, ** $p < 0.01$ versus solvent treatment.

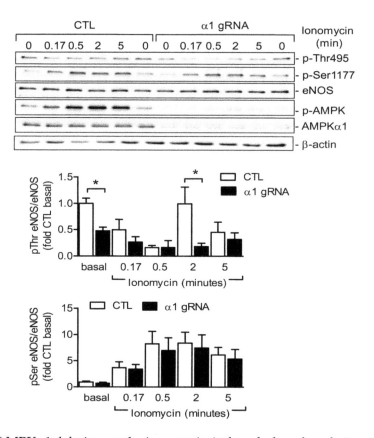

Figure 6. Effect of AMPKα1 deletion on the ionomycin-induced phosphorylation of eNOS on Thr495 and Ser1177. The AMPKα1 subunit was deleted in cultured human cells using the "clustered regularly interspaced short palindromic repeats" (CRISPR)/CRISPR-associated protein 9 (Cas9) system and AMPKα1-specific guide RNAs (gRNAs), and was stimulated with ionomycin (100 nmol/L) for up to 5 min. Representative Western blots are shown of six independent experiments. Bar graphs summarize the evaluation of p-eNOS to total eNOS ($n = 6$); * $p < 0.05$ versus control guide RNA.

3. Discussion

The results of the present study revealed that global deletion of the AMPKα1 or AMPKα2 subunits in healthy animals had no major impact on the relaxant function of isolated endothelium-intact murine aortae or carotid arteries. A small decrease in the contractile response to phenylephrine was apparent in global AMPKα1-deficient arteries, which was much more pronounced following endothelial-specific deletion of the AMPKα1 subunit. Moreover, the attenuated contractile response observed in arteries from AMPKα1$^{\Delta EC}$ mice was sensitive to L-NAME and removal of the endothelium, indicating that an increase in NO production by the endothelium underlies the effects observed. Mechanistically, the activity of AMPKα1 could be linked to the phosphorylation of eNOS on the inhibitory Thr495 site.

There are numerous reports describing the effects on AMPK activators on the phosphorylation of eNOS on Ser1177, which suggests that AMPK acts as an eNOS activator [13,40]. However, most studies linking AMPK with eNOS Ser1177 relied on compound C to inhibit AMPK, or AICAR, phenformin, or resveratrol to activate AMPK. This is a concern as the specificity of these pharmacological tools is questionable, with AMPK-dependent and -independent effects being attributed to both activators and inhibitors [41]. In the present study, the previously reported AMPK activators, amurensin G [16] and resveratrol [34], were studied together with the reportedly more specific small-molecule activators of AMPK, PT-1 [42] and 991 [43], in vascular reactivity studies. No endothelial-specific and AMPKα1-dependent effects were detected using any of the substances tested. Most studies using genetically modified models that reported effects on vascular reactivity focused on global AMPKα-deficient mice, and the defects were usually attributed to vascular smooth-muscle cells [44,45]. Effects on vascular function in global AMPKα1$^{-/-}$ mice were only observed in exercising or angiotensin-II-treated mice. The protective effects of voluntary exercise on vascular function were attributed to AMPKα1 via an effect on eNOS [26,46]. However, the only study to investigate changes in vascular reactivity in vessels from mice lacking AMPKα subunits specifically in endothelial cells linked changes in blood pressure with a carybdotoxin-sensitive potassium channel and endothelial-cell hyperpolarization [28].

The majority of reports describing AMPK-mediated effects on vascular function in disease models, as well as in healthy mice, focused on the AMPKα2 subunit, which suppresses reduced nicotinamide adenine dinucleotide phosphate (NADPH) oxidase activity and the production of reactive oxygen species to inhibit 26S proteasomal activity [47]. One consequence of this was the stabilization of GTP cyclohydrolase, the key sepiapterin biosynthetic enzyme that generates the essential eNOS cofactor, tetrahydrobiopterin [29]. It was, therefore, somewhat surprising that no major alterations in NO-mediated relaxation due to ACh could be detected in arteries from animals constitutively lacking the AMPKα2 subunit. Moreover, endothelial-specific deletion of AMPKα2 also failed to affect vascular NO production.

The enzyme eNOS can be phosphorylated on serine, threonine, and tyrosine residues, findings which highlight the potential role of phosphorylation in regulating eNOS activity. There are numerous putative phosphorylation sites, but most is known about the functional consequences of phosphorylation of a serine residue (human eNOS sequence: Ser1177) in the reductase domain and a threonine residue (human eNOS sequence Thr495) within the calmodulin (CaM)-binding domain. Maximal eNOS activity is linked with the simultaneous dephosphorylation of Thr495 and phosphorylation of Ser1177 [39,48].

In unstimulated cultured endothelial cells, Ser1177 is not phosphorylated, but it is rapidly phosphorylated after the application of fluid shear stress [35], VEGF [49] or bradykinin [39]. The kinases involved in this process vary with the stimuli applied. For example, while shear stress elicits the phosphorylation of Ser1177 by protein kinase A (PKA), insulin, estrogen, and VEGF mainly phosphorylate eNOS in endothelial cells via protein kinase B (Akt) [50]. The bradykinin-, Ca^{2+} ionophore-, and thapsigargin-induced phosphorylation of Ser1177 is mediated by Ca^{2+}/calmodulin-dependent kinase II (CaMKII) [39]. Thr495, on the other hand, is constitutively phosphorylated in all endothelial cells investigated to date, and it is a negative regulatory site, i.e., phosphorylation is associated with a decrease in enzyme activity [38,39]. The constitutively active kinase that phosphorylates eNOS Thr495 is most probably protein kinase C (PKC) [38], even though there is some confusion regarding the specific isoform(s) involved. AMPK can, however, also phosphorylate Thr495 [37]. The results of this study clearly indicate a role for endothelial cell AMPKα1 in the negative regulation of NO production and vascular tone, and as such, are in line with a previous study that reported an increased NO component to total relaxation in the mesenteric arteries of AMPKα1$^{\Delta EC}$ mice compared to wild type [28], this correlated to an enhanced eNOS Thr495 phosphorylation in mesenteric arteries compared to the aorta in wild type mice [51]. Our study also goes further to demonstrate that, in in vitro kinase assays, AMPKα1 clearly phosphorylated eNOS on Thr495, an effect that was prevented by the mutation of Thr495 to Ala or Asp. Also, in AMPKα1-depleted human endothelial cells, basal eNOS phosphorylation on Thr495 was decreased and its re-phosphorylation in response to agonist stimulation was significantly delayed, an effect that can account for the increase in NO generation by AMPKα1-deficient endothelial cells. At this stage, it is not possible to rule out a role for AMPK in the regulation of Ser1177 phosphorylation, as the higher basal phosphorylation of this residue in the transduced cells studied may have masked AMPK-dependent effects. However, the functional studies using vessels from AMPKα1 knockout mice clearly hint at an inhibitory rather than a stimulatory effect of AMPK on eNOS activity. The link between eNOS Thr495 phosphorylation and NO production can be explained by interference with the binding of CaM to the CaM-binding domain. Indeed, in endothelial cells stimulated with agonists such as bradykinin, histamine, or a Ca^{2+} ionophore, substantially more CaM binds to eNOS when Thr495 is dephosphorylated [39]. Analysis of the crystal structure of the eNOS CaM-binding domain with CaM indicates that the phosphorylation of eNOS Thr495 not only causes electrostatic repulsion of nearby glutamate residues within CaM, but may also affect eNOS Glu498, and thus, induce a conformational change within eNOS itself [52]. AMPK activation was also linked with the phosphorylation of eNOS on Ser1177 in isolated endothelial cells [13,37,53], but contrasted somewhat with the lack of effect on endothelium-dependent vascular reactivity [27]. In the present study, only a small increase in Ser1177 phosphorylation was detected in vitro using different cellular sources of eNOS (i.e., HEK cells or human endothelial cells).

In cultured endothelial cells, we found thiopental to be an effective AMPK activator and could demonstrate that AMPKα1 phosphorylates eNOS on Thr495, an observation that fits well with an earlier report [37]. This phosphorylation step is generally associated with eNOS inhibition due to the decreased binding of Ca^{2+}/calmodulin to the enzyme [39], and implies that the activation of AMPK in isolated vessels would act to decrease relaxation and increase vascular tone, which is exactly the response that was observed in the vascular reactivity studies.

In addition to direct phosphorylation, there are various signaling pathways described for AMPK to influence eNOS activity. AMPK was previously reported to prevent the estradiol-induced phosphorylation of eNOS by preventing the association of eNOS with heat-shock protein 90 (Hsp90), which is generally required for kinase binding to the eNOS signalosome [54]. Any link between AMPK and Hsp90 was not addressed in the current study given the clear effect of AMPKα1 on eNOS phosphorylation in vitro. Direct effects on eNOS activity may not be the only way via which AMPK activation can affect NO signaling. Indeed, AMPKα1 activation could affect the bioavailability of NO by improving mitochondrial function and stimulating the transcriptional regulation of anti-inflammatory enzymes, such as superoxide dismutase 2, to alter the production of reactive oxygen species [55].

In summary, endothelial AMPKα subunits have no direct activating effect on eNOS in vivo. Rather, since AMPKα1 phosphorylates eNOS on the inhibitory Thr495 site, AMPK activation attenuates NO production. No link between AMPKα2 and phenylephrine- or ACh-induced changes in vascular tone were detected. Moreover, while some of AMPK activators tested did affect vascular tone, the effects were independent of the endothelial-specific deletion of AMPKα1.

4. Materials and Methods

4.1. Materials

The antibodies used were directed against p-Ser1177 (Cell signaling, Cat. No. 9571) and p-Thr495 eNOS (Cell signaling, Cat. No. 9574), eNOS (BD Transduction, 610296), p-Thr172 AMPK (Cell signaling, Cat. No. 2535), AMPKα2 (Cell signaling, Cat. No. 2757), β-actin (Sigma, Cat. No. A5441), Flag (Sigma, Cat. No. F3165), and c-Myc (Santa Cruz, Cat. No. SC-40). The AMPKα1 antibody was generated by Eurogentec by injecting rabbits with the AMPKα1-specific peptide H_2N–CRA RHT LDE LNPQKS KHQ–$CONH_2$. All other substances were obtained from Sigma-Aldrich (Munich, Germany). $^{32}P\gamma$-ATP was obtained from Hartmann Analytics (Braunschweig, Germany).

4.2. Animals

AMPKα1$^{-/-}$ or AMPKα2$^{-/-}$ mice (kindly provided by Benoit Viollet, Paris via the European Mouse Mutant Archive, Munich, Germany) were bred heterozygous and housed at the Goethe University Hospital and knockouts or their respective wild-type littermates were used. AMPKα1$^{flox/flox}$ and α2$^{flox/flox}$ mice with *loxP* sites flanking a coding exon (provided by Benoit Viollet) were crossed with transgenic mouse lines overexpressing Cre recombinase under control of the vascular endothelial (VE)-cadherin promoter to generate the appropriate endothelial-specific AMPKα deletion; Cre$^{+/-}$ mice are referred throughout as AMPKα1$^{\Delta EC}$ and AMPKα2$^{\Delta EC}$ mice and Cre$^{-/-}$ mice are referred as their respective WT littermates. The investigation conforms to the Guide for the Care and Use of Laboratory Animals published by the European Commission Directive 86/609/EEC. For the isolation of tissues, mice were euthanized with 4% isoflurane in air and subsequent exsanguination.

4.3. Vascular Reactivity Measurements

Aortae and carotid arteries were prepared free of adhering tissue and cut into 2.0-mm segments. Aortic rings were mounted in standard 10-mL organ bath chambers, stretched to 1 *g* tension and responses were measured in *g*. Carotid artery rings were mounted in 5-mL wire myograph chambers (DMT, Aarhus, Denmark), stretched to 90% of their diameter at 100 mmHg, and responses were measured in mN/mm segment length. Contractile responses to a high K$^+$ buffer (80 mmol/L KCl) or cumulatively increasing concentrations of phenylephrine were assessed. Relaxation to cumulatively increasing concentrations of ACh, resveratrol (Sigma, Munich, Germany), 2-chloro-5-[[5-[[5-(4,5-dimethyl-2-nitrophenyl)-2-furanyl]methylene]-4,5-dihydro-4-oxo-2-thiazolyl] amino]benzoic acid (PT-1; Tocris, Biotechne, Wiesbaden, Germany), amurensin G (kindly provided by K.W. Kang, Seoul, Korea), 5-((6-chloro-5-(1-methyl-1H-indol-5-yl)-1H-benzo [d]imidazol-2-yl)oxy)-2-methylbenzoic acid (991; SpiroChem AG, Switzerland), or SNP was assessed in segments pre-contracted with phenylephrine to 80% of their maximal contraction due to KCl in the presence and absence of L-NAME. Relaxation was expressed as the percentage of phenylephrine precontraction. Removal of the endothelium was performed by intraluminal application of CHAPS (0.5%, 30 s) into the aortae.

4.4. Cell Culture

Human endothelial cells: Human umbilical vein endothelial cells were isolated and cultured as previously described [56] and used up to passage 2. The use of human material in this study conforms to the principles outlined in the Declaration of Helsinki, and the isolation of endothelial cells was approved in written form by the ethics committee of Goethe University. For lentiviral and adenoviral transduction, human umbilical vein endothelial cells (Promocell, Heidelberg, Germany) were used and cultured up to passage 8 in endothelial growth medium 2 (Promocell, Heidelberg, Germany).

Murine pulmonary endothelial cells: Mouse lungs were freshly processed as previously described [18].

4.5. Adenoviral Transduction of Human Umbilical Vein Endothelial Cells

Adenoviral particles expressing the C-terminal Flag-tagged human full-length eNOS were used to transduce cultured umbilical vein endothelial cells as described previously [57].

4.6. In Vitro Kinase Assay

The eNOS wild-type or mutant proteins with C-terminal myc or Flag tags were overexpressed by transfection in HEK cells or adenoviral transduction in human umbilical vein endothelial cells, and after two days, cells were lysed and eNOS was immunoprecipitated by c-myc or Flag immunoprecipitation (IP). The immunoprecipitated proteins were used as a substrate for kinase assays with purified AMPKα1/β1/γ1 subunits (Merck Millipore, Darmstadt, Germany, Cat No. 1480) [20]. The lysates were separated by SDS-PAGE and blotted with antibodies specific for the phosphorylation sites of eNOS. Alternatively, ^{32}PγATP was used to radioactively label the protein. Proteins were separated by SDS-PAGE, and the gel was exposed to X-ray film after drying.

4.7. CRISPR/Cas9-Mediated Knock-Down of AMPKα1

Human umbilical vein endothelial cells (Promocell, Heidelberg, Germany) were transduced with lentiviral particles mediating the expression of Cas9 (Lenti-Cas9-2A-Blast was provided by Jason Moffat (Addgene plasmid # 73310)) and selected by blasticidin (10 μg/mL). Thereafter, a second lentiviral transduction with guide RNAs directed against AMPKα1 (Addgene numbers 76253 and 75254 provided by David Root, Cambridge, MA, USA) was performed, and puromycin (1 μg/mL) was used to select for double-transduced cells. The efficiency of the knockdown was analyzed by Western blotting.

4.8. Immunoblotting

Cells were lysed in Triton X-100 lysis buffer (Tris/HCl pH 7.5; 50 mmol/L; NaCl, 150 mmol/L; ethyleneglycoltetraacetic acid (EGTA), 2 mmol/L; ethylenediaminetetraacetic acid (EDTA) 2 mmol/L; Triton X-100, 1% (v/v); NaF, 25 mmol/L; $Na_4P_2O_7$, 10 mmol/L; 2 μg/mL each of leupeptin, pepstatin A, antipain, aprotinin, chymostatin, and trypsin inhibitor, and phenylmethylsulfonyl fluoride (PMSF), 40 μg/mL). Detergent-soluble proteins were heated with SDS-PAGE sample buffer and separated by SDS-PAGE, and specific proteins were detected by immunoblotting. To assess the phosphorylation of proteins, either equal amounts of protein from each sample were loaded twice and one membrane incubated with the phospho-specific antibody and the other with an antibody recognizing total protein, or blots were reprobed with the appropriate antibody.

4.9. Statistical Analyses

Data are expressed as mean ± standard error of the mean (SEM). Statistical evaluation was done using Student's t-test for unpaired data or ANOVA for repeated measures where appropriate. Values of $p < 0.05$ were considered statistically significant.

Supplementary Materials:
Figure S1. Vascular function in carotid arteries from wild-type (WT) and AMPKα1 $^{-/-}$ mice. (A) Contraction induced by KCl (80 mmol/L), (B) concentration response curves to phenylephrine (PE), and relaxation curves to (C) acetylcholine (ACh) or (D) sodium nitroprusside (SNP) in PE-contracted vessels. The graphs summarize data obtained from 7 animals in each group. Figure S2. Endothelial cell specific deletion of AMPKα1. (A) AMPKα1 expression in freshly isolated pulmonary endothelial cells from AMPKα1^{ΔEC} or Cre$^{-/-}$ (wild-type; WT) mice. (B) Expression of eNOS, AMPKα1 and AMPKα2 in aortic ring lysates from WT or AMPKα1^{ΔEC} (ΔEC) mice. (A) The blots presented are representative of 12 additional experiments using 2 mice per group. Figure S3. Effect of endothelial specific deletion of AMPKα2 on vascular reactivity of aortic rings (A) Dose dependent contraction to PE of wild-type (open symbols) or AMPKα2^{ΔEC} mice (closed symbols). (B) Relaxation curves of aortic rings to acetylcholine (ACh) after PE constriction of wild-type (open symbols) or AMPKα2^{ΔEC} mice (closed symbols). (C) Dose-dependent relaxation to SNP. The graphs summarize data obtained from 6 animals in each group. Figure S4. Effect of AMPK activators on the relaxation of aortic rings. (A,B) Concentration

dependent effects of resveratrol (A) and amurensin G (B) on vascular tone in phenylephrine preconstricted aortic rings from wild-type (WT) and AMPKα1$^{\Delta EC}$ (α1$^{\Delta EC}$) mice; $n = 6$ animals in each group. (C,D) Time-dependent effects of PT-1 (30 µmol/L) and 991 (30 µmol/L) on vascular tone in phenylephrine preconstricted aortic rings from wild-type (WT) and AMPKα1$^{\Delta EC}$ (α1$^{\Delta EC}$) mice; $n = 4$ animals in each group. (E) Effects of the AMPK activators on the phosphorylation of AMPK (on Thr172) and ACC (Ser79) in endothelial cells isolated from aortic rings from wild-type mice. Experiments were performed in the absence (Basal) and presence of 991 (30 µmol/L), AICAR (0.5 mmol/L) or PT-1 (30 µmol/L) for 60 min. Comparable results were obtained in 3 additional independent experiments.

Author Contributions: N.Z., A.E.L., H.S., V.R., and B.F. performed the experiments and interpreted the data. I.F. and B.F. planned the study and wrote the manuscript.

Acknowledgments: The authors are indebted to Isabel Winter, Katharina Bruch, Mechtild Pipenbrock-Gyamfi, and Katharina Herbig for expert technical assistance.

Abbreviations

AMPK	AMP-activated protein kinase
eNOS	endothelial nitric-oxide synthase
AICAR	5-aminoimidazole-4-carboxamide ribonucleotide
VEGF	vascular endothelial growth factor
PPAR	peroxisome proliferator-activated receptor
SNP	sodium nitroprusside
ACh	acetylcholine
PE	phenylephrine
L-NAME	N$^{\omega}$-nitro-L-arginine methyl ester
CRISPR	clustered regularly interspaced short palindromic repeats
Cas9	CRISPR-associated protein 9

References

1. Musi, N.; Fujii, N.; Hirshman, M.F.; Ekberg, I.; Froberg, S.; Ljungqvist, O.; Thorell, A.; Goodyear, L.J. AMP-activated protein kinase (AMPK) is activated in muscle of subjects with type 2 diabetes during exercise. *Diabetes* **2001**, *50*, 921–927. [CrossRef] [PubMed]

2. Laderoute, K.R.; Amin, K.; Calaoagan, J.M.; Knapp, M.; Le, T.; Orduna, J.; Foretz, M.; Viollet, B. 5′-AMP-activated protein kinase (AMPK) is induced by low-oxygen and glucose deprivation conditions found in solid-tumor microenvironments. *Mol. Cell. Biol.* **2006**, *26*, 5336–5347. [CrossRef] [PubMed]

3. Herzig, S.; Shaw, R.J. AMPK: Guardian of metabolism and mitochondrial homeostasis. *Nat. Rev. Mol. Cell Biol.* **2018**, *19*, 121–135. [CrossRef] [PubMed]

4. Fisslthaler, B.; Fleming, I. Activation and signaling by the AMP-activated protein kinase in endothelial cells. *Circ. Res.* **2009**, *105*, 114–127. [CrossRef] [PubMed]

5. Bess, E.; Fisslthaler, B.; Fromel, T.; Fleming, I. Nitric oxide-induced activation of the AMP-activated protein kinase α2 subunit attenuates IκB kinase activity and inflammatory responses in endothelial cells. *PLoS ONE* **2011**, *6*, e20848. [CrossRef] [PubMed]

6. Cacicedo, J.M.; Yagihashi, N.; Keaney, J.F.; Ruderman, N.B.; Ido, Y. AMPK inhibits fatty acid-induced increases in NF-κB transactivation in cultured human umbilical vein endothelial cells. *Biochem. Biophys. Res. Commun.* **2004**, *324*, 1204–1209. [CrossRef] [PubMed]

7. Nagata, D.; Mogi, M.; Walsh, K. AMP-activated protein kinase (AMPK) signaling in endothelial cells is essential for angiogenesis in response to hypoxic stress. *J. Biol. Chem.* **2003**, *278*, 31000–31006. [CrossRef] [PubMed]

8. Yu, J.W.; Deng, Y.P.; Han, X.; Ren, G.F.; Cai, J.; Jiang, G.J. Metformin improves the angiogenic functions of endothelial progenitor cells via activating AMPK/eNOS pathway in diabetic mice. *Cardiovasc. Diabetol.* **2016**, *15*, 88. [CrossRef] [PubMed]

9. Reihill, J.A.; Ewart, M.A.; Hardie, D.G.; Salt, I.P. AMP-activated protein kinase mediates VEGF-stimulated endothelial NO production. *Biochem. Biophys. Res. Commun.* **2007**, *354*, 1084–1088. [CrossRef] [PubMed]

10. Cheng, K.K.; Lam, K.S.; Wang, Y.; Huang, Y.; Carling, D.; Wu, D.; Wong, C.; Xu, A. Adiponectin-induced endothelial nitric oxide synthase activation and nitric oxide production are mediated by APPL1 in endothelial cells. *Diabetes* **2007**, *56*, 1387–1394. [CrossRef] [PubMed]

11. Boyle, J.G.; Logan, P.J.; Ewart, M.A.; Reihill, J.A.; Ritchie, S.A.; Connell, J.M.; Cleland, S.J.; Salt, I.P. Rosiglitazone stimulates nitric oxide synthesis in human aortic endothelial cells via AMP-activated protein kinase. *J. Biol. Chem.* **2008**, *283*, 11210–11217. [CrossRef] [PubMed]

12. Rossoni, L.; Wareing, M.; Wenceslau, C.; Al-Abri, M.; Cobb, C.; Austin, C. Acute simvastatin increases endothelial nitric oxide synthase phosphorylation via AMP-activated protein kinase and reduces contractility of isolated rat mesenteric resistance arteries. *Clin. Sci. (Lond.)* **2011**, *121*, 449–458. [CrossRef] [PubMed]

13. Morrow, V.A.; Foufelle, F.; Connell, J.M.; Petrie, J.R.; Gould, G.W.; Salt, I.P. Direct activation of AMP-activated protein kinase stimulates nitric-oxide synthesis in human aortic endothelial cells. *J. Biol. Chem.* **2003**, *278*, 31629–31639. [CrossRef] [PubMed]

14. Davis, B.J.; Xie, Z.; Viollet, B.; Zou, M.H. Activation of the AMP-activated kinase by antidiabetes drug metformin stimulates nitric oxide synthesis in vivo by promoting the association of heat shock protein 90 and endothelial nitric oxide synthase. *Diabetes* **2006**, *55*, 496–505. [CrossRef] [PubMed]

15. Zou, M.H.; Kirkpatrick, S.S.; Davis, B.J.; Nelson, J.S.; Wiles, W.G.; Schlattner, U.; Neumann, D.; Brownlee, M.; Freeman, M.B.; Goldman, M.H. Activation of the AMP-activated protein kinase by the anti-diabetic drug metformin in vivo. Role of mitochondrial reactive nitrogen species. *J. Biol. Chem.* **2004**, *279*, 43940–43951. [CrossRef] [PubMed]

16. Hien, T.T.; Oh, W.K.; Quyen, B.T.; Dao, T.T.; Yoon, J.H.; Yun, S.Y.; Kang, K.W. Potent vasodilation effect of amurensin G is mediated through the phosphorylation of endothelial nitric oxide synthase. *Biochem. Pharmacol.* **2012**, *84*, 1437–1450. [CrossRef] [PubMed]

17. Li, X.; Dai, Y.; Yan, S.; Shi, Y.; Li, J.; Liu, J.; Cha, L.; Mu, J. Resveratrol lowers blood pressure in spontaneously hypertensive rats via calcium-dependent endothelial NO production. *Clin. Exp. Hypertens.* **2016**, *38*, 287–293. [CrossRef] [PubMed]

18. Fleming, I.; Fisslthaler, B.; Dixit, M.; Busse, R. Role of PECAM-1 in the shear-stress-induced activation of Akt and the endothelial nitric oxide synthase (eNOS) in endothelial cells. *J. Cell Sci.* **2005**, *118*, 4103–4111. [CrossRef] [PubMed]

19. Dixit, M.; Bess, E.; Fisslthaler, B.; Hartel, F.V.; Noll, T.; Busse, R.; Fleming, I. Shear stress-induced activation of the AMP-activated protein kinase regulates FoxO1a and angiopoietin-2 in endothelial cells. *Cardiovasc. Res.* **2008**, *77*, 160–168. [CrossRef] [PubMed]

20. Fisslthaler, B.; Fleming, I.; Keseru, B.; Walsh, K.; Busse, R. Fluid shear stress and NO decrease the activity of the hydroxy-methylglutaryl coenzyme A reductase in endothelial cells via the AMP-activated protein kinase and FoxO1. *Circ. Res.* **2007**, *100*, e12–e21. [CrossRef] [PubMed]

21. Thors, B.; Halldorsson, H.; Jonsdottir, G.; Thorgeirsson, G. Mechanism of thrombin mediated eNOS phosphorylation in endothelial cells is dependent on ATP levels after stimulation. *Biochim. Biophys. Acta* **2008**, *1783*, 1893–1902. [CrossRef] [PubMed]

22. Bradley, E.A.; Eringa, E.C.; Stehouwer, C.D.; Korstjens, I.; van Nieuw Amerongen, G.P.; Musters, R.; Sipkema, P.; Clark, M.G.; Rattigan, S. Activation of AMP-activated protein kinase by 5-aminoimidazole-4-carboxamide-1-beta-D-ribofuranoside in the muscle microcirculation increases nitric oxide synthesis and microvascular perfusion. *Arterioscler. Thromb. Vasc. Biol.* **2010**, *30*, 1137–1142. [CrossRef] [PubMed]

23. Schneider, H.; Schubert, K.M.; Blodow, S.; Kreutz, C.P.; Erdogmus, S.; Wiedenmann, M.; Qiu, J.; Fey, T.; Ruth, P.; Lubomirov, L.T.; et al. AMPK Dilates Resistance Arteries via Activation of SERCA and BKCa Channels in Smooth Muscle. *Hypertension* **2015**, *66*, 108–116. [CrossRef] [PubMed]

24. Schubert, K.M.; Qiu, J.; Blodow, S.; Wiedenmann, M.; Lubomirov, L.T.; Pfitzer, G.; Pohl, U.; Schneider, H. The AMP-Related Kinase (AMPK) Induces Ca^{2+}-Independent Dilation of Resistance Arteries by Interfering With Actin Filament Formation. *Circ. Res.* **2017**, *121*, 149–161. [CrossRef] [PubMed]

25. Davis, B.; Rahman, A.; Arner, A. AMP-activated kinase relaxes agonist induced contractions in the mouse aorta via effects on PKC signaling and inhibits NO-induced relaxation. *Eur. J. Pharmacol.* **2012**, *695*, 88–95. [CrossRef] [PubMed]

26. Schuhmacher, S.; Foretz, M.; Knorr, M.; Jansen, T.; Hortmann, M.; Wenzel, P.; Oelze, M.; Kleschyov, A.L.; Daiber, A.; Keaney, J.F.; et al. α1 AMP-activated protein kinase preserves endothelial function during chronic angiotensin II treatment by limiting Nox2 upregulation. *Arterioscler. Thromb. Vasc. Biol.* **2011**, *31*, 560–566. [CrossRef] [PubMed]

27. Goirand, F.; Solar, M.; Athea, Y.; Viollet, B.; Mateo, P.; Fortin, D.; Leclerc, J.; Hoerter, J.; Ventura-Clapier, R.; Garnier, A. Activation of AMP kinase alpha1 subunit induces aortic vasorelaxation in mice. *J. Physiol.* **2007**, *581*, 1163–1171. [CrossRef] [PubMed]

28. Enkhjargal, B.; Godo, S.; Sawada, A.; Suvd, N.; Saito, H.; Noda, K.; Satoh, K.; Shimokawa, H. Endothelial AMP-activated protein kinase regulates blood pressure and coronary flow responses through hyperpolarization mechanism in mice. *Arterioscler. Thromb. Vasc. Biol.* **2014**, *34*, 1505–1513. [CrossRef] [PubMed]

29. Wang, S.; Xu, J.; Song, P.; Viollet, B.; Zou, M.H. In vivo activation of AMP-activated protein kinase attenuates diabetes-enhanced degradation of GTP cyclohydrolase I. *Diabetes* **2009**, *58*, 1893–1901. [CrossRef] [PubMed]

30. Stahmann, N.; Woods, A.; Carling, D.; Heller, R. Thrombin activates AMP-activated protein kinase in endothelial cells via a pathway involving Ca^{2+}/calmodulin-dependent protein kinase kinase beta. *Mol. Cell. Biol.* **2006**, *26*, 5933–5945. [CrossRef] [PubMed]

31. Stahmann, N.; Woods, A.; Spengler, K.; Heslegrave, A.; Bauer, R.; Krause, S.; Viollet, B.; Carling, D.; Heller, R. Activation of AMP-activated protein kinase by vascular endothelial growth factor mediates endothelial angiogenesis independently of nitric-oxide synthase. *J. Biol. Chem.* **2010**, *285*, 10638–10652. [CrossRef] [PubMed]

32. Fleming, I.; Bauersachs, J.; Fisslthaler, B.; Busse, R. Calcium-independent activation of endothelial nitric oxide synthase in response to tyrosine phosphatase inhibitors and fluid shear stress. *Circ. Res.* **1998**, *82*, 686–695. [CrossRef] [PubMed]

33. Schneider, J.C.; El, K.D.; Chereau, C.; Lanone, S.; Huang, X.L.; De Buys Roessingh, A.S.; Mercier, J.C.; Dall'Ava-Santucci, J.; Dinh-Xuan, A.T. Involvement of Ca^{2+}/calmodulin-dependent protein kinase II in endothelial NO production and endothelium-dependent relaxation. *Am. J. Physiol. Heart Circ. Physiol.* **2003**, *284*, H2311–H2319. [CrossRef] [PubMed]

34. Zang, M.; Xu, S.; Maitland-Toolan, K.A.; Zuccollo, A.; Hou, X.; Jiang, B.; Wierzbicki, M.; Verbeuren, T.J.; Cohen, R.A. Polyphenols stimulate AMP-activated protein kinase, lower lipids, and inhibit accelerated atherosclerosis in diabetic LDL receptor-deficient mice. *Diabetes* **2006**, *55*, 2180–2191. [CrossRef] [PubMed]

35. Dimmeler, S.; Fleming, I.; Fisslthaler, B.; Hermann, C.; Busse, R.; Zeiher, A.M. Activation of nitric oxide synthase in endothelial cells by Akt-dependent phosphorylation. *Nature* **1999**, *399*, 601–605. [CrossRef] [PubMed]

36. Fulton, D.; Mcgiff, J.C.; Quilley, J. Pharmacological evaluation of an epoxide as the putative hyperpolarizing factor mediating the nitric oxide-independent vasodilator effect of bradykinin in the rat heart. *J. Pharm. Exp. Ther.* **1999**, *287*, 497–503.

37. Chen, Z.P.; Mitchelhill, K.I.; Michell, B.J.; Stapleton, D.; Rodriguez-Crespo, I.; Witters, L.A. AMP-activated protein kinase phosphorylation of endothelial no synthase. *FEBS Lett.* **1999**, 285–289. [CrossRef]

38. Michell, B.J.; Chen, Z.p.; Tiganis, T.; Stapleton, D.; Katsis, F.; Power, D.A.; Sim, A.T.; Kemp, B.E. Coordinated control of endothelial nitric-oxide synthase phosphorylation by protein kinase C and the cAMP-dependent protein kinase. *J. Biol. Chem.* **2001**, *276*, 17625–17628. [CrossRef] [PubMed]

39. Fleming, I.; Fisslthaler, B.; Dimmeler, S.; Kemp, B.E.; Busse, R. Phosphorylation of Thr(495) regulates Ca(2+)/calmodulin-dependent endothelial nitric oxide synthase activity. *Circ. Res.* **2001**, *88*, E68–E75. [CrossRef] [PubMed]

40. Gaskin, F.S.; Kamada, K.; Yusof, M.; Korthuis, R.J. 5′-AMP-activated protein kinase activation prevents postischemic leukocyte-endothelial cell adhesive interactions. *Am. J. Physiol. Heart Circ. Physiol.* **2007**, *292*, H326–H332. [CrossRef] [PubMed]

41. Huang, Y.; Smith, C.A.; Chen, G.; Sharma, B.; Miner, A.S.; Barbee, R.W.; Ratz, P.H. The AMP-dependent protein kinase (AMPK) activator A-769662 causes arterial relaxation by reducing cytosolic free calcium independently of an increase in AMPK phosphorylation. *Front. Pharmacol.* **2017**, *8*. [CrossRef] [PubMed]

42. Jensen, T.E.; Ross, F.A.; Kleinert, M.; Sylow, L.; Knudsen, J.R.; Gowans, G.J.; Hardie, D.G.; Richter, E.A. PT-1 selectively activates AMPKγ1 complexes in mouse skeletal muscle, but activates all three + subunit

complexes in cultured human cells by inhibiting the respiratory chain. *Biochem. J.* **2015**, *467*, 461–472. [CrossRef] [PubMed]

43. Bultot, L.; Jensen, T.E.; Lai, Y.C.; Madsen, A.L.B.; Collodet, C.; Kviklyte, S.; Deak, M.; Yavari, A.; Foretz, M.; Ghaffari, S.; et al. Benzimidazole derivative small-molecule 991 enhances AMPK activity and glucose uptake induced by AICAR or contraction in skeletal muscle. *Am. J. Physiol. Endocrinol. Metab.* **2016**, *311*, E706–E719. [CrossRef] [PubMed]

44. Wang, S.; Liang, B.; Viollet, B.; Zou, M.H. Inhibition of the AMP-activated protein kinase-alpha2 accentuates agonist-induced vascular smooth muscle contraction and high blood pressure in mice. *Hypertension* **2011**, *57*, 1010–1017. [CrossRef] [PubMed]

45. Sun, G.Q.; Li, Y.B.; Du, B.; Meng, Y. Resveratrol via activation of AMPK lowers blood pressure in DOCA-salt hypertensive mice. *Clin. Exp. Hypertens.* **2015**, *37*, 616–621. [CrossRef] [PubMed]

46. Kröller-Schön, S.; Jansen, T.; Hauptmann, F.; Schüler, A.; Heeren, T.; Hausding, M.; Oelze, M.; Viollet, B.; Keaney, J.F.; Wenzel, P.; et al. α1AMP-activated protein kinase mediates vascular protective effects of exercise. *Arterioscler. Thromb. Vasc. Biol.* **2012**, *32*, 1634–1641. [CrossRef] [PubMed]

47. Wang, S.; Zhang, M.; Liang, B.; Xu, J.; Xie, Z.; Liu, C.; Viollet, B.; Yan, D.; Zou, M.H. AMPKalpha2 deletion causes aberrant expression and activation of NAD(P)H oxidase and consequent endothelial dysfunction in vivo: Role of 26S proteasomes. *Circ. Res.* **2010**, *106*, 1117–1128. [CrossRef] [PubMed]

48. Mount, P.F.; Kemp, B.E.; Power, D.A. Regulation of endothelial and myocardial NO synthesis by multi-site eNOS phosphorylation. *J. Mol. Cell. Cardiol.* **2007**, *42*, 271–279. [CrossRef] [PubMed]

49. Michell, B.J.; Griffiths, J.E.; Mitchelhill, K.I.; Rodriguez-Crespo, I.; Tiganis, T.; Bozinovski, S.; de Montellano, P.R.O.; Kemp, B.E.; Pearson, R.B. The Akt kinase signals directly to endothelial nitric oxide synthase. *Curr. Biol.* **1999**, *9*, 845–848. [CrossRef]

50. Gallis, B.; Corthals, G.L.; Goodlett, D.R.; Ueba, H.; Kim, F.; Presnell, S.R.; Figeys, D.; Harrison, D.G.; Berk, B.C.; Aebersold, R.; et al. Identification of flow-dependent endothelial nitric-oxide synthase phosphorylation sites by mass spectrometry and regulation of phosphorylation and nitric oxide production by the phosphatidylinositol 3-Kinase inhibitor LY294002. *J. Biol. Chem.* **1999**, *274*, 30101–30108. [CrossRef] [PubMed]

51. Ohashi, J.; Sawada, A.; Nakajima, S.; Noda, K.; Takaki, A.; Shimokawa, H. Mechanisms for enhanced endothelium-derived hyperpolarizing factor-mediated responses in microvessels in mice. *Circ. J.* **2012**, *76*, 1768–1779. [CrossRef] [PubMed]

52. Aoyagi, M.; Arvai, A.S.; Tainer, J.A.; Getzoff, E.D. Structural basis for endothelial nitric oxide synthase binding to calmodulin. *EMBO J.* **2003**, *22*, 766–775. [CrossRef] [PubMed]

53. Chen, Z.; Peng, I.C.; Sun, W.; Su, M.I.; Hsu, P.H.; Fu, Y.; Zhu, Y.; DeFea, K.; Pan, S.; Tsai, M.D.; et al. AMP-activated protein kinase functionally phosphorylates endothelial nitric oxide synthase Ser633. *Circ. Res.* **2009**, *104*, 496–505. [CrossRef] [PubMed]

54. Schulz, E.; Anter, E.; Zou, M.H.; Keaney, J.F., Jr. Estradiol-mediated endothelial nitric oxide synthase association with heat shock protein 90 requires adenosine monophosphate-dependent protein kinase. *Circulation* **2005**, *111*, 3473–3480. [CrossRef] [PubMed]

55. Zippel, N.; Malik, R.A.; Fromel, T.; Popp, R.; Bess, E.; Strilic, B.; Wettschureck, N.; Fleming, I.; Fisslthaler, B. Transforming growth factor-beta-activated kinase 1 regulates angiogenesis via AMP-activated protein kinase-α1 and redox balance in endothelial cells. *Arterioscler. Thromb. Vasc. Biol.* **2013**, *33*, 2792–2799. [CrossRef] [PubMed]

56. Busse, R.; Lamontagne, D. Endothelium-derived bradykinin is responsible for the increase in calcium produced by angiotensin-converting enzyme inhibitors in human endothelial cells. *Naunyn Schmiedebergs Arch. Pharmacol.* **1991**, *344*, 126–129. [CrossRef] [PubMed]

57. Michaelis, U.R.; Falck, J.R.; Schmidt, R.; Busse, R.; Fleming, I. Cytochrome P4502C9-derived epoxyeicosatrienoic acids induce the expression of cyclooxygenase-2 in endothelial cells. *Arterioscler. Thromb. Vasc. Biol.* **2005**, *25*, 321–326. [CrossRef] [PubMed]

AMP-Activated Protein Kinase and Host Defense against Infection

tttttttt

Prashanta Silwal [1,2], Jin Kyung Kim [1,2,3], Jae-Min Yuk [4] and Eun-Kyeong Jo [1,2,3,*]

[1] Department of Microbiology, Chungnam National University School of Medicine, Daejeon 35015, Korea; pst.ktz@gmail.com (P.S.); pcjlovesh6@naver.com (J.K.K.)
[2] Infection Control Convergence Research Center, Chungnam National University School of Medicine, Daejeon 35015, Korea
[3] Department of Medical Science, Chungnam National University School of Medicine, Daejeon 35015, Korea
[4] Department of Infection Biology, Chungnam National University School of Medicine, Daejeon 35015, Korea; yjaemin0@cnu.ac.kr
* Correspondence: hayoungj@cnu.ac.kr

Abstract: 5′-AMP-activated protein kinase (AMPK) plays diverse roles in various physiological and pathological conditions. AMPK is involved in energy metabolism, which is perturbed by infectious stimuli. Indeed, various pathogens modulate AMPK activity, which affects host defenses against infection. In some viral infections, including hepatitis B and C viral infections, AMPK activation is beneficial, but in others such as dengue virus, Ebola virus, and human cytomegaloviral infections, AMPK plays a detrimental role. AMPK-targeting agents or small molecules enhance the antiviral response and contribute to the control of microbial and parasitic infections. In addition, this review focuses on the double-edged role of AMPK in innate and adaptive immune responses to infection. Understanding how AMPK regulates host defenses will enable development of more effective host-directed therapeutic strategies against infectious diseases.

Keywords: AMPK; infection; mycobacteria; host defense

1. Introduction

5′-AMP-activated protein kinase (AMPK) is an intracellular serine/threonine kinase and a key energy sensor that is activated under conditions of metabolic stress [1–3]. It governs a variety of biological processes for the maintenance of energy homeostasis in response to metabolic stresses such as adenosine triphosphate (ATP) depletion [2]. Due to its critical function in metabolic homeostasis, much research has focused on the roles of AMPK in metabolic diseases and cancers [4–6]. However, much less is known about the function of AMPK in infection [7]. Due to the energetic demands of infected cells, most infections by intracellular pathogens are associated with activation of host AMPK, presumably to promote microbial proliferation [8]. AMPK functions as a modulator of host defenses against intracellular bacterial, viral, and parasitic infections [9–12]. Indeed, numerous viruses have the ability to trigger metabolic changes, thereby modulating AMPK activity and substrate selection [13], and AMPK signaling could facilitate or inhibit intracellular viral replication depending on the virus infection [14].

This review focuses on the double-edged role of AMPK in the regulation of host antimicrobial defenses in infections of viruses, bacteria, and parasites. In this review, we describe the existing evidence for the defensive and inhibitory roles of AMPK and the mechanisms underlying its regulation of innate and inflammatory responses. Finally, we describe AMPK-targeting agents that enhance host defenses against infection or control harmful inflammation.

2. Overview of AMPK

5′-AMP-activated protein kinase (AMPK), a serine/threonine kinase, is a key player in bioenergetic homeostasis to preserve cellular ATP [1]. AMPK is activated in response to an increased cellular adenosine monophosphate (AMP)/ATP or adenosine diphosphate (ADP)/ATP ratio, thus promoting catabolic pathways and suppressing biosynthetic pathways [1–3]. Mammalian AMPK exists as a heterotrimeric complex comprising a catalytic subunit α (α1 and α2), a scaffolding β subunit (β1 and β2), and a regulatory γ subunit (γ1, γ2, and γ3) (Figure 1A) [15]. Multiple isoforms of AMPK are encoded by distinct genes of the subunit isotypes, depending on the cell/tissue or species [16]. The AMPK subunit composition and ligand-induced activities of each AMPK isoform complex can differ among cell types, although the α1, β1, and γ1 isoforms are ubiquitously expressed [16,17].

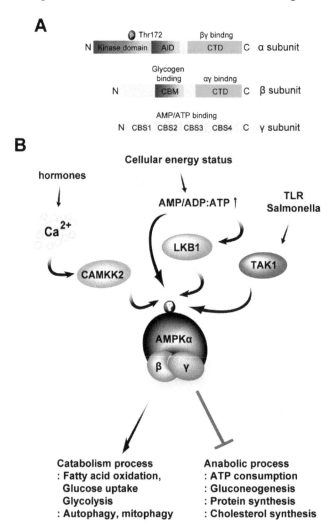

Figure 1. Domain structures of the 5′ AMP-activated protein kinase (AMPK) subunits and the mechanisms that regulate activation of AMPK signaling pathways. (**A**) Conserved domain structure of AMPK subunits consisting of a catalytic α subunit, scaffolding β subunit, and regulatory γ subunit. AID, autoinhibitory domain; CBM, carbohydrate-binding module; CBS, cystathionine-beta-synthase; CTD, C-terminal domain. (**B**) AMPK is activated by the upstream kinases LKB1, CAMKK2 and TAK1 associated with the canonical pathway (triggered by an increased cellular AMP/ATP ratio) or the non-canonical pathway (triggered by an increased intracellular Ca^{2+} concentration or infection/TLR activation). Activated AMPK modulates cellular homeostasis, such as energy metabolism and autophagy, and mitochondrial homeostasis. (black arrow indicate activation/increase; bar-headed red arrow indicates inhibition/decrease). CAMKK2, calcium/calmodulin-dependent kinase kinase 2; LKB1, liver kinase B1; TAK1, Transforming growth factor-β-activated kinase 1; TLR, Toll-like receptor.

The AMPK α subunit contains a kinase domain at the N terminus, which is activated by phosphorylation of Thr-172 by the major upstream liver kinase B1 (LKB1) [18,19]. In contrast to LKB1, the upstream Ca^{2+}-calmodulin-dependent kinase kinase (CaMKK) activates AMPK in response to an increased intracellular Ca^{2+} concentration in the absence of significant changes in ATP/ADP/AMP levels [20]. The regulatory β subunit of AMPK contains a glycogen-binding domain that can sense the structural state of glycogen [21]. Four consecutive cystathionine-β-synthase domains in the regulatory γ subunit are essential for binding to adenosine nucleotides to form an active αβγ complex (Figure 1B) [22,23].

Different AMPK isoforms may have distinct biological functions in different physiological and pathological systems. AMPK governs the cellular energy status by acting as a crucial regulator of energy homeostasis in response to various metabolic stresses, including starvation, hypoxia, and muscle contraction. AMPK activity can be altered by numerous factors, including hormones, cytokines, and nutrients, as well as diverse pathological changes such as metabolic disturbances [24,25]. Because AMPK is important in the adaptation to energy stress, dysregulation of or decreased AMPK activation is implicated in the development of metabolic disorders associated with insulin resistance [6]. In addition to its primary role in the regulation of energy metabolism, AMPK signaling plays a critical role in host–microbial interactions [7]. Furthermore, infections by several viruses result in dysregulation or stimulation of AMPK activity [13]. In mycobacterial infections, AMPK activation promotes activation of host defenses in macrophages and in vivo [12,26]. However, much less is known about the function of AMPK in innate host defenses compared with that in the regulation of metabolism and its mitochondrial function.

3. Multifaceted Role of AMPK in Antimicrobial Responses

Viruses have evolved strategies to manipulate the AMPK signaling pathway to escape host defenses. Indeed, several pathogens can modulate the activity of AMPK/mTOR to obtain sufficient energy for their growth and proliferation [8]. In this review, we discuss microbial manipulation of AMPK activity to affect host defenses against infections. Figure 2 summarizes the multiple roles of AMPK in the viral and bacterial infections addressed in this review. The detailed mechanisms and outcomes of host–pathogen interactions in terms of AMPK modulation are described in Tables 1–4.

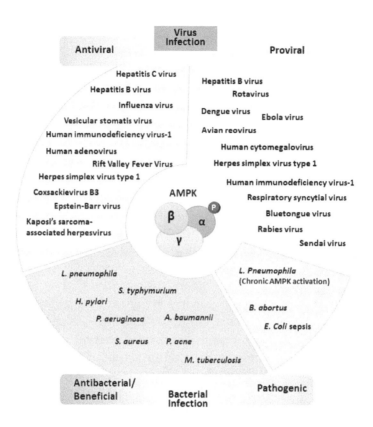

Figure 2. Multifaceted roles of AMPK in viral and bacterial infections. A variety of viruses and bacteria modulate host AMPK activity to promote their growth in host cells. Activation of the AMPK signaling pathway has been implicated in both beneficial antiviral (**left upper**) and detrimental proviral (**right upper**) responses. In addition, AMPK activation promotes the host response to infections by various bacteria (**left lower**) but, in some cases, promotes a detrimental response (**right lower**). The detailed mechanisms by which AMPK activation/inhibition affects infection outcomes are listed in Tables 1–4.

3.1. Roles of AMPK in Viral Infections

3.1.1. Beneficial Effects of AMPK on Virus Infections

Hepatitis C virus (HCV) is a major etiologic agent of chronic liver disease worldwide. HCV infection inhibits AMPKα phosphorylation and signaling [27], and the AMPK agonist metformin suppresses HCV replication in an autophagy-independent manner [28]. Moreover, HCV core protein increases the levels of reactive oxygen species (ROS) and alters the NAD/NADH ratio to decrease the activity and expression of sirtuin 1 (SIRT1) and AMPK, thereby altering the metabolic profile of hepatocytes. This mechanism is implicated in the pathogenesis of hepatic metabolic diseases [29]. As AMPK is a crucial regulator of lipid and glucose metabolism, pharmacological restoration of AMPK activity inhibits lipid accumulation and viral replication in HCV-infected cells [27]. In addition, metformin enhances type I interferon (IFN) signaling by activating AMPK, resulting in inhibition of HCV replication [30]. AMPK inhibition resulted in the downregulation of type I IFN signaling and rescue of the metformin-mediated decrease in the HCV core protein level [30]. Moreover, the AMPK activator 5-aminoimidazole-4-carboxamide 1-β-D-ribofuranoside (AICAR) inhibits HCV replication by activating AMPK signaling, although the anti-HCV effect of metformin is independent of AMPK activation [31]. In chronic HCV infection, the expression of Sucrose-non-fermenting protein kinase 1/AMP-activated protein kinase-related protein kinase (SNARK), an AMPK-related kinase, is increased to promote transforming growth factor β signaling, which is critical for hepatic fibrogenesis [32]. A more recent study showed that HCV-mediated ROS production triggers AMPK activation to attenuate lipid synthesis and promote fatty acid β-oxidation in HCV-infected cells [33]. These data suggest that HCV inhibits AMPK activation to promote its replication, and that the restoration of AMPK activity may be an effective therapeutic modality for HCV infection that acts by metabolic reprogramming or modulation of type I IFN production in host cells [27,28,30,33].

In hepatitis B viral (HBV) infection, AMPK can promote or inhibit viral replication. The detrimental effects of AMPK are described in the following section. Xie et al. reported that AMPK, which is activated by HBV-induced ROS accumulation, suppresses HBV replication [9]. Mechanistically, AMPK activation leads to HBV-mediated autophagic activation, which enhances autolysosome-dependent degradation to restrict viral proliferation [9]. AMPK activity is also involved in the defense against vesicular stomatitis virus, the causal agent of an influenza-like illness, by activating stimulator of IFN genes (STING) [10]. Treatment of mouse macrophages or fibroblasts with an AMPK inhibitor suppressed the production of type I IFN and TNF-α in response to a STING-dependent ligand or agonist, suggesting a role for AMPK in STING signaling [10]. AMPK plays a role in the excessive inflammatory cytokine/chemokine levels in Mint3/Apba3 depletion models of severe pneumonia due to influenza virus [34]. Indeed, food-derived polyphenols, such as epigallocatechin gallate and curcumin, are useful for controlling viral and bacterial infections [35]. Although a review of AMPK-modulating polyphenols is beyond our scope, we highlight the therapeutic promise of polyphenols against infection. For example, curcumin from *Curcuma longa* inhibits influenza A viral infection in vitro and in vivo, at least in part by activating AMPK [36]. The polyphenol epigallocatechin gallate attenuates Tat-induced human immunodeficiency virus (HIV)-1 transactivation by activating AMPK [37]. Further studies should examine the ability of food-derived polyphenols to activate AMPK signaling to control viral replication in host cells.

Human adenovirus type 36, which is associated with obesity, inhibits fatty acid oxidation and AMPK activity and increases accumulation of lipid droplets in infected cells [38]. The AMPK signaling pathway and its upstream regulator LKB1 repress replication of the bunyavirus Rift Valley Fever virus (RVFV), a re-emerging human pathogen [39]. The mechanisms of the antiviral effects of AMPK on RVFV and other viruses are mediated by AMPK inhibition of fatty acid synthesis [39]. Pharmacologic activation of AMPK suppresses RVFV infection and reduces lipid levels by inhibiting fatty acid biosynthesis [39]. In addition, the AMPK/Sirt1 activators resveratrol and quercetin significantly reduce the viral titer and gene expression, as well as increase the viability of infected neurons, in herpes simplex virus type 1 (HSV-1) infection [40]. Moreover, coxsackievirus B3 (CVB3) infection triggers AMPK activation, which suppresses viral replication in HeLa and primary myocardial cells [41]. The AMPK agonists AICAR and metformin suppress CVB3 replication and attenuate lipid accumulation by inhibiting lipid biosynthesis [41]. Thus, regulation of fatty acid metabolism by AMPK signaling is an essential component of cell autonomous immune responses [39].

Latent membrane protein 1 (LMP1) of Epstein-Barr virus (EBV) inactivates LKB1/AMPK, whereas AMPK activation by AICAR abrogated LMP1-mediated proliferation and transformation of nasopharyngeal epithelial cells, suggesting therapeutic potential for EBV-associated nasopharyngeal carcinoma [42]. Moreover, constitutive activation of AMPK inhibited lytic replication of Kaposi's sarcoma-associated herpesvirus in primary human umbilical vein endothelial cells [43]. These data suggest that AMPK suppresses cell transformation and infection-related tumorigenesis in a context-dependent manner. The roles of AMPK in viral infection are listed in Table 1.

3.1.2. Detrimental Effects of AMPK on Virus Infections

Several viruses manipulate AMPK signaling to promote their replication. Genome-scale RNA interference screening of host factors in rotaviral infection identified AMPK as a critical factor in the initiation of a rotavirus-favorable environment [44]. In dengue viral infections, the 3-hydroxy-3-methylglutaryl-CoA reductase (HMGGR) activity elevated by AMPK inactivation resulted in generation of a cholesterol-rich environment in the endoplasmic reticulum, which promoted formation of viral replication complexes [45]. Also, dengue viral infection stimulates AMPK activation to induce proviral lipophagy, thereby enhancing fatty acid β-oxidation and viral replication [46].

Table 1. Beneficial Effects of AMPK in viral infection.

Pathogen	Small Molecules/Chemicals	Agonist/Antagonist	Involvement of AMPK	Outcome (In Vitro/In Vivo)	Ref.
Hepatitis C virus (HCV)	HCV	-	HCV infection inhibit AMPKα phosphorylation and Akt-TSC-mTORC1 pathway	AMPK inhibition is required for HCV replication (in vitro)	[27,28]
	AICAR, Metformin, A769662	Agonist	Restoration of AMPKα activity	Antiviral effects (in vitro)	[27,28]
	Metformin	Agonist	Type I interferon signaling through AMPK pathway activation	Inhibits HCV replication (in vitro)	[30]
	AICAR	Agonist	AMPK activation (Indirect effects counteracted by compound C)	Suppression of HCV replication (in vitro)	[31]
Hepatitis B virus (HBV)	HBV	-	ROS-dependent AMPK activation in HBV-producing cells	Negatively regulates HBV production	[9]
	AICAR constitutive active AMPKα	Agonist	AMPK activation, autophagic flux activation	Inhibits HBV production (in vitro)	[9]
	Compound C dominant-negative AMPKα	Antagonist	AMPK inhibition	Enhances HBV production (in vitro and in vivo)	[9]
Vesicular stomatitis virus (VSV)	AICAR	Agonist	STING-dependent signaling activation	Type I IFN production and antiviral responses (in vitro)	[10]
	Compound C	Antagonist	Inhibition of STING-dependent signaling	Suppression of IFN-β production (in vitro)	[10]
	Mint3 depletion	-	AMPK activation	Attenuates severe pneumonia by influenza infection (in vivo)	[34]
Influenza virus	AICAR	Agonist	AMPK activation in Mint3 depletion model	Decreases inflammatory cytokine production in Mint3-deficient macrophages (in vitro)	[34]
	Curcumin	Activator	AMPK activation	Inhibits influenza A virus infection (in vitro and in vivo)	[36]
Human immunodeficiency virus-1	Epigallocatechin gallate	Activator	AMPK activation	Attenuation of Tat-induced human immunodeficiency virus 1 (HIV-1) transactivation	[37]
Human adenovirus	Adenovirus	-	Inhibit AMPK activity/signaling	Virus induces lipid droplets, presumably associated with obesity (in vitro)	[38]
Rift Valley Fever Virus (RVFV)	A769662, 2-deoxy-D-glucose (2-DG)	Agonist	LKB1/AMPK signaling activation; Inhibition of fatty acid synthesis	Restriction of viral infection (in vitro)	[39]
Herpes simplex virus type 1 (HSV-1)	AICAR, Resveratrol, Quercetin	Activator/agonist	AMPK/Sirt1 activation	Reduces viral titer and the expression of viral genes (in vitro)	[40]

Table 1. *Cont.*

Pathogen	Small Molecules/Chemicals	Agonist/Antagonist	Involvement of AMPK	Outcome (In Vitro/In Vivo)	Ref.
Coxsackievirus B3 (CVB3)	-	-	AMPK activation by CVB3	Restriction of viral replication; reversed by siRNA against AMPK	[41]
	AICAR, A769662, Metformin	Agonist	AMPK activation	Restriction of viral replication; improve the survival rate of infected mice (in vitro and in vivo)	[41]
Epstein-Barr virus (EBV)	LMP1 of EBV	-	LKB1-AMPK inactivation	AMPK inactivation leads to proliferation and transformation of epithelial cells associated with EBV infection (in vitro)	[42]
	AICAR	Agonist	AMPK activation	Inhibition of proliferation of nasopharyngeal epithelial cells (in vitro)	[42]
Kaposi's sarcoma-associated herpesvirus (KSHV)	AICAR, Metformin, Constitutive active AMPK	Agonist	AMPK as a KSHV restriction factor	Inhibits the expression of viral lytic genes and virion production (in vitro)	[43]
	Compound C, Knockdown of AMPKα1	Antagonist	AMPK inhibition	Enhances viral lytic gene expression and virion production (in vitro)	[43]

In HBV infection, the HBV X protein activates AMPK, and inhibition of AMPK reduces HBV replication in rat primary hepatocytes [47]. Inhibition of AMPK led to activation of mTORC1, which is required for inhibition of HBV replication in the presence of low AMPK activity [47]. The crosstalk between AMPK and mTORC1 may enable development of therapeutics that suppress HBV replication and thus also hepatocellular carcinoma (HCC) development [47]. However, as described in Section 3.1.1, AMPK activation by AICAR inhibits extracellular HBV production in HepG2 cells [9]. The discrepancy may be attributed to the use of different cell lines in the two studies [9,47]. Further work should address the role of AMPK in HBV infection in vitro and in vivo. In infection by Zaire Ebolavirus (EBOV), the expression levels of the γ2 subunit of AMPK are correlated with EBOV transduction in host cells. In mouse embryonic fibroblasts treated with a small-molecule inhibitor of AMPK (compound C), it was shown that AMPK activity is required for EBOV replication in host cells and EBOV glycoprotein-mediated entry/uptake [48]. In addition, Avian reoviral infection upregulates AMPK phosphorylation, which leads to activation of mitogen-activated protein kinase (MAPK) p38 in Vero cells, which enhances viral replication [49]. The nonstructural protein p17 of avian reovirus positively regulates AMPK activity, inducing autophagy and increasing viral replication [50].

In HSV-1 infection, the activated AMPK/Sirt1 axis inhibits host-cell apoptosis during early-stage infection, which promotes viral latency and protects neurons [51]. However, during the later stages of infection, HSV-1 induces apoptosis of host cells concomitantly with Sirt1 activation [51]. In HIV-infected cocaine abusers, AMPK signaling plays a role in energy deficit and neuronal dysfunction, which are associated with the development of neuroAIDS [52]. These data suggest that differential regulation of AMPK signaling is a determinant of the viral infection course.

Using a kinome-profiling approach, AMPK and related kinases were found to be effectors of human cytomegalovirus (HCMV) replication [53,54]. HCMV infection induces AMPK and CaMKK2 (upstream activator of AMPK)-dependent remodeling of core metabolism, both of which are required for optimal yield and replication of HCMV [53,54]. Notably, inhibition of AMPK activity by short-interfering RNA-mediated AMPK knockdown or an AMPK antagonist (compound C) prevents viral gene expression, providing valuable insight into the mechanisms of HCMV infection [53]. In addition, the AMPK activation-dependent modulation by HCMV of host-cell metabolism is associated with HCMV replication [55]. HCMV-mediated AMPK activation is dependent on CaMKK, and inhibition of AMPK activity abrogated HCMV replication and DNA synthesis [55]. Furthermore, the cardiac glycoside digitoxin induces phosphorylation of AMPK/ULK1, whereas it suppresses mTOR activity to increase autophagic flux and inhibit HCMV replication [56]. Moreover, HCMV induces production of the host protein viperin [57], which is required for AMPK activation, transcriptional activation of GLUT4 and lipogenic enzymes, and lipid synthesis [58]. The enhanced lipid synthesis promotes formation of the viral envelope and production of HCMV virions [58].

In infection, host-cell autophagy plays an important role in host defense and virus survival. Several viruses can manipulate or subvert the autophagic machinery to favor viral replication. For example, respiratory syncytial virus activates autophagy via the AMPK/mTOR signaling pathway to enhance its replication by inhibiting host-cell apoptosis [59]. Bluetongue virus, a double-stranded segmented RNA virus, also induces host-cell autophagy by activating AMPK [60]. Moreover, AMPK is an upstream regulator of rabies virus-induced incomplete autophagy to provide the scaffolds for viral replication [61]. In Sendai viral infection, AMPK activity is required for autophagic initiation to promote viral replication [62]. In oncogenic EBV infection, the increased cell survival caused by AMPK-mediated autophagic activation maintains early hyperproliferation of infected cells [63]. These data suggest AMPK activity to be a therapeutic target for the development of novel antiviral agents.

Importantly, type I IFN, a critical effector in the antiviral response, attenuates AMPK phosphorylation and increases the intracellular ATP level [64]. In addition, IFN-β-mediated glycolytic metabolism is important for the acute phase of the antiviral response to CVB3 [64]. The antiviral cytokine IFN-β regulates host-cell metabolism to enhance glucose uptake and ATP generation, which promote the antiviral response [64]. In infection by snakehead vesiculovirus, miR-214 targeting AMPK suppressed viral replication and upregulated IFN-α expression [65]. Thus, regulation of AMPK activity by the host–pathogen interaction mediates diverse metabolic effects, which modulate viral replication and the host defense response. The beneficial and detrimental effects of AMPK on viral infections are summarized in Tables 1 and 2, respectively.

3.2. Bacterial Infections and AMPK Activation

Intracellular pathogens manipulate the AMPK signaling pathway to alter their metabolic environment to favor bacterial survival or pathogenesis. Mitochondrial dysfunction triggers AMPK signaling, thus enhancing the proliferation of *Legionella pneumophila*, a respiratory pathogen, in *Dictyostelium* cells [66]. Inhibition of AMPK activation reversed the increased *Legionella* proliferation in host cells with mitochondrial disease [66]. However, the AMPK activator metformin triggers mitochondrial ROS generation and activates the AMPK signaling pathway to enhance the host response to *L. pneumophila* in macrophages and promote survival in a murine model of *L. pneumophila* pneumonia [67]. Thus, the role of AMPK activation in bacterial infections differs depending on the host species.

Salmonella typhimurium degrades SIRT1/AMPK to evade host xenophagy [68]. In addition, cytosolic *Salmonella* is ubiquitinated and targeted for xenophagy by AMPK activation [69,70]. AMPK activation by AICAR induces autophagy and colocalization of *Salmonella*-containing vacuoles with LC3 autophagosomes [68], whereas inhibition of AMPK by compound C increases bacterial replication by suppressing autophagy [69]. In *Salmonella*-infected cells, AMPK activation is mediated through toll-like receptor-activated TGF-β-activated kinase 1 (TAK1) [69] which is a direct upstream kinase of AMPK in addition to LKB1 and CaMKK2 [71]. In *Brucella abortus* infection, AMPK activation enhances intracellular growth of *B. abortus* by inhibiting nicotinamide adenine dinucleotide phosphate (NADPH) oxidase-mediated ROS generation [72]. In models of *Escherichia coli* sepsis, ATP-induced pyroptosis is blocked by piperine, a phytochemical present in black pepper (*Piper nigrum* Linn) [73]. The inhibitory effects of piperine on pyroptosis and systemic inflammation are mediated by regulation of the AMPK signaling pathway, as shown by suppression of ATP-mediated AMPK activation by piperine treatment in vitro and in vivo [73]. These data suggest that AMPK plays multiple roles in bacterial infections.

AMPK activation promotes host defenses against infections by several microbes. Transcriptomic and proteomic analyses of a *Caenorhabditis elegans* model indicated that AMPKs function as regulators and mediators of the immune response to infection by, for example, *Bacillus thuringiensis* [74]. Several small-molecule AMPK activators exert protective effects against *Helicobacter pylori*-induced apoptosis of gastric epithelial cells [70,75]. The AMPK agonists A-769662 and resveratrol, as well as AMPKα overexpression, inhibit apoptosis in *H. pylori*-infected gastric epithelial cells [76]. The AMPK activator compound 13 ameliorates *H. pylori*-induced apoptosis of gastric epithelial cells by modulating ROS levels via the AMPK-heme oxygenase-1 axis [76]. Blockade of AMPK signaling significantly abrogates the protective effect of compound 13 against *H. pylori* within gastric epithelial cells [76].

Metformin and AICAR repress infection by *Pseudomonas aeruginosa*, an important opportunistic pathogen, of airway epithelial cells by inhibiting bacterial growth and increasing transepithelial electrical resistance [77]. AMPKα1 depletion increased the susceptibility to *Staphylococcus aureus* endophthalmitis in mice [78], suggesting a protective role for AMPK in bacterial retinal inflammation. Moreover, AMPK activation by AICAR enhances its anti-inflammatory effects, phagocytosis, and bactericidal activity against *S. aureus* infection of various phagocytic cells including microglia, macrophages, and neutrophils [78]. Epigallocatechin gallate, a polyphenol in green tea, inhibits the viability of *Propionibacterium acnes*, a pathogen associated with acne, and exerted an antilipogenic effect in SEB-1 sebocytes by activating the AMPK/sterol regulatory element-binding protein pathway [35]. Moreover, *Acinetobacter baumannii*, an emerging opportunistic pathogen, activates autophagy via the AMPK/ERK/mTOR pathway to promote an antimicrobial response to intracellular *A. baumannii* [79]. The roles of AMPK in bacterial infection are summarized in Table 3.

Table 2. Detrimental Effects of AMPK in viral infection.

Pathogen	Small Molecules/Chemicals	Agonist/Antagonist	Involvement of AMPK	Outcome (In Vitro/In Vivo)	Ref.
Rotavirus	RNAi	-	AMPK-mediated glycolysis, fatty acid oxidation and autophagy	Development of a rotavirus replication-permissive environment (in vitro)	[44]
	AICAR, Metformin	Agonist	AMPK activation (AICAR, directly; Metformin, indirectly)	Upregulation of the proportion of viral infected cells (in vitro)	[44]
	Dorsomorphin	Inhibitor	Inhibition of AMPK activity	Reduces the number of infected cells (in vitro)	[44]
Dengue virus	Virus infection	-	Elevates 3-hydroxy-3-methylglutaryl-CoA reductase activity through AMPK inactivation	Promotes the formation of viral replicative complexes (in vitro)	[45]
	Metformin, A769662	Agonist	AMPK activation	Antiviral effects (in vitro)	[45]
	Compound C	Antagonist	AMPK inhibition	Augments the viral genome copies (in vitro)	[45]
	Virus infection	-	AMPK activation; induction of lipophagy	Increases viral replication (in vivo)	[46]
	Compound C siRNA against AMPKα1	Antagonist	Inhibition of proviral lipophagy	Decreases viral replication (in vivo)	[46]
Hepatitis B virus (HBV)	HBx protein	-	Decreased ATP, activates AMPK in rat primary hepatocytes	AMPK inhibition decreases HBV replication (in vitro)	[47]
	Compound C	Antagonist	Activates mTORC1	Reduces HBV replication (in vitro)	[47]
Ebola virus	Compound C	Antagonist	Less permissive to Ebola virus infection (Similar effects in AMPKα1- or AMPKα2-deleted mouse embryonic fibroblasts)	Inhibits EBOV replication in Vero cells (in vitro)	[48]
Avian reovirus	Virus infection	-	Upregulates AMPK phosphorylation leading to p38 MAPK activation	Increases virus replication (in vitro)	[49]
	P17 protein	-	P17 protein activates AMPK to induce autophagy	Increases virus replication (in vitro)	[50]
	AICAR	Agonist	AMPK activation (Indirect effects through p38 MAPK)	Increases virus replication (in vitro)	[49]
	Compound C	Antagonist	AMPK inhibition	Decreases virus replication (in vitro)	[49]
Herpes simplex virus type 1 (HSV-1)	HSV-1		In early infection, AMPK is down-regulated, and then recovered gradually	AMPK/Sirt1 axis inhibits host apoptosis in early infection (in vitro)	[51]
Human immunodeficiency virus-1 (HIV1)	Cocaine		Induces AMPK upregulation; AMPK plays a role in energy deficit and metabolic dysfunction	Cocaine exposure during HIV infection accelerates neuronal dysfunction (in vitro)	[52]

Table 2. Cont.

Pathogen	Small Molecules/Chemicals	Agonist/Antagonist	Involvement of AMPK	Outcome (In Vitro/In Vivo)	Ref.
Human cytomegalovirus (HCMV)	RNAi	-	AMPK may activate numerous metabolic pathways during HCMV infection	siRNA to AMPKα reduces HCMV replication (in vitro)	[53,54]
	HCMV	-	Upregulation of host AMPK	Favors viral replication (in vitro)	[53,54]
	Compound C	Antagonist	Interferes with normal accumulation of viral proteins and alters the core metabolism	Compound C inhibits the viral production of HCMV (in vitro); blocks the immediate early phase of viral replication (in vitro)	[53,54]
	RNAi to AMPK	-	Blocks glycolytic activation in HCMV-infected cells	RNA-based inhibition of AMPK attenuates HCMV replication (in vitro)	[55]
	Digitoxin	Activator	Digitoxin modulates AMPK-ULK1 and mTOR activity to increase autophagic flux	Viral inhibition (in vitro)	[56]
	Digitoxin + AICAR	-	Combination reduces autophagy	Viral replication (in vitro)	[56]
	HCMV	-	Induces targeting host protein viperin to mitochondria; viperin is required for AMPK activation and regulate lipid metabolism	Viperin-dependent lipogenesis promotes viral replication and production by infected host cells (in vitro)	[57]
Respiratory syncytial virus (RSV)	RSV	-	RSV induces autophagy through ROS and AMPK activation	RSV-induced autophagy favors viral replication (in vitro)	[59]
	Compound C	Antagonist	Inhibition of AMPK and autophagy	Compound C reduces viral gene and protein expression, and total viral titers (in vitro)	
Bluetongue virus	Bluetongue virus	-	Induces autophagy through activation of AMPK	Favors viral replication (in vitro)	
	Compound C siRNA to AMPK	Antagonist	Inhibits BTV1-induced autophagy	AMPK inhibition decreases viral titers (in vitro)	[60]
Rabies virus	Rabies virus		Incomplete autophagy induction via CASP2-AMPK-MAPK1/3/11-AKT1-mTOR pathways	Enhances viral replication (in vitro)	[61]
Sendai virus	Sendai virus	-	Induces host protein TDRD7, an inhibitor of autophagy-inducing AMPK	Host autophagy and viral replication is inhibited by TDRD7 (in vitro)	[62]
	Compound C, shRNA to AMPK	Antagonist	Inhibition of AMPK activity; inhibits viral protein	AMPK activity is required for viral replication (in vitro)	
Snakehead vesiculo-virus	Snakehead vesiculo-virus	-	Downregulates miR-214, which targets AMPK	AMPK upregulation promotes viral replication through reduction of IFN-α expression (in vitro)	[65]

Table 3. The roles of AMPK in bacterial infection.

Pathogen	Small Molecules/Chemicals	Agonist/Antagonist	Involvement of AMPK	Outcome (In Vitro/In Vivo)	Ref.
L. pneumophila	-	-	Chronic AMPK activation involved in host susceptibility to infection (Direct effects by AMPKα antisense)	Bacterial multiplication in host cells with mitochondrial dysfunction	[66]
	Metformin	Agonist	Bactericidal effects are mediated by mitochondrial ROS production (Indirect)	Antimicrobial responses (in vitro and in vivo)	[67]
S. typhimurium	-	-	*S. typhimurium* exhibits virulence through lysosomal degradation of SIRT1 and AMPK to impair autophagy	Bacterial evasion from autophagic clearance (in vitro)	[68]
	AICAR	Agonist	Upregulation of autophagy	Increased colocalization of salmonella containing vacuole with LC3 (in vitro)	[68]
	-	-	AMPK activation via TAK1; autophagy initiation by ULK1 phosphorylation	Autophagy activation (in vitro)	[69]
	Compound C	Antagonist	AMPK inhibition	Increased bacterial replication by suppression of autophagy (in vitro)	[69]
B. abortus	-	-	AMPK activation via inositol-requiring enzyme 1 (IRE1)	Promote intracellular growth of *B. abortus* (in vitro)	[72]
	Compound C	Antagonist	AMPK inhibition; activation of NADPH oxidase-mediated ROS production	Suppression of intracellular growth (in vitro)	[72]
E. coli	Piperine	Antagonist	Inhibits ATP-induced pyroptosis by suppressing AMPK activation	Inhibition of pyroptosis; Attenuation of systemic inflammation (in vitro and in vivo)	[73]
	ATP Metformin	Agonist	AMPK activation; increases pyroptosis by inflammasome activation	Activation of pyroptosis (in vitro)	[73]
B. thuringiensis	-	-	AMPK identified by transcriptome and proteome data analysis in vivo (Indirect)	Potentially related to regulation of immune defense (Not determined)	[74]
H. pylori	-	-	TAK1-mediated AMPK activation	Protects gastric epithelial cells from *H. pylori*-induced apoptosis (in vitro)	[75]
	A-769662 Resveratrol	Agonist	Inhibits *H. pylori*-induced apoptosis (Direct effects by overexpression of AMPKα)	Alleviates *H. pylori*-induced gastric epithelial cell apoptosis (in vitro)	[76]
	Compound 13	Agonist	Inhibits *H. pylori*-induced apoptosis through AMPK-heme oxygenase-1 signaling	Alleviates *H. pylori*-induced gastric epithelial cell apoptosis (in vitro)	[76]
	Compound C	Antagonist	Inhibitory effects upon compound 13-mediated anti-*H. pylori* activities (Direct effects by AMPKα1 shRNAs)	Aggravates *H. pylori*-induced gastric epithelial cell apoptosis (in vitro)	[76]
P. aeruginosa	AICAR Metformin	Agonist	Counteracts the bacterial effects on the reduction of transepithelial electrical resistance (Indirect effects)	Inhibits hyperglycemia-induced bacterial growth; Improve airway epithelial barrier function (in vitro)	[77]

Table 3. *Cont.*

Pathogen	Small Molecules/Chemicals	Agonist/Antagonist	Involvement of AMPK	Outcome (In Vitro/In Vivo)	Ref.
S. aureus	AICAR	Agonist	AMPK activation	Reduces bacterial burden and intraocular inflammation; Increases bacterial killing in macrophages (in vitro and in vivo)	[78]
	Compound C	Antagonist	Downregulates AMPK activity (Direct effects by AMPKα1 knockout mice)	Counteracts AICAR-mediated anti-inflammatory effects (in vivo); Increases susceptibility towards *S. aureus* endophthalmitis (in vivo)	[78]
P. acne	Epigallocatechin gallate	-	Activates AMPK-sterol regulatory element-binding proteins pathway activation	Antilipogenic effects in SEB-1 sebocytes (in vitro)	[35]
A. baumannii	-	-	Activates autophagy through Beclin-1-dependent AMPK/ERK/mTOR pathway (Indirect effects by different *A. baumannii* strains)	Autophagy may promote antimicrobial responses (in vivo)	[79]
	Metformin	Agonist	AMPK activation; Increased mtROS production; Increases phago-lysosomal fusion (Direct effects upon bacterial growth in vitro)	Inhibition of intracellular growth of *M. tuberculosis* (drug-resistant strain; in vitro); Increases the efficacy of conventional TB drugs in vivo	[80]
	AICAR	Agonist	AMPK-PPARGC1A signaling-mediated autophagy activation; Enhancement of phagosomal maturation (Direct effects by shRNA against AMPKα)	Upregulation of antimicrobial responses (in vitro and in vivo)	[12]
Mycobacterium tuberculosis	Compound C	Antagonist	Counteracts the effects by AICAR upon intracellular inhibition of *M. tuberculosis* growth	Downregulation of antimicrobial responses (in vitro)	[12]
	Vitamin D (1,25-D3)	-	Induces autophagy through LL-37 and AMPK activation (Indirect effects upon LL-37 function)	Promotes autophagy and antimicrobial response in human monocytes/macrophages (in vitro)	[81]
	Phenylbutyrate Vitamin D	-	Induces LL-37-mediated autophagy (Indirect effects; AMPK is involved in LL-37-mediated autophagy)	Improves intracellular killing of *M. tuberculosis* (in vitro)	[82]
	Gamma-aminobutyric acid (GABA)	Agonist	Induces autophagy (Direct effects by shRNA against AMPK)	Promotes antimicrobial effects against *M. tuberculosis* (in vitro and in vivo)	[83]
	Ohmyungsamycins	-	Activates AMPK and autophagy; Intracellular inhibition of bacterial growth; Amelioration of inflammation (Indirect effects upon host autophagy)	Promotes antimicrobial effects against *M. tuberculosis* (in vitro and in vivo)	[26]
	Compound C	Antagonist	Blocks the secretion of neutrophil Matrix metalloproteinase-8 (MMP-8)	Neutrophil MMP-8 secretion is related to matrix destruction in human pulmonary TB (in vitro and in human TB lung specimens)	[84]

3.3. Roles of AMPK in Mycobacterial Infection

The seminal study by Singhal et al. addressed the effect of metformin as an adjunctive therapy for tuberculosis. Importantly, metformin suppressed the intracellular growth of *Mycobacterium tuberculosis* (Mtb) in vitro, including drug-resistant strains, by activating the AMPK signaling pathway [80]. In vivo, metformin attenuated the immunopathology and enhanced the immune response and showed a synergistic effect with conventional anti-TB drugs in Mtb-infected mice [80]. The microbicidal effect of metformin in macrophages is due, at least in part, to mitochondrial ROS generation, which is associated with AMPK signaling [80]. Type 2 diabetes mellitus (DM) is re-emerging as a risk factor for human tuberculosis, thus candidate host directed therapeutic targets for tuberculosis combined with DM should be identified [85]. Human cohort studies showed that metformin treatment for DM is associated with a decreased prevalence of latent tuberculosis compared with alternative DM treatments, suggesting metformin to be a candidate HDT for tuberculosis patients with type 2 DM [80,85]. Indeed, Mtb infection inhibits AMPK phosphorylation but increases mTOR kinase activation in macrophages [12]. The AMPK activator AICAR via autophagic activation enhances phagosomal maturation and antimicrobial responses in macrophages in Mtb infection [12]. In human monocytes/macrophages, vitamin D-mediated antimicrobial responses are mediated by the antimicrobial peptide LL-37 via AMPK activation [81]. In addition, LL-37-induced autophagy by phenylbutyrate, alone or in combination with vitamin D, promotes intracellular killing of Mtb in human macrophages via AMPK- and PtdIns3K-dependent pathways [82]. Recent findings revealed the role of gamma-aminobutyric acid (GABA) in AMPK activation to enhance the autophagy and the antimicrobial responses [83]. Silencing of AMPK by a lentiviral short hairpin RNA (shRNA) specific to AMPK reduces GABA-induced autophagic activation as well as phagosomal maturation during Mtb infection [83].

MicroRNAs are small non-coding RNAs involved in the regulation of diverse physiological and pathological processes, including Mtb infection. Mycobacterial infection of macrophages upregulates miR-33 and miR-33*, which target and suppress AMPKα [86]. Interestingly, miR-33/miR-33* regulates autophagy by suppressing AMPK-dependent activation of the transcription of autophagy- and lysosome-related genes and promoting accumulation of lipid bodies in Mtb infection [86]. Mtb infection increases the expression of MIR144*/has-miR-144-5p, which targets DNA damage regulated autophagy modulator 2 (DRAM2), to inhibit the antimicrobial responses to Mtb infection in human monocytes/macrophages. In contrast, autophagic activators enhance production of the autophagy-related protein DRAM2 by activating the AMPK signaling pathway; this contributes to host defenses against Mtb in human macrophages [87].

Although AMPK may play a protective role in tuberculosis, it has also been reported to exert an immunopathological effect by driving the secretion of neutrophil matrix metalloproteinase-8 (MMP-8), resulting in matrix destruction and cavitation, which enhance the spread of Mtb [84]. Neutrophil-derived MMP-8 secretion is upregulated in Mtb infection and neutrophils from AMPK-deficient patients express lower levels of MMP-8, suggesting a key role for MMP-8 in tuberculosis immunopathology [84]. Because the pathogenesis of tuberculosis is complex, further information on the function of AMPK in the immune response to Mtb infection is needed for development of improved therapeutic strategies [88]. The roles of AMPK in mycobacterial infection are listed in Table 3.

3.4. Roles of AMPK in Parasite Infections

The immune response to parasitic helminths involves M2-type cells, CD4(+) Th2 cells, and group 2 innate lymphoid cells. AMPK activation regulates type 2 immune responses and ameliorates lung injury in response to hookworm infections [89]. Mice deficient in AMPK α1 subunit exhibited impaired type 2 responses, an increased intestinal worm burden, and exacerbated lung injury [89]. In *Leishmania*-infected macrophages, *Leishmania infantum* causes a metabolic switch to enhance oxidative phosphorylation by activating LKB1/AMPK and SIRT1 [90]. Impairment of metabolic reprogramming by SIRT1 or AMPK suppresses intracellular growth of the parasite, suggesting a role for AMPK/SIRT1 in intracellular proliferation of *L. infantum* [90]. In *Schistosoma japonicum* egg antigen (SEA)-mediated autophagy, which is modulated by IL-7 and the AMPK signaling pathway, ameliorate liver pathology, suggesting AMPK to be a therapeutic target factor for schistosomiasis [91].

Notably, host AMPK activity is decreased by hepatic *Plasmodium* infection. Activation of the AMPK signaling pathway by AMPK agonists, including salicylate, suppresses the intracellular replication of malaria parasites, including that of the human pathogen *Plasmodium falciparum* [92]. These data suggest that host AMPK signaling is a therapeutic target for hepatic *Plasmodium* infection [92]. In addition, resveratrol protects cardiac function and reduces lipid peroxidation and trypanosomal burden in the heart by activating AMPK, suggesting a role for AMPK in Chagas heart disease [93]. The roles of AMPK in parasitic infections are listed in Table 4.

Table 4. The role of AMPK in parasitic infection.

Pathogen	Small Molecules/Chemicals	Agonist/Antagonist	Involvement of AMPK (Direct/Indirect)	Outcome (In Vitro/In Vivo)	Ref.
Hookworm *Nippostrongylus brasiliensis*	-		AMPKα1 deficiency inhibit IL-13 and CCL17, and defective type 2 immune resistance (Direct effects using by AMPKα1 knockout mice)	AMPKα1 suppresses lung injury and drives M2 polarization during infection	[89]
L. infantum	-	-	Infection leads to a metabolic switch to activate AMPK through the SIRT1-LKB1 axis (Direct effects using by AMPKα1 knockout mice)	Ablation of AMPK promotes parasite clearance in vitro and in vivo	[90]
S. japonicum			Infection-driven IL-7-IL-7R signaling inhibits autophagy; IL-7 inhibits macrophage autophagy via AMPK	Anti-autophagic IL-7 increases liver pathology (in vivo)	[91]
	Metformin	Agonist	Decreases the autophagosome formation in macrophages	in vitro	
	Compound C siAMPKα	Antagonist	Increases autophagosome formation in macrophages (Direct effects by siAMPKα)	in vitro	
P. falciparum	-	-	AMPK activity is suppressed upon infection	Decreases *Plasmodium* hepatic growth	[92]
	Salicylate Metformin A769662	Agonist	AMPK activation impairs the intracellular replication of malaria	Antimalarial interventions (in vitro and in vivo)	
Trypanosoma cruzi	Resveratrol, Metformin	Agonist	AMPK activation reduces heart oxidative stress (Indirect effects)	Reduces heart parasite burden; Protects heart function in Chagas heart disease (in vivo)	[93]

3.5. AMPK in Fungal Infection

A recent phosphoproteomic analysis of *Cryptococcus neoformans* (Cn) infection showed that AMPK activation is triggered by fungal phagocytosis and is required for autophagic induction. Interestingly, AMPK depletion in monocytes promoted host resistance to fungal infection in mouse models, suggesting that AMPK represses the immune response to *Cryptococcus* infection [94].

4. Roles of AMPK in Innate and Adaptive Immune Responses

The roles of AMPK in modulation of the mitochondrial network and energy metabolism, which are associated with the immune response, have been investigated [95]. Here, we briefly review recent data on AMPK regulation of innate and adaptive immune responses in infection and inflammation. Figure 3 summarizes the regulatory roles and mechanisms of a variety of small-molecule AMPK activators in terms of innate immune and inflammatory responses.

The data must be interpreted cautiously, as the small-molecule activators (e.g., AICAR, metformin, and compound C) function via off-target mechanisms, such as AMPK-independent pathways or inhibition of protein kinases other than AMPK [31,96,97]. The beneficial effects of these compounds remain to be fully determined. Thus, selective compounds such as MK-8722 [98] and SC4 [99] and the selective inhibitor SBI-0206965 [100] should be considered for future AMPK-targeted treatment strategies.

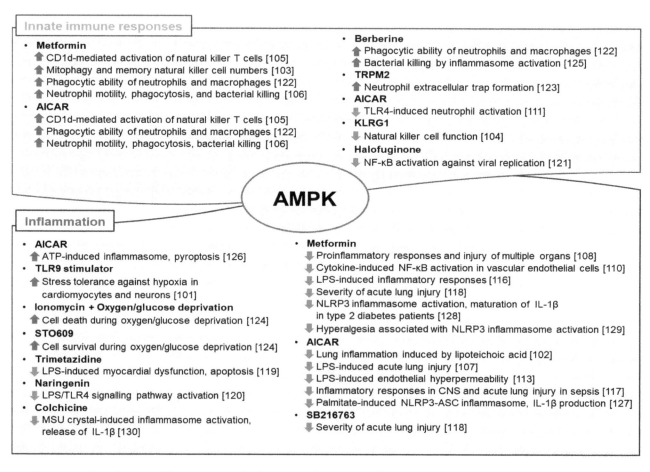

Figure 3. Regulatory effects and underlying mechanisms of small-molecule AMPK activators on the innate immune and inflammatory responses. (red upward arrows indicate activation/increase and blue downward arrows indicate inhibition/decrease)

4.1. Role of AMPK in Regulation of the Innate Immune Response

AMPK is involved in regulation of the innate immune response. For example, the innate-immune stimulator toll-like receptor (TLR) 9 inhibits energy substrates (intracellular ATP levels) and activates AMPK, which enhances stress tolerance in cardiomyocytes and neurons, while stimulation by the TLR9 ligand induces inflammation [101]. The AMPK activator AICAR suppresses the lung inflammation induced by lipoteichoic acid, a major component of the cell wall of Gram-positive bacteria [102]. Natural killer (NK) cells are crucial in the innate immune response to viral infections and transformed cells. Activation of the AMPK signaling pathway or inhibition of mTOR is associated with enhanced mitophagy and an increased number of memory NK cells in antiviral responses [103]. In contrast, increased expression of the inhibitory killer cell lectin-like receptor G1 in aged humans is related to AMPK activation, which has been implicated in disruption of NK cell function [104]. In addition, AMPK activation contributes to CD1d-mediated activation of NK T cells, an important cell type in the innate immune response [105]. The findings above suggest that AMPK plays a pleiotropic role in the regulation of the innate immune response depending on the stimulus and cell type in question.

4.2. AMPK Regulation of Local and Systemic Inflammation

AMPK activators enhance neutrophil chemotaxis, phagocytosis, and bacterial killing to protect against peritonitis-induced sepsis [106]. Indeed, AMPK activators including metformin inhibit injurious inflammatory responses, including neutrophil proinflammatory responses and injury to multiple organs such as the lung, liver, and kidney [107–109]. Pharmacologic activation of AMPK by metformin, berberine, or AICAR dampens excessive TLR4/NF-κB signaling, M2-type macrophage polarization, and the production of proinflammatory mediators in vitro and in models of sepsis [110–115]. The anti-inflammatory effect of metformin in mice with lipopolysachharide (LPS)-induced septic shock and in ob/ob mice is mediated at least in part by AMPK activation [116]. In septic mice, AMPK activation by AICAR or metformin reduces the severity of sepsis-induced lung injury, enhances AMPK phosphorylation in the brain, and attenuates the inflammatory response [117,118].

Treatment with trimetazidine protects against LPS-induced myocardial dysfunction, exerts an anti-apoptotic effect, and attenuates the inflammatory response due to its effect on the SIRT1/AMPK pathway [119]. Moreover, the flavonoid naringenin dampens inflammation in vitro and protects against murine endotoxemia in vivo; these effects are mediated by AMPK/ATF3-dependent inhibition of the TLR4 signaling pathway [120]. In severe acute HBV infection, halofuginone, a plant alkaloid, inhibits viral replication by activating AMPK-mediated anti-inflammatory responses [121]. AMPK activation also enhances the phagocytic capacities of neutrophils and macrophages [122]. Transient receptor potential melastatin 2, an oxidant sensor cation channel, promotes extracellular trap formation by neutrophils via the AMPK/p38 MAPK pathway, enhancing their antimicrobial activity [123]. AMPK activation not only modulates the acute inflammatory response but also promotes neutrophil-dependent bacterial uptake and killing [106].

In perinatal hypoxic–ischemic encephalopathy, prolonged activation of AMPK signaling suppresses the response to oxygen/glucose deprivation and promotes neonatal hypoxic–ischemic injury [124]. Although AMPK inhibition increases neuronal survival, blockade of AMPK prior to oxygen/glucose deprivation increases cell damage and death [124]. Therefore, the clinical implications of AMPK activation are complex, and further preclinical and clinical data are needed to enable therapeutic use of AMPK activators in patients with acute or chronic inflammation.

4.3. Role of AMPK in Inflammasome Activation

AMPK is implicated in modulation of NLRP3 inflammasome activation. The bactericidal activity of the isoquinoline alkaloid berberine exerts a bactericidal effect by augmenting inflammasome activation via AMPK signaling [125]. However, metformin increases mortality of mice with bacteremia, likely via an AMPK-mediated increase in ATP-induced inflammasome activation and pyroptosis [126].

AMPK is implicated in the inhibition of palmitate-induced inflammasome activation [127]. The AMPK activator AICAR inhibits palmitate-induced activation of the NLRP3 inflammasome and IL-1β secretion by suppressing ROS generation [127]. In addition, NLRP3 inflammasome activation and production of IL-1β are upregulated in the peripheral mononuclear cells of drug-naïve type-2 diabetic patients, suggesting a role of the inflammasome in the pathogenesis of type-2 diabetes [128]. Interestingly, AMPK activation is responsible for the significantly reduced mature IL-1β level in peripheral myeloid cells from type-2 diabetic patients after two months of metformin therapy [128]. In a model of hyperalgesia, which is associated with NLRP3 inflammasome activation, metformin attenuated the clinical symptoms and improved the biochemical parameters, whereas blockade of AMPK activation by compound C provoked hyperalgesia and increased the levels of IL-1β and IL-18 [129]. Furthermore, pharmacological activation of AMPK inhibits the monosodium urate (MSU) crystal-induced inflammatory response, suggesting a role for AMPK in gouty inflammation. Moreover, colchicine, an inhibitor of microtubule assembly used to treat gouty arthritis, enhances AMPKα-mediated phosphorylation, thereby inhibiting inflammasome activation and IL-1β release [130]. Further studies on the efficacy of AMPK activators against inflammasome-associated diseases are thus warranted.

4.4. Role of AMPK in the Regulation of the Adaptive Immune Responses

AMPKα1 is a key regulator of the adaptive immune response, particularly T helper (Th1) 1 and Th17 cell differentiation and the T-cell responses to viral and bacterial infections [131]. In addition, in models of simian immunodeficiency viral infection, AMPK activation is associated with the virus-specific CD8(+) cytotoxic T-lymphocyte population and control of Simian Immunodeficiency Virus (SIV) [132]. The mechanism(s) by which AMPK signaling activates innate and adaptive immune responses and controls excessive inflammation must be determined if the potential of AMPK-targeted therapy is to be realized.

5. Conclusions

Although much research has focused on the role of AMPK in the regulation of mitochondrial and metabolic homeostasis, several issues remain to be addressed. Further work should focus on the mechanism(s) by which AMPK modulates host defenses against infections in vivo. Several pathogens modulate the host metabolic environment to promote their survival and replication. Because of its role in regulating mitochondrial metabolism, dynamics, and biogenesis, AMPK signaling can provide energy to the pathogen and/or host, benefitting either. Stimulation of AMPK activity enhances host defenses against diverse viruses, bacteria, and parasites, notably Mtb. Moreover, AMPK links the innate and adaptive immune responses to infection. However, the molecular mechanisms underlying AMPK regulation of innate and adaptive immunity are unclear. AMPK-targeted small molecules have potential as antimicrobial agents as well as metabolic drugs. Further work is needed to enable development of therapeutics that target AMPK to control inflammation and promote host defenses against infection. This work should focus on elucidating the mechanisms by which AMPK and/or AMPK-targeting compounds modulate host defenses against infection.

Abbreviations

ADP	Adenosine diphosphate
AICAR	5-aminoimidazole-4-carboxamide 1-β-D-ribofuanoside
AMP	Adenosine monophosphate
AMPK	5′-AMP-activated protein kinase
ATF3	Activating transcription factor 3
ATP	Adenosine triphosphate
CaMKK	Ca2+/calmodulin-dependent protein kinase kinases
CCL17	Chemokine ligand 17
CVB3	Coxsackie virus B3
DM	Diabetes mellitus
EBOV	Zaire Ebolavirus
EBV	Epstein-Barr Virus
GABA	Gamma-aminobutyric acid
HBV	Hepatitis B virus

GABA   Gamma-aminobutyric acid
HBV   Hepatitis B virus
HCC   Hepatocellular carcinoma
HCMV   Human cytomegalovirus
HCV   Hepatitis C virus
HIV   Human immunodeficiency virus
HMGCR   3-hydroxy-3-methylglutaryl-CoA reductase
HSV-1   Herpex Simplex Virus Type 1
IFN   Interferons
IL-17   Interleukin-17
IL-1β   Interleukin-1β
IL-37   Interleukin-37
IL-7   Interleukin-7
KSHV   Kaposi's sarcoma-associated herpesvirus
LKB1   Liver kinase B1
LMP1   Latent membrane protein 1
MAPK   Mitogen-activated protein kinase
MMP-8   Matrix metalloproteinase-8
mTOR   Mammalian target of rapamycin
mTORC1   Mammalian target of rapamycin complex 1
NADPH   Nicotinamide adenine dinucleotide phosphate
NF-κB   Nuclear factor kappa-light-chain-enhancer of activated B cells
PPARGC1A   Peroxisome proliferator-activated receptor-gamma, coactivator 1α
ROS   Reactive oxygen species
RVFV   Rift Valley Fever Virus
SIRT1   Sirtuin 1
SIV   Simian immunodeficiency virus
SNARK   Sucrose-non-fermenting protein kinase 1/AMP-activated protein kinase-related protein kinase
STING   Stimulator of IFN genes
TAK1   Transforming growth factor (TGF)-β-activated kinase 1
TLR   Toll-like receptor
TSC   Tuberous sclerosis complex

References

1. Hardie, D.G.; Ross, F.A.; Hawley, S.A. AMPK: A nutrient and energy sensor that maintains energy homeostasis. *Nat. Rev. Mol. Cell Biol.* **2012**, *13*, 251–262. [CrossRef] [PubMed]
2. Gowans, G.J.; Hardie, D.G. AMPK: A cellular energy sensor primarily regulated by AMP. *Biochem. Soc. Trans.* **2014**, *42*, 71–75. [CrossRef] [PubMed]
3. Gowans, G.J.; Hawley, S.A.; Ross, F.A.; Hardie, D.G. AMP is a true physiological regulator of AMP-activated protein kinase by both allosteric activation and enhancing net phosphorylation. *Cell Metab.* **2013**, *18*, 556–566. [CrossRef] [PubMed]
4. Shackelford, D.B.; Shaw, R.J. The LKB1-AMPK pathway: Metabolism and growth control in tumour suppression. *Nat. Rev. Cancer* **2009**, *9*, 563–575. [CrossRef] [PubMed]
5. Faubert, B.; Boily, G.; Izreig, S.; Griss, T.; Samborska, B.; Dong, Z.; Dupuy, F.; Chambers, C.; Fuerth, B.J.; Viollet, B.; et al. AMPK is a negative regulator of the Warburg effect and suppresses tumor growth in vivo. *Cell Metab.* **2013**, *17*, 113–124. [CrossRef] [PubMed]
6. Hardie, D.G. AMP-activated protein kinase: A cellular energy sensor with a key role in metabolic disorders and in cancer. *Biochem. Soc. Trans.* **2011**, *39*, 1–13. [CrossRef] [PubMed]
7. Moreira, D.; Silvestre, R.; Cordeiro-da-Silva, A.; Estaquier, J.; Foretz, M.; Viollet, B. AMP-activated Protein Kinase as a Target for Pathogens: Friends or Foes? *Curr. Drug Targets* **2016**, *17*, 942–953. [CrossRef] [PubMed]

8.   Brunton, J.; Steele, S.; Ziehr, B.; Moorman, N.; Kawula, T. Feeding uninvited guests: MTOR and AMPK set the table for intracellular pathogens. *PLoS Pathog.* **2013**, *9*, e1003552. [CrossRef] [PubMed]

9.   Xie, N.; Yuan, K.; Zhou, L.; Wang, K.; Chen, H.N.; Lei, Y.; Lan, J.; Pu, Q.; Gao, W.; Zhang, L.; et al. PRKAA/AMPK restricts HBV replication through promotion of autophagic degradation. *Autophagy* **2016**, *12*, 1507–1520. [CrossRef] [PubMed]

10.   Prantner, D.; Perkins, D.J.; Vogel, S.N. AMP-activated Kinase (AMPK) Promotes Innate Immunity and Antiviral Defense through Modulation of Stimulator of Interferon Genes (STING) Signaling. *J. Biol. Chem.* **2017**, *292*, 292–304. [CrossRef] [PubMed]

11.   Shin, D.M.; Yang, C.S.; Lee, J.Y.; Lee, S.J.; Choi, H.H.; Lee, H.M.; Yuk, J.M.; Harding, C.V.; Jo, E.K. *Mycobacterium tuberculosis* lipoprotein-induced association of TLR2 with protein kinase C zeta in lipid rafts contributes to reactive oxygen species-dependent inflammatory signalling in macrophages. *Cell Microbiol.* **2008**, *10*, 1893–1905. [CrossRef] [PubMed]

12.   Yang, C.S.; Kim, J.J.; Lee, H.M.; Jin, H.S.; Lee, S.H.; Park, J.H.; Kim, S.J.; Kim, J.M.; Han, Y.M.; Lee, M.S.; et al. The AMPK-PPARGC1A pathway is required for antimicrobial host defense through activation of autophagy. *Autophagy* **2014**, *10*, 785–802. [CrossRef] [PubMed]

13.   Mankouri, J.; Harris, M. Viruses and the fuel sensor: The emerging link between AMPK and virus replication. *Rev. Med. Virol.* **2011**, *21*, 205–212. [CrossRef] [PubMed]

14.   Mesquita, I.; Moreira, D.; Sampaio-Marques, B.; Laforge, M.; Cordeiro-da-Silva, A.; Ludovico, P.; Estaquier, J.; Silvestre, R. AMPK in Pathogens. *EXS* **2016**, *107*, 287–323. [CrossRef] [PubMed]

15.   Kim, J.; Yang, G.; Kim, Y.; Kim, J.; Ha, J. AMPK activators: Mechanisms of action and physiological activities. *Exp. Mol. Med.* **2016**, *48*, e224. [CrossRef] [PubMed]

16.   Wu, J.; Puppala, D.; Feng, X.; Monetti, M.; Lapworth, A.L.; Geoghegan, K.F. Chemoproteomic analysis of intertissue and interspecies isoform diversity of AMP-activated protein kinase (AMPK). *J. Biol. Chem.* **2013**, *288*, 35904–35912. [CrossRef] [PubMed]

17.   Rajamohan, F.; Reyes, A.R.; Frisbie, R.K.; Hoth, L.R.; Sahasrabudhe, P.; Magyar, R.; Landro, J.A.; Withka, J.M.; Caspers, N.L.; Calabrese, M.F.; et al. Probing the enzyme kinetics, allosteric modulation and activation of alpha1- and alpha2-subunit-containing AMP-activated protein kinase (AMPK) heterotrimeric complexes by pharmacological and physiological activators. *Biochem. J.* **2016**, *473*, 581–592. [CrossRef] [PubMed]

18.   Woods, A.; Johnstone, S.R.; Dickerson, K.; Leiper, F.C.; Fryer, L.G.; Neumann, D.; Schlattner, U.; Wallimann, T.; Carlson, M.; Carling, D. LKB1 is the upstream kinase in the AMP-activated protein kinase cascade. *Curr. Biol.* **2003**, *13*, 2004–2008. [CrossRef] [PubMed]

19.   Carling, D.; Aguan, K.; Woods, A.; Verhoeven, A.J.; Beri, R.K.; Brennan, C.H.; Sidebottom, C.; Davison, M.D.; Scott, J. Mammalian AMP-activated protein kinase is homologous to yeast and plant protein kinases involved in the regulation of carbon metabolism. *J. Biol. Chem.* **1994**, *269*, 11442–11448. [PubMed]

20.   Hawley, S.A.; Pan, D.A.; Mustard, K.J.; Ross, L.; Bain, J.; Edelman, A.M.; Frenguelli, B.G.; Hardie, D.G. Calmodulin-dependent protein kinase kinase-beta is an alternative upstream kinase for AMP-activated protein kinase. *Cell. Metab.* **2005**, *2*, 9–19. [CrossRef] [PubMed]

21.   McBride, A.; Ghilagaber, S.; Nikolaev, A.; Hardie, D.G. The glycogen-binding domain on the AMPK beta subunit allows the kinase to act as a glycogen sensor. *Cell. Metab.* **2009**, *9*, 23–34. [CrossRef] [PubMed]

22.   Viana, R.; Towler, M.C.; Pan, D.A.; Carling, D.; Viollet, B.; Hardie, D.G.; Sanz, P. A conserved sequence immediately N-terminal to the Bateman domains in AMP-activated protein kinase gamma subunits is required for the interaction with the beta subunits. *J. Biol. Chem.* **2007**, *282*, 16117–16125. [CrossRef] [PubMed]

23.   Xiao, B.; Heath, R.; Saiu, P.; Leiper, F.C.; Leone, P.; Jing, C.; Walker, P.A.; Haire, L.; Eccleston, J.F.; Davis, C.T.; et al. Structural basis for AMP binding to mammalian AMP-activated protein kinase. *Nature* **2007**, *449*, 496–500. [CrossRef] [PubMed]

24.   Steinberg, G.R.; Kemp, B.E. AMPK in Health and Disease. *Physiol. Rev.* **2009**, *89*, 1025–1078. [CrossRef] [PubMed]

25.   Viollet, B.; Horman, S.; Leclerc, J.; Lantier, L.; Foretz, M.; Billaud, M.; Giri, S.; Andreelli, F. AMPK inhibition in health and disease. *Crit. Rev. Biochem. Mol. Biol.* **2010**, *45*, 276–295. [CrossRef] [PubMed]

26.   Kim, T.S.; Shin, Y.H.; Lee, H.M.; Kim, J.K.; Choe, J.H.; Jang, J.C.; Um, S.; Jin, H.S.; Komatsu, M.; Cha, G.H.; et al. Ohmyungsamycins promote antimicrobial responses through autophagy activation via AMP-activated protein kinase pathway. *Sci. Rep.* **2017**, *7*, 3431. [CrossRef] [PubMed]

27. Mankouri, J.; Tedbury, P.R.; Gretton, S.; Hughes, M.E.; Griffin, S.D.; Dallas, M.L.; Green, K.A.; Hardie, D.G.; Peers, C.; Harris, M. Enhanced hepatitis C virus genome replication and lipid accumulation mediated by inhibition of AMP-activated protein kinase. *Proc. Natl. Acad. Sci. USA* **2010**, *107*, 11549–11554. [CrossRef] [PubMed]

28. Huang, H.; Kang, R.; Wang, J.; Luo, G.; Yang, W.; Zhao, Z. Hepatitis C virus inhibits AKT-tuberous sclerosis complex (TSC), the mechanistic target of rapamycin (MTOR) pathway, through endoplasmic reticulum stress to induce autophagy. *Autophagy* **2013**, *9*, 175–195. [CrossRef] [PubMed]

29. Yu, J.W.; Sun, L.J.; Liu, W.; Zhao, Y.H.; Kang, P.; Yan, B.Z. Hepatitis C virus core protein induces hepatic metabolism disorders through down-regulation of the SIRT1-AMPK signaling pathway. *Int. J. Infect. Dis.* **2013**, *17*, e539–e545. [CrossRef] [PubMed]

30. Tsai, W.L.; Chang, T.H.; Sun, W.C.; Chan, H.H.; Wu, C.C.; Hsu, P.I.; Cheng, J.S.; Yu, M.L. Metformin activates type I interferon signaling against HCV via activation of adenosine monophosphate-activated protein kinase. *Oncotarget* **2017**, *8*, 91928–91937. [CrossRef] [PubMed]

31. Nakashima, K.; Takeuchi, K.; Chihara, K.; Hotta, H.; Sada, K. Inhibition of hepatitis C virus replication through adenosine monophosphate-activated protein kinase-dependent and -independent pathways. *Microbiol. Immunol.* **2011**, *55*, 774–782. [CrossRef] [PubMed]

32. Goto, K.; Lin, W.; Zhang, L.; Jilg, N.; Shao, R.X.; Schaefer, E.A.; Zhao, H.; Fusco, D.N.; Peng, L.F.; Kato, N.; et al. The AMPK-related kinase SNARK regulates hepatitis C virus replication and pathogenesis through enhancement of TGF-beta signaling. *J. Hepatol.* **2013**, *59*, 942–948. [CrossRef] [PubMed]

33. Douglas, D.N.; Pu, C.H.; Lewis, J.T.; Bhat, R.; Anwar-Mohamed, A.; Logan, M.; Lund, G.; Addison, W.R.; Lehner, R.; Kneteman, N.M. Oxidative Stress Attenuates Lipid Synthesis and Increases Mitochondrial Fatty Acid Oxidation in Hepatoma Cells Infected with Hepatitis C Virus. *J. Biol. Chem.* **2016**, *291*, 1974–1990. [CrossRef] [PubMed]

34. Uematsu, T.; Fujita, T.; Nakaoka, H.J.; Hara, T.; Kobayashi, N.; Murakami, Y.; Seiki, M.; Sakamoto, T. Mint3/Apba3 depletion ameliorates severe murine influenza pneumonia and macrophage cytokine production in response to the influenza virus. *Sci. Rep.* **2016**, *6*, 37815. [CrossRef] [PubMed]

35. Yoon, J.Y.; Kwon, H.H.; Min, S.U.; Thiboutot, D.M.; Suh, D.H. Epigallocatechin-3-gallate improves acne in humans by modulating intracellular molecular targets and inhibiting *P. acnes*. *J. Investig. Dermatol.* **2013**, *133*, 429–440. [CrossRef] [PubMed]

36. Han, S.; Xu, J.; Guo, X.; Huang, M. Curcumin ameliorates severe influenza pneumonia via attenuating lung injury and regulating macrophage cytokines production. *Clin. Exp. Pharmacol. Physiol.* **2018**, *45*, 84–93. [CrossRef] [PubMed]

37. Zhang, H.S.; Wu, T.C.; Sang, W.W.; Ruan, Z. EGCG inhibits Tat-induced LTR transactivation: Role of Nrf2, AKT, AMPK signaling pathway. *Life Sci.* **2012**, *90*, 747–754. [CrossRef] [PubMed]

38. Wang, Z.Q.; Yu, Y.; Zhang, X.H.; Floyd, E.Z.; Cefalu, W.T. Human adenovirus 36 decreases fatty acid oxidation and increases de novo lipogenesis in primary cultured human skeletal muscle cells by promoting Cidec/FSP27 expression. *Int. J. Obes.* **2010**, *34*, 1355–1364. [CrossRef] [PubMed]

39. Moser, T.S.; Schieffer, D.; Cherry, S. AMP-activated kinase restricts Rift Valley fever virus infection by inhibiting fatty acid synthesis. *PLoS Pathog.* **2012**, *8*, e1002661. [CrossRef] [PubMed]

40. Leyton, L.; Hott, M.; Acuna, F.; Caroca, J.; Nunez, M.; Martin, C.; Zambrano, A.; Concha, M.I.; Otth, C. Nutraceutical activators of AMPK/Sirt1 axis inhibit viral production and protect neurons from neurodegenerative events triggered during HSV-1 infection. *Virus Res.* **2015**, *205*, 63–72. [CrossRef] [PubMed]

41. Xie, W.; Wang, L.; Dai, Q.; Yu, H.; He, X.; Xiong, J.; Sheng, H.; Zhang, D.; Xin, R.; Qi, Y.; et al. Activation of AMPK restricts coxsackievirus B3 replication by inhibiting lipid accumulation. *J. Mol. Cell Cardiol.* **2015**, *85*, 155–167. [CrossRef] [PubMed]

42. Lo, A.K.; Lo, K.W.; Ko, C.W.; Young, L.S.; Dawson, C.W. Inhibition of the LKB1-AMPK pathway by the Epstein-Barr virus-encoded LMP1 promotes proliferation and transformation of human nasopharyngeal epithelial cells. *J. Pathol.* **2013**, *230*, 336–346. [CrossRef] [PubMed]

43. Cheng, F.; He, M.; Jung, J.U.; Lu, C.; Gao, S.J. Suppression of Kaposi's Sarcoma-Associated Herpesvirus Infection and Replication by 5′-AMP-Activated Protein Kinase. *J. Virol.* **2016**, *90*, 6515–6525. [CrossRef] [PubMed]

44. Green, V.A.; Pelkmans, L. A Systems Survey of Progressive Host-Cell Reorganization during Rotavirus Infection. *Cell Host Microbe* **2016**, *20*, 107–120. [CrossRef] [PubMed]

45. Soto-Acosta, R.; Bautista-Carbajal, P.; Cervantes-Salazar, M.; Angel-Ambrocio, A.H.; Del Angel, R.M. DENV up-regulates the HMG-CoA reductase activity through the impairment of AMPK phosphorylation: A potential antiviral target. *PLoS Pathog.* **2017**, *13*, e1006257. [CrossRef] [PubMed]

46. Jordan, T.X.; Randall, G. Dengue Virus Activates the AMP Kinase-mTOR Axis to Stimulate a Proviral Lipophagy. *J. Virol.* **2017**, *91*. [CrossRef] [PubMed]

47. Bagga, S.; Rawat, S.; Ajenjo, M.; Bouchard, M.J. Hepatitis B virus (HBV) X protein-mediated regulation of hepatocyte metabolic pathways affects viral replication. *Virology* **2016**, *498*, 9–22. [CrossRef] [PubMed]

48. Kondratowicz, A.S.; Hunt, C.L.; Davey, R.A.; Cherry, S.; Maury, W.J. AMP-activated protein kinase is required for the macropinocytic internalization of ebolavirus. *J. Virol.* **2013**, *87*, 746–755. [CrossRef] [PubMed]

49. Ji, W.T.; Lee, L.H.; Lin, F.L.; Wang, L.; Liu, H.J. AMP-activated protein kinase facilitates avian reovirus to induce mitogen-activated protein kinase (MAPK) p38 and MAPK kinase 3/6 signalling that is beneficial for virus replication. *J. Gen. Virol.* **2009**, *90*, 3002–3009. [CrossRef] [PubMed]

50. Chi, P.I.; Huang, W.R.; Lai, I.H.; Cheng, C.Y.; Liu, H.J. The p17 nonstructural protein of avian reovirus triggers autophagy enhancing virus replication via activation of phosphatase and tensin deleted on chromosome 10 (PTEN) and AMP-activated protein kinase (AMPK), as well as dsRNA-dependent protein kinase (PKR)/eIF2alpha signaling pathways. *J. Biol. Chem.* **2013**, *288*, 3571–3584. [CrossRef] [PubMed]

51. Martin, C.; Leyton, L.; Arancibia, Y.; Cuevas, A.; Zambrano, A.; Concha, M.I.; Otth, C. Modulation of the AMPK/Sirt1 axis during neuronal infection by herpes simplex virus type 1. *J. Alzheimers Dis.* **2014**, *42*, 301–312. [CrossRef] [PubMed]

52. Samikkannu, T.; Atluri, V.S.; Nair, M.P. HIV and Cocaine Impact Glial Metabolism: Energy Sensor AMP-activated protein kinase Role in Mitochondrial Biogenesis and Epigenetic Remodeling. *Sci. Rep.* **2016**, *6*, 31784. [CrossRef] [PubMed]

53. Terry, L.J.; Vastag, L.; Rabinowitz, J.D.; Shenk, T. Human kinome profiling identifies a requirement for AMP-activated protein kinase during human cytomegalovirus infection. *Proc. Natl. Acad. Sci. USA* **2012**, *109*, 3071–3076. [CrossRef] [PubMed]

54. Hutterer, C.; Wandinger, S.K.; Wagner, S.; Muller, R.; Stamminger, T.; Zeittrager, I.; Godl, K.; Baumgartner, R.; Strobl, S.; Marschall, M. Profiling of the kinome of cytomegalovirus-infected cells reveals the functional importance of host kinases Aurora A, ABL and AMPK. *Antivir. Res.* **2013**, *99*, 139–148. [CrossRef] [PubMed]

55. McArdle, J.; Moorman, N.J.; Munger, J. HCMV targets the metabolic stress response through activation of AMPK whose activity is important for viral replication. *PLoS Pathog.* **2012**, *8*, e1002502. [CrossRef] [PubMed]

56. Mukhopadhyay, R.; Venkatadri, R.; Katsnelson, J.; Arav-Boger, R. Digitoxin Suppresses Human Cytomegalovirus Replication via Na$^+$, K$^+$/ATPase alpha1 Subunit-Dependent AMP-Activated Protein Kinase and Autophagy Activation. *J. Virol.* **2018**, *92*. [CrossRef] [PubMed]

57. Seo, J.Y.; Yaneva, R.; Hinson, E.R.; Cresswell, P. Human cytomegalovirus directly induces the antiviral protein viperin to enhance infectivity. *Science* **2011**, *332*, 1093–1097. [CrossRef] [PubMed]

58. Seo, J.Y.; Cresswell, P. Viperin regulates cellular lipid metabolism during human cytomegalovirus infection. *PLoS Pathog.* **2013**, *9*, e1003497. [CrossRef] [PubMed]

59. Li, M.; Li, J.; Zeng, R.; Yang, J.; Liu, J.; Zhang, Z.; Song, X.; Yao, Z.; Ma, C.; Li, W.; et al. Respiratory Syncytial Virus Replication Is Promoted by Autophagy-Mediated Inhibition of Apoptosis. *J. Virol.* **2018**, *92*. [CrossRef] [PubMed]

60. Lv, S.; Xu, Q.Y.; Sun, E.C.; Zhang, J.K.; Wu, D.L. Dissection and integration of the autophagy signaling network initiated by bluetongue virus infection: Crucial candidates ERK1/2, Akt and AMPK. *Sci. Rep.* **2016**, *6*, 23130. [CrossRef] [PubMed]

61. Liu, J.; Wang, H.; Gu, J.; Deng, T.; Yuan, Z.; Hu, B.; Xu, Y.; Yan, Y.; Zan, J.; Liao, M.; et al. BECN1-dependent CASP2 incomplete autophagy induction by binding to rabies virus phosphoprotein. *Autophagy* **2017**, *13*, 739–753. [CrossRef] [PubMed]

62. Subramanian, G.; Kuzmanovic, T.; Zhang, Y.; Peter, C.B.; Veleeparambil, M.; Chakravarti, R.; Sen, G.C.; Chattopadhyay, S. A new mechanism of interferon's antiviral action: Induction of autophagy, essential for paramyxovirus replication, is inhibited by the interferon stimulated gene, TDRD7. *PLoS Pathog.* **2018**, *14*, e1006877. [CrossRef] [PubMed]

63. McFadden, K.; Hafez, A.Y.; Kishton, R.; Messinger, J.E.; Nikitin, P.A.; Rathmell, J.C.; Luftig, M.A. Metabolic stress is a barrier to Epstein-Barr virus-mediated B-cell immortalization. *Proc. Natl. Acad. Sci. USA* **2016**, *113*, E782–E790. [CrossRef] [PubMed]

64. Burke, J.D.; Platanias, L.C.; Fish, E.N. Beta interferon regulation of glucose metabolism is PI3K/Akt dependent and important for antiviral activity against coxsackievirus B3. *J. Virol.* **2014**, *88*, 3485–3495. [CrossRef] [PubMed]

65. Zhang, C.; Feng, S.; Zhang, W.; Chen, N.; Hegazy, A.M.; Chen, W.; Liu, X.; Zhao, L.; Li, J.; Lin, L.; et al. MicroRNA miR-214 Inhibits Snakehead Vesiculovirus Replication by Promoting IFN-alpha Expression via Targeting Host Adenosine 5'-Monophosphate-Activated Protein Kinase. *Front. Immunol.* **2017**, *8*, 1775. [CrossRef] [PubMed]

66. Francione, L.; Smith, P.K.; Accari, S.L.; Taylor, P.E.; Bokko, P.B.; Bozzaro, S.; Beech, P.L.; Fisher, P.R. *Legionella pneumophila* multiplication is enhanced by chronic AMPK signalling in mitochondrially diseased Dictyostelium cells. *Dis. Model Mech.* **2009**, *2*, 479–489. [CrossRef] [PubMed]

67. Kajiwara, C.; Kusaka, Y.; Kimura, S.; Yamaguchi, T.; Nanjo, Y.; Ishii, Y.; Udono, H.; Standiford, T.J.; Tateda, K. Metformin Mediates Protection against *Legionella* Pneumonia through Activation of AMPK and Mitochondrial Reactive Oxygen Species. *J. Immunol.* **2018**, *200*, 623–631. [CrossRef] [PubMed]

68. Ganesan, R.; Hos, N.J.; Gutierrez, S.; Fischer, J.; Stepek, J.M.; Daglidu, E.; Kronke, M.; Robinson, N. *Salmonella* Typhimurium disrupts Sirt1/AMPK checkpoint control of mTOR to impair autophagy. *PLoS Pathog.* **2017**, *13*, e1006227. [CrossRef] [PubMed]

69. Liu, W.; Jiang, Y.; Sun, J.; Geng, S.; Pan, Z.; Prinz, R.A.; Wang, C.; Sun, J.; Jiao, X.; Xu, X. Activation of TGF-beta-activated kinase 1 (TAK1) restricts *Salmonella* Typhimurium growth by inducing AMPK activation and autophagy. *Cell Death Dis.* **2018**, *9*, 570. [CrossRef] [PubMed]

70. Tattoli, I.; Sorbara, M.T.; Philpott, D.J.; Girardin, S.E. Bacterial autophagy: The trigger, the target and the timing. *Autophagy* **2012**, *8*, 1848–1850. [CrossRef] [PubMed]

71. Neumann, D. Is TAK1 a Direct Upstream Kinase of AMPK? *Int. J. Mol. Sci.* **2018**, *19*, 2412. [CrossRef] [PubMed]

72. Liu, N.; Li, Y.; Dong, C.; Xu, X.; Wei, P.; Sun, W.; Peng, Q. Inositol-Requiring Enzyme 1-Dependent Activation of AMPK Promotes *Brucella abortus* Intracellular Growth. *J. Bacteriol.* **2016**, *198*, 986–993. [CrossRef] [PubMed]

73. Liang, Y.D.; Bai, W.J.; Li, C.G.; Xu, L.H.; Wei, H.X.; Pan, H.; He, X.H.; Ouyang, D.Y. Piperine Suppresses Pyroptosis and Interleukin-1beta Release upon ATP Triggering and Bacterial Infection. *Front. Pharmacol.* **2016**, *7*, 390. [CrossRef] [PubMed]

74. Yang, W.; Dierking, K.; Esser, D.; Tholey, A.; Leippe, M.; Rosenstiel, P.; Schulenburg, H. Overlapping and unique signatures in the proteomic and transcriptomic responses of the nematode *Caenorhabditis elegans* toward pathogenic *Bacillus thuringiensis*. *Dev. Comp. Immunol.* **2015**, *51*, 1–9. [CrossRef] [PubMed]

75. Lv, G.; Zhu, H.; Zhou, F.; Lin, Z.; Lin, G.; Li, C. AMP-activated protein kinase activation protects gastric epithelial cells from *Helicobacter pylori*-induced apoptosis. *Biochem. Biophys. Res. Commun.* **2014**, *453*, 13–18. [CrossRef] [PubMed]

76. Zhao, H.; Zhu, H.; Lin, Z.; Lin, G.; Lv, G. Compound 13, an alpha1-selective small molecule activator of AMPK, inhibits *Helicobacter pylori*-induced oxidative stresses and gastric epithelial cell apoptosis. *Biochem. Biophys. Res. Commun.* **2015**, *463*, 510–517. [CrossRef] [PubMed]

77. Patkee, W.R.; Carr, G.; Baker, E.H.; Baines, D.L.; Garnett, J.P. Metformin prevents the effects of *Pseudomonas aeruginosa* on airway epithelial tight junctions and restricts hyperglycaemia-induced bacterial growth. *J. Cell Mol. Med.* **2016**, *20*, 758–764. [CrossRef] [PubMed]

78. Kumar, A.; Giri, S.; Kumar, A. 5-Aminoimidazole-4-carboxamide ribonucleoside-mediated adenosine monophosphate-activated protein kinase activation induces protective innate responses in bacterial endophthalmitis. *Cell Microbiol.* **2016**, *18*, 1815–1830. [CrossRef] [PubMed]

79. Wang, Y.; Zhang, K.; Shi, X.; Wang, C.; Wang, F.; Fan, J.; Shen, F.; Xu, J.; Bao, W.; Liu, M.; et al. Critical role of bacterial isochorismatase in the autophagic process induced by *Acinetobacter baumannii* in mammalian cells. *FASEB J.* **2016**, *30*, 3563–3577. [CrossRef] [PubMed]

80. Singhal, A.; Jie, L.; Kumar, P.; Hong, G.S.; Leow, M.K.; Paleja, B.; Tsenova, L.; Kurepina, N.; Chen, J.; Zolezzi, F.; et al. Metformin as adjunct antituberculosis therapy. *Sci. Transl. Med.* **2014**, *6*, 263ra159. [CrossRef] [PubMed]

81. Yuk, J.M.; Shin, D.M.; Lee, H.M.; Yang, C.S.; Jin, H.S.; Kim, K.K.; Lee, Z.W.; Lee, S.H.; Kim, J.M.; Jo, E.K. Vitamin D3 induces autophagy in human monocytes/macrophages via cathelicidin. *Cell Host Microbe* **2009**, *6*, 231–243. [CrossRef] [PubMed]

82. Rekha, R.S.; Rao Muvva, S.S.; Wan, M.; Raqib, R.; Bergman, P.; Brighenti, S.; Gudmundsson, G.H.; Agerberth, B. Phenylbutyrate induces LL-37-dependent autophagy and intracellular killing of *Mycobacterium tuberculosis* in human macrophages. *Autophagy* **2015**, *11*, 1688–1699. [CrossRef] [PubMed]

83. Kim, J.K.; Kim, Y.S.; Lee, H.M.; Jin, H.S.; Neupane, C.; Kim, S.; Lee, S.H.; Min, J.J.; Sasai, M.; Jeong, J.H.; et al. GABAergic signaling linked to autophagy enhances host protection against intracellular bacterial infections. *Nat. Commun.* **2018**, *9*, 4184. [CrossRef] [PubMed]

84. Ong, C.W.; Elkington, P.T.; Brilha, S.; Ugarte-Gil, C.; Tome-Esteban, M.T.; Tezera, L.B.; Pabisiak, P.J.; Moores, R.C.; Sathyamoorthy, T.; Patel, V.; et al. Neutrophil-Derived MMP-8 Drives AMPK-Dependent Matrix Destruction in Human Pulmonary Tuberculosis. *PLoS Pathog.* **2015**, *11*, e1004917. [CrossRef] [PubMed]

85. Restrepo, B.I. Metformin: Candidate host-directed therapy for tuberculosis in diabetes and non-diabetes patients. *Tuberculosis (Edinb)* **2016**, *101S*, S69–S72. [CrossRef] [PubMed]

86. Ouimet, M.; Koster, S.; Sakowski, E.; Ramkhelawon, B.; van Solingen, C.; Oldebeken, S.; Karunakaran, D.; Portal-Celhay, C.; Sheedy, F.J.; Ray, T.D.; et al. *Mycobacterium tuberculosis* induces the miR-33 locus to reprogram autophagy and host lipid metabolism. *Nat. Immunol.* **2016**, *17*, 677–686. [CrossRef] [PubMed]

87. Kim, J.K.; Lee, H.M.; Park, K.S.; Shin, D.M.; Kim, T.S.; Kim, Y.S.; Suh, H.W.; Kim, S.Y.; Kim, I.S.; Kim, J.M.; et al. MIR144* inhibits antimicrobial responses against *Mycobacterium tuberculosis* in human monocytes and macrophages by targeting the autophagy protein DRAM2. *Autophagy* **2017**, *13*, 423–441. [CrossRef] [PubMed]

88. Dorhoi, A.; Kaufmann, S.H. Pathology and immune reactivity: Understanding multidimensionality in pulmonary tuberculosis. *Semin. Immunopathol.* **2016**, *38*, 153–166. [CrossRef] [PubMed]

89. Nieves, W.; Hung, L.Y.; Oniskey, T.K.; Boon, L.; Foretz, M.; Viollet, B.; Herbert, D.R. Myeloid-Restricted AMPKalpha1 Promotes Host Immunity and Protects against IL-12/23p40-Dependent Lung Injury during Hookworm Infection. *J. Immunol.* **2016**, *196*, 4632–4640. [CrossRef] [PubMed]

90. Moreira, D.; Rodrigues, V.; Abengozar, M.; Rivas, L.; Rial, E.; Laforge, M.; Li, X.; Foretz, M.; Viollet, B.; Estaquier, J.; et al. *Leishmania infantum* modulates host macrophage mitochondrial metabolism by hijacking the SIRT1-AMPK axis. *PLoS Pathog.* **2015**, *11*, e1004684. [CrossRef] [PubMed]

91. Zhu, J.; Zhang, W.; Zhang, L.; Xu, L.; Chen, X.; Zhou, S.; Xu, Z.; Xiao, M.; Bai, H.; Liu, F.; et al. IL-7 suppresses macrophage autophagy and promotes liver pathology in *Schistosoma japonicum*-infected mice. *J. Cell. Mol. Med.* **2018**, *22*, 3353–3363. [CrossRef] [PubMed]

92. Ruivo, M.T.G.; Vera, I.M.; Sales-Dias, J.; Meireles, P.; Gural, N.; Bhatia, S.N.; Mota, M.M.; Mancio-Silva, L. Host AMPK Is a Modulator of Plasmodium Liver Infection. *Cell. Rep.* **2016**, *16*, 2539–2545. [CrossRef] [PubMed]

93. Vilar-Pereira, G.; Carneiro, V.C.; Mata-Santos, H.; Vicentino, A.R.; Ramos, I.P.; Giarola, N.L.; Feijo, D.F.; Meyer-Fernandes, J.R.; Paula-Neto, H.A.; Medei, E.; et al. Resveratrol Reverses Functional Chagas Heart Disease in Mice. *PLoS Pathog.* **2016**, *12*, e1005947. [CrossRef] [PubMed]

94. Pandey, A.; Ding, S.L.; Qin, Q.M.; Gupta, R.; Gomez, G.; Lin, F.; Feng, X.; Fachini da Costa, L.; Chaki, S.P.; Katepalli, M.; et al. Global Reprogramming of Host Kinase Signaling in Response to Fungal Infection. *Cell Host Microbe* **2017**, *21*, 637–649. [CrossRef] [PubMed]

95. Andris, F.; Leo, O. AMPK in lymphocyte metabolism and function. *Int. Rev. Immunol.* **2015**, *34*, 67–81. [CrossRef] [PubMed]

96. Vincent, E.E.; Coelho, P.P.; Blagih, J.; Griss, T.; Viollet, B.; Jones, R.G. Differential effects of AMPK agonists on cell growth and metabolism. *Oncogene* **2015**, *34*, 3627–3639. [CrossRef] [PubMed]

97. Bain, J.; Plater, L.; Elliott, M.; Shpiro, N.; Hastie, C.J.; McLauchlan, H.; Klevernic, I.; Arthur, J.S.; Alessi, D.R.; Cohen, P. The selectivity of protein kinase inhibitors: A further update. *Biochem. J.* **2007**, *408*, 297–315. [CrossRef] [PubMed]

98. Myers, R.W.; Guan, H.P.; Ehrhart, J.; Petrov, A.; Prahalada, S.; Tozzo, E.; Yang, X.; Kurtz, M.M.; Trujillo, M.; Gonzalez Trotter, D.; et al. Systemic pan-AMPK activator MK-8722 improves glucose homeostasis but induces cardiac hypertrophy. *Science* **2017**, *357*, 507–511. [CrossRef] [PubMed]

99. Ngoei, K.R.W.; Langendorf, C.G.; Ling, N.X.Y.; Hoque, A.; Varghese, S.; Camerino, M.A.; Walker, S.R.; Bozikis, Y.E.; Dite, T.A.; Ovens, A.J.; et al. Structural Determinants for Small-Molecule Activation of Skeletal Muscle AMPK alpha2beta2gamma1 by the Glucose Importagog SC4. *Cell Chem. Biol.* **2018**, *25*, 728–737. [CrossRef] [PubMed]

100. Dite, T.A.; Langendorf, C.G.; Hoque, A.; Galic, S.; Rebello, R.J.; Ovens, A.J.; Lindqvist, L.M.; Ngoei, K.R.W.; Ling, N.X.Y.; Furic, L.; et al. AMP-activated protein kinase selectively inhibited by the type II inhibitor SBI-0206965. *J. Biol. Chem.* **2018**, *293*, 8874–8885. [CrossRef] [PubMed]

101. Shintani, Y.; Kapoor, A.; Kaneko, M.; Smolenski, R.T.; D'Acquisto, F.; Coppen, S.R.; Harada-Shoji, N.; Lee, H.J.; Thiemermann, C.; Takashima, S.; et al. TLR9 mediates cellular protection by modulating energy metabolism in cardiomyocytes and neurons. *Proc. Natl. Acad. Sci. USA* **2013**, *110*, 5109–5114. [CrossRef] [PubMed]

102. Hoogendijk, A.J.; Pinhancos, S.S.; van der Poll, T.; Wieland, C.W. AMP-activated protein kinase activation by 5-aminoimidazole-4-carbox-amide-1-beta-D-ribofuranoside (AICAR) reduces lipoteichoic acid-induced lung inflammation. *J. Biol. Chem.* **2013**, *288*, 7047–7052. [CrossRef] [PubMed]

103. O'Sullivan, T.E.; Johnson, L.R.; Kang, H.H.; Sun, J.C. BNIP3- and BNIP3L-Mediated Mitophagy Promotes the Generation of Natural Killer Cell Memory. *Immunity* **2015**, *43*, 331–342. [CrossRef] [PubMed]

104. Muller-Durovic, B.; Lanna, A.; Covre, L.P.; Mills, R.S.; Henson, S.M.; Akbar, A.N. Killer Cell Lectin-like Receptor G1 Inhibits NK Cell Function through Activation of Adenosine 5′-Monophosphate-Activated Protein Kinase. *J. Immunol.* **2016**, *197*, 2891–2899. [CrossRef] [PubMed]

105. Webb, T.J.; Carey, G.B.; East, J.E.; Sun, W.; Bollino, D.R.; Kimball, A.S.; Brutkiewicz, R.R. Alterations in cellular metabolism modulate CD1d-mediated NKT-cell responses. *Pathog. Dis.* **2016**, *74*. [CrossRef] [PubMed]

106. Park, D.W.; Jiang, S.; Tadie, J.M.; Stigler, W.S.; Gao, Y.; Deshane, J.; Abraham, E.; Zmijewski, J.W. Activation of AMPK enhances neutrophil chemotaxis and bacterial killing. *Mol. Med.* **2013**, *19*, 387–398. [CrossRef] [PubMed]

107. Zhao, X.; Zmijewski, J.W.; Lorne, E.; Liu, G.; Park, Y.J.; Tsuruta, Y.; Abraham, E. Activation of AMPK attenuates neutrophil proinflammatory activity and decreases the severity of acute lung injury. *Am. J. Physiol. Lung Cell Mol. Physiol.* **2008**, *295*, L497–L504. [CrossRef] [PubMed]

108. Bergheim, I.; Luyendyk, J.P.; Steele, C.; Russell, G.K.; Guo, L.; Roth, R.A.; Arteel, G.E. Metformin prevents endotoxin-induced liver injury after partial hepatectomy. *J. Pharmacol. Exp. Ther.* **2006**, *316*, 1053–1061. [CrossRef] [PubMed]

109. Escobar, D.A.; Botero-Quintero, A.M.; Kautza, B.C.; Luciano, J.; Loughran, P.; Darwiche, S.; Rosengart, M.R.; Zuckerbraun, B.S.; Gomez, H. Adenosine monophosphate-activated protein kinase activation protects against sepsis-induced organ injury and inflammation. *J. Surg. Res.* **2015**, *194*, 262–272. [CrossRef] [PubMed]

110. Hattori, Y.; Suzuki, K.; Hattori, S.; Kasai, K. Metformin inhibits cytokine-induced nuclear factor κB activation via AMP-activated protein kinase activation in vascular endothelial cells. *Hypertension* **2006**, *47*, 1183–1188. [CrossRef] [PubMed]

111. Zmijewski, J.W.; Lorne, E.; Zhao, X.; Tsuruta, Y.; Sha, Y.; Liu, G.; Siegal, G.P.; Abraham, E. Mitochondrial respiratory complex I regulates neutrophil activation and severity of lung injury. *Am. J. Respir. Crit. Care Med.* **2008**, *178*, 168–179. [CrossRef] [PubMed]

112. Sag, D.; Carling, D.; Stout, R.D.; Suttles, J. Adenosine 5′-monophosphate-activated protein kinase promotes macrophage polarization to an anti-inflammatory functional phenotype. *J. Immunol.* **2008**, *181*, 8633–8641. [CrossRef] [PubMed]

113. Xing, J.; Wang, Q.; Coughlan, K.; Viollet, B.; Moriasi, C.; Zou, M.H. Inhibition of AMP-activated protein kinase accentuates lipopolysaccharide-induced lung endothelial barrier dysfunction and lung injury in vivo. *Am. J. Pathol.* **2013**, *182*, 1021–1030. [CrossRef] [PubMed]

114. Jeong, H.W.; Hsu, K.C.; Lee, J.W.; Ham, M.; Huh, J.Y.; Shin, H.J.; Kim, W.S.; Kim, J.B. Berberine suppresses proinflammatory responses through AMPK activation in macrophages. *Am. J. Physiol. Endocrinol. Metab.* **2009**, *296*, E955–E964. [CrossRef] [PubMed]

115. Vaez, H.; Rameshrad, M.; Najafi, M.; Barar, J.; Barzegari, A.; Garjani, A. Cardioprotective effect of metformin in lipopolysaccharide-induced sepsis via suppression of toll-like receptor 4 (TLR4) in heart. *Eur. J. Pharmacol.* **2016**, *772*, 115–123. [CrossRef] [PubMed]

116. Kim, J.; Kwak, H.J.; Cha, J.Y.; Jeong, Y.S.; Rhee, S.D.; Kim, K.R.; Cheon, H.G. Metformin suppresses lipopolysaccharide (LPS)-induced inflammatory response in murine macrophages via activating transcription factor-3 (ATF-3) induction. *J. Biol. Chem.* **2014**, *289*, 23246–23255. [CrossRef] [PubMed]

117. Mulchandani, N.; Yang, W.L.; Khan, M.M.; Zhang, F.; Marambaud, P.; Nicastro, J.; Coppa, G.F.; Wang, P. Stimulation of Brain AMP-Activated Protein Kinase Attenuates Inflammation and Acute Lung Injury in Sepsis. *Mol. Med.* **2015**, *21*, 637–644. [CrossRef] [PubMed]

118. Liu, Z.; Bone, N.; Jiang, S.; Park, D.W.; Tadie, J.M.; Deshane, J.; Rodriguez, C.A.; Pittet, J.F.; Abraham, E.; Zmijewski, J.W. AMP-Activated Protein Kinase and Glycogen Synthase Kinase 3beta Modulate the Severity of Sepsis-Induced Lung Injury. *Mol. Med.* **2016**, *21*, 937–950. [CrossRef] [PubMed]

119. Chen, J.; Lai, J.; Yang, L.; Ruan, G.; Chaugai, S.; Ning, Q.; Chen, C.; Wang, D.W. Trimetazidine prevents macrophage-mediated septic myocardial dysfunction via activation of the histone deacetylase sirtuin 1. *Br. J. Pharmacol.* **2016**, *173*, 545–561. [CrossRef] [PubMed]

120. Liu, X.; Wang, N.; Fan, S.; Zheng, X.; Yang, Y.; Zhu, Y.; Lu, Y.; Chen, Q.; Zhou, H.; Zheng, J. The citrus flavonoid naringenin confers protection in a murine endotoxaemia model through AMPK-ATF3-dependent negative regulation of the TLR4 signalling pathway. *Sci. Rep.* **2016**, *6*, 39735. [CrossRef] [PubMed]

121. Zhan, W.; Kang, Y.; Chen, N.; Mao, C.; Kang, Y.; Shang, J. Halofuginone ameliorates inflammation in severe acute hepatitis B virus (HBV)-infected SD rats through AMPK activation. *Drug Des. Dev. Ther.* **2017**, *11*, 2947–2955. [CrossRef] [PubMed]

122. Bae, H.B.; Zmijewski, J.W.; Deshane, J.S.; Tadie, J.M.; Chaplin, D.D.; Takashima, S.; Abraham, E. AMP-activated protein kinase enhances the phagocytic ability of macrophages and neutrophils. *FASEB J.* **2011**, *25*, 4358–4368. [CrossRef] [PubMed]

123. Li, L.; Xu, S.; Guo, T.; Gong, S.; Zhang, C. Effect of dapagliflozin on intestinal flora in MafA-deficient mice. *Curr. Pharm. Des.* **2018**. [CrossRef] [PubMed]

124. Rousset, C.I.; Leiper, F.C.; Kichev, A.; Gressens, P.; Carling, D.; Hagberg, H.; Thornton, C. A dual role for AMP-activated protein kinase (AMPK) during neonatal hypoxic-ischaemic brain injury in mice. *J. Neurochem.* **2015**, *133*, 242–252. [CrossRef] [PubMed]

125. Li, C.G.; Yan, L.; Jing, Y.Y.; Xu, L.H.; Liang, Y.D.; Wei, H.X.; Hu, B.; Pan, H.; Zha, Q.B.; Ouyang, D.Y.; et al. Berberine augments ATP-induced inflammasome activation in macrophages by enhancing AMPK signaling. *Oncotarget* **2017**, *8*, 95–109. [CrossRef] [PubMed]

126. Zha, Q.B.; Wei, H.X.; Li, C.G.; Liang, Y.D.; Xu, L.H.; Bai, W.J.; Pan, H.; He, X.H.; Ouyang, D.Y. ATP-Induced Inflammasome Activation and Pyroptosis Is Regulated by AMP-Activated Protein Kinase in Macrophages. *Front. Immunol.* **2016**, *7*, 597. [CrossRef] [PubMed]

127. Wen, H.; Gris, D.; Lei, Y.; Jha, S.; Zhang, L.; Huang, M.T.; Brickey, W.J.; Ting, J.P. Fatty acid-induced NLRP3-ASC inflammasome activation interferes with insulin signaling. *Nat. Immunol.* **2011**, *12*, 408–415. [CrossRef] [PubMed]

128. Lee, H.M.; Kim, J.J.; Kim, H.J.; Shong, M.; Ku, B.J.; Jo, E.K. Upregulated NLRP3 inflammasome activation in patients with type 2 diabetes. *Diabetes* **2013**, *62*, 194–204. [CrossRef] [PubMed]

129. Bullon, P.; Alcocer-Gomez, E.; Carrion, A.M.; Marin-Aguilar, F.; Garrido-Maraver, J.; Roman-Malo, L.; Ruiz-Cabello, J.; Culic, O.; Ryffel, B.; Apetoh, L.; et al. AMPK Phosphorylation Modulates Pain by Activation of NLRP3 Inflammasome. *Antioxid. Redox. Signal.* **2016**, *24*, 157–170. [CrossRef] [PubMed]

130. Wang, Y.; Viollet, B.; Terkeltaub, R.; Liu-Bryan, R. AMP-activated protein kinase suppresses urate crystal-induced inflammation and transduces colchicine effects in macrophages. *Ann. Rheum. Dis.* **2016**, *75*, 286–294. [CrossRef] [PubMed]

131. Blagih, J.; Coulombe, F.; Vincent, E.E.; Dupuy, F.; Galicia-Vazquez, G.; Yurchenko, E.; Raissi, T.C.; van der Windt, G.J.; Viollet, B.; Pearce, E.L.; et al. The energy sensor AMPK regulates T cell metabolic adaptation and effector responses in vivo. *Immunity* **2015**, *42*, 41–54. [CrossRef] [PubMed]

132. Iseda, S.; Takahashi, N.; Poplimont, H.; Nomura, T.; Seki, S.; Nakane, T.; Nakamura, M.; Shi, S.; Ishii, H.; Furukawa, S.; et al. Biphasic CD8+ T-Cell Defense in Simian Immunodeficiency Virus Control by Acute-Phase Passive Neutralizing Antibody Immunization. *J. Virol.* **2016**, *90*, 6276–6290. [CrossRef] [PubMed]

6

AMPK: Regulation of Metabolic Dynamics in the Context of Autophagy

Isaac Tamargo-Gómez [1,2] **and Guillermo Mariño** [1,2,*]

[1] Instituto de Investigación Sanitaria del Principado de Asturias, 33011 Oviedo, Spain; isaactamargo13@gmail.com

[2] Departamento de Biología Funcional, Universidad de Oviedo, 33011 Oviedo, Spain

[*] Correspondence: marinoguillermo@uniovi.es

Abstract: Eukaryotic cells have developed mechanisms that allow them to link growth and proliferation to the availability of energy and biomolecules. AMPK (adenosine monophosphate-activated protein kinase) is one of the most important molecular energy sensors in eukaryotic cells. AMPK activity is able to control a wide variety of metabolic processes connecting cellular metabolism with energy availability. Autophagy is an evolutionarily conserved catabolic pathway whose activity provides energy and basic building blocks for the synthesis of new biomolecules. Given the importance of autophagic degradation for energy production in situations of nutrient scarcity, it seems logical that eukaryotic cells have developed multiple molecular links between AMPK signaling and autophagy regulation. In this review, we will discuss the importance of AMPK activity for diverse aspects of cellular metabolism, and how AMPK modulates autophagic degradation and adapts it to cellular energetic status. We will explain how AMPK-mediated signaling is mechanistically involved in autophagy regulation both through specific phosphorylation of autophagy-relevant proteins or by indirectly impacting in the activity of additional autophagy regulators.

Keywords: AMPK; autophagy; metabolism; mTOR; ULK

1. Introduction

Eukaryotic cells are able to adapt to adverse fluctuations in the cellular environment. In order to do so, cells have developed molecular sensors, which react to circumstances that may perturb cell homeostasis. AMPK (adenosine monophosphate-activated protein kinase) is one of the main cellular sensors able to link a variety of cellular functions and processes to energy availability. AMPK is an evolutionarily conserved protein kinase, present both in unicellular organisms, such as baker's yeast, and also in more complex multicellular eukaryotes, as mammals. In 1981, the role for AMPK yeast ortholog, SNF1 (Sucrose Non-fermenting 1) as the main energy sensor in this organism was described [1]. This kinase is responsible for activating alternate metabolic pathways when the main carbon or nitrogen sources change, thus adapting cellular metabolism to oscillations in the cellular environment. In higher eukaryotes, AMPK is activated when AMP (adenosine monophosphate):ATP (adenosine triphosphate) and/or ADP (adenosine diphosphate):ATP ratios increase [1,2]. Once activated, AMPK maintains energy homeostasis by two complementary actions, an inhibition of ATP-consuming anabolic processes coupled with an activation of ATP-generating catabolic processes [3]. Due to its importance for cellular energy homeostasis, the activity of AMPK is tightly controlled by multiple upstream regulators, which contributes to link cellular metabolism with oscillating parameters in the cellular environment and also with changes in cellular nutritional and energetic requirements [4]. Moreover, AMPK signaling pathways are involved in numerous physiological processes apart from their main metabolic functions, such as cytoskeleton remodeling and transcriptional control or regulation of essential cellular processes, such as apoptosis or autophagy [5].

Autophagy is the cellular process by which organelles, proteins, and different macromolecules are delivered to the lysosomes for degradation [6]. This process can be classified into at least three different pathways: Macroautophagy (which we will refer to as autophagy), microautophagy, and chaperone-mediated autophagy, which mainly differ in the way in which autophagic cargo is transferred to the lysosome [7–9]. Autophagy plays essential roles in all eukaryotic cells and has been implicated in multiple processes, such as cell differentiation, cell death, and regulation of innate and adaptive immune responses or antigen presentation, among many other processes in high eukaryotes [10,11]. Despite all these functions, autophagy's most evolutionarily conserved role, from yeast to mammals, is to sustain energy balance in the cell by providing ATP and building blocks (lipids, amino acids, nucleotides, etc.) out of the degradation of non-essential or damaged cellular structures. Although autophagy regulation is complex and a variety of signaling cascades and regulatory mechanisms modulate autophagic activity, AMPK is probably the most conserved autophagy inducer through evolution. AMPK activity is linked to autophagic degradation in almost all eukaryotic cells.

In this review, we describe the main metabolic regulatory functions of AMPK, including its prominent role in autophagy regulation. We will discuss how AMPK activity is key to the coordination of catabolic pathways which produce energy and micro-molecules with anabolic processes, which use energy and micro-molecules to synthesize new macro-molecules which may be essential to sustain cell viability when cells face significant alterations in the intracellular/extracellular environment.

2. AMPK: Structure and Activation Mechanism

Evolutionarily, AMPK is a highly conserved serine/threonine protein kinase, and a member of the AMPK-related kinase family, which is comprised of thirteen kinases in the human genome. In mammalian cells, AMPK exists as a heterotrimeric complex formed by a catalytic α subunit and regulatory β and γ subunits [12]. There are multiple isoforms for each subunit encoded by different genes, *PRKAA* (5'-AMP-activated protein kinase catalytic subunit alpha), *PRKAB* (5'-AMP-activated protein kinase subunit beta) and *PRKAG* (5'-AMP-activated protein kinase subunit gamma). In humans, there are seven genes encoding AMPK subunits: Two isoforms for the α subunit (α1 and α2), encoded by the genes *PRKAA1* and *PRKAA2*, two isoforms of the β subunit (β1 and β2), encoded by *PRKAB1*, and three isoforms of the γ subunit (γ1, γ2 and γ3), encoded by *PRKAG1*, *PRKAG2* and *PRKAG3*, respectively [13]. Each AMPK complex is composed by one α-subunit, one β-subunit, and one γ-subunit (Figure 1). The fact that all combinations are possible leads to twelve different conformations in which α, β and γ subunits may constitute a functional AMPK complex, which are normally associated with a specific tissue or a determined cell type, or determined subcellular localizations inside cells [14].

The α-subunit presents a serine/threonine kinase domain at the N-terminal region and a critical residue, Thr172. This conserved residue can be phosphorylated by several different upstream kinases, which constitutes the main mechanism by which AMPK activity is regulated at the short term. Different groups have shown that the LKB1 (Liver kinase B1) kinase is able to phosphorylate Thr172 in response to a variety of signals [15,16]. Furthermore, Thr172 can be phosphorylated by CAMKK2 (Calcium/calmodulin-dependent protein kinase kinase 2) kinase in response to calcium flux, independently of LKB1 [17]. Several other studies have suggested that the MAPKKK family member TAK1/MAP3K7 (Transforming growth factor beta-activated kinase 1)/(Mitogen-activated protein kinase 7) might also phosphorylate Thr172 [18,19]. Apart from the direct regulation of AMPK activity by phosphorylation, its γ-subunit acts a sensor that enables AMPK to respond to changes in AMP/ATP or ADP/ATP levels [20]. Moreover, AMP alone is able to directly modulate AMPK activity through three distinct mechanisms. First, AMP may stimulate the phosphorylation of Thr172 by acting directly on upstream kinases [21]. Second, through allosteric modulation, AMP allows AMPK to be a more attractive substrate for its upstream kinases [22]. Third, AMP could inhibit Thr172 dephosphorylation by protecting it from phosphatases activity, or also increase AMPK activity once Thr172 is phosphorylated, also in an allosteric fashion [23,24].

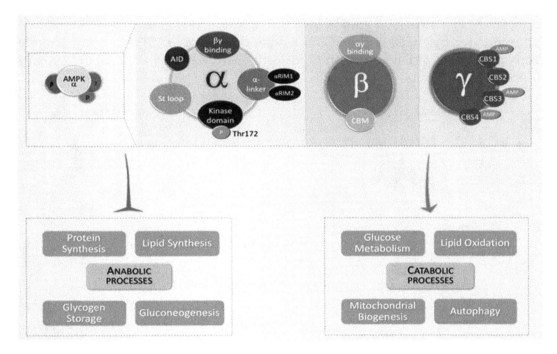

Figure 1. The domain structure of AMPK (adenosine monophosphate-activated protein kinase) heterotrimer. Functional AMPK complexes consist of one catalytic and two regulatory subunits. When activated, AMPK acts by decreasing energy-consuming anabolic processes (lipid synthesis, glycogen storage, gluconeogenesis, and protein synthesis) and increasing energy-providing catabolic processes that provide ATP (glucose metabolism, lipid oxidation, mitochondrial biogenesis and autophagy).

3. AMPK as an Energy-Sensing Kinase for Metabolic Regulation

AMPK is one of the main energy-sensing kinases in eukaryotic cells, able to regulate a wide variety of metabolic processes either by directly acting on metabolically-relevant proteins or by indirectly influencing gene expression [4,25]. During energy stress, AMPK directly activates metabolic enzymes and regulates different energy-consuming/producing pathways, (i.e., lipid or glucose metabolism, or mitochondrial biogenesis) in order to maintain an adequate energy balance. Moreover, AMPK can inhibit the activity of several transcription factors involved in anabolic routes, such as lipid, protein, and carbohydrate biosynthesis, to minimize ATP consumption. By contrast, AMPK promotes the activity of numerous transcription factors implicated in catabolism pathways to stimulate ATP production such as glucose uptake and metabolism (Figure 2) [26].

3.1. Lipid Metabolism

AMPK is a major regulator of cellular lipid metabolism. Once activated, AMPK is able to reduce the activity of essential enzymes for lipid synthesis and related processes, in order to couple their activity to cell energy levels. One of the best-studied AMPK functions in this context is the inhibitory phosphorylation of ACC1 (acetyl-CoA carboxylase 1) and ACC2 (acetyl-CoA carboxylase 2), which catalyze the first step in lipid synthesis. Several studies using animal models have shown that AMPK-mediated phosphorylation of Ser79 in ACC1 and Ser221 in ACC2 are mechanistically involved in the regulation of lipid homeostasis by AMPK [27]. Similarly, AMPK inhibits HMGCR (3-hydroxy-3-methylglutaryl-coenzyme A reductase), which catalyzes an essential step in cholesterol synthesis [28]. Conversely, AMPK positively promotes triglyceride conversion to fatty acids by stimulating lipases such as ATGL (adipocyte triglyceride lipase) and HSL (hormone-sensitive lipase) [29]. When cellular energy is low, free fatty acids can be imported into mitochondria for β-oxidation. This process requires the activity of the acyl-transferases form the CPT1 (Carnitine palmitoyl-transferase 1) family [30]. AMPK indirectly participates in the regulation of CPT1 because malonyl-CoA generated by ACC1 and ACC2 is a potent inhibitor of CPT1 activity. Therefore, AMPK activity directly decreases lipid synthesis and indirectly increases fatty acid import into mitochondria for β-oxidation [31,32].

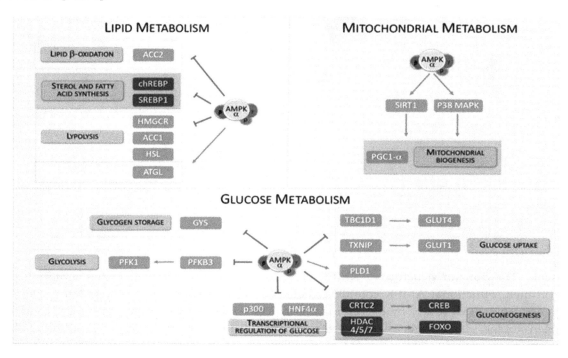

Figure 2. AMPK regulates different metabolic key targets. The metabolic pathways modulated by AMPK can be classified into three general categories: Lipid metabolism, mitochondrial metabolism, and glucose metabolism. The arrow indicates key targets for AMPK involved in these three metabolic categories. Transcriptional regulators are shown in dark squares.

In addition to its direct activity, AMPK is able to modulate lipid metabolism by regulating the activity of several transcription factors involved in lipid synthesis and associated processes. Specifically, AMPK phosphorylation inhibits the transcriptional activity of SREBP1 (sterol regulatory element binding protein 1), ChREBP (carbohydrate-responsive element binding protein) or HNF4α (hepatocyte nuclear factor 4α) [33–36]. Thus, by inhibiting phosphorylation of these transcription factors, AMPK negatively regulates lipogenic processes through the modulation of lipid-specific transcriptional programs.

3.2. Glucose Metabolism

In parallel to its inhibitory role for lipid synthesis, AMPK activity contributes to ATP generation through the modulation of a variety catabolic and anabolic pathways involved in glucose metabolism [37]. AMPK promotes glucose uptake by its inhibitory phosphorylation on TBC1D1 (TBC domain family member 1) and TXNIP (thioredoxin-interacting protein). These two factors respectively inhibit translocation of glucose transporters GLUT1 (Glucose transporter 1) and GLUT4 (Glucose transporter 4) to the plasma membrane. Thus, a high AMPK activity is associated to an increased presence of GLUT1 and GLUT4 glucose transporters in the plasma membrane [38,39]. Consistently, AMPK positively regulates glycolysis by phosphorylating PFKFB3 (6-phosphofructo-2-kinase/fructose-2,6-biphosphatase) [40]. Moreover, AMPK also inhibits glucose conversion into glycogen by inhibitory phosphorylation on different isoforms of GYS (glycogen synthase) [41,42]. Paradoxically, AMPK is involved in the regulation of glycogen supercompensation in skeletal muscle. Under conditions of prolonged physical exercise, sustained activation of AMPK boosts glycogen synthesis, especially in skeletal muscle. This is produced as a result of an increase in glucose uptake, which leads to an accumulation of intracellular G6P (glucose 6 phosphate) that allosterically activates the GYS, thus bypassing the inhibitory action of AMPK on this enzyme [43,44]. In parallel to its direct action on specific enzymes, AMPK also modulates glucose metabolism in a transcriptional fashion. This is the case for gluconeogenesis, which is activated during fasting or reduced glucose intake to maintain blood glucose levels. After re-feeding, a surge in insulin levels leads to phosphorylation of liver AMPK by either AKT (RAC-alpha serine/threonine-protein kinase) or LKB1 kinases, which in turn leads to transcriptional inhibition of gluconeogenesis key genes. This effect is partially achieved by AMPK-dependent phosphorylation and nuclear exclusion of CRTC2 (cyclic-AMP-regulated transcriptional co-activator 2) and HDACs (class IIA histone deacetylases), which are all essential co-factors for the transcription of gluconeogenic genes [45,46]. Thus, AMPK can influence glucose metabolism either by direct regulation of specific proteins or by transcriptional regulation of key genes involved in glucose metabolism [47].

3.3. Mitochondrial Biogenesis

AMPK activity is also important for maintaining mitochondrial function. In situations of energy imbalance, AMPK contributes to mitochondrial biogenesis in order to increase ATP production. During this process, cells increase their individual mitochondrial mass, which requires an increase in the expression of mitochondrial protein genes [48]. One of the main regulators of this process is PGC1-α (peroxisome proliferator-activated receptor-gamma coactivator), which transcriptionally controls the expression of a wide variety of mitochondrial genes. Overexpression of PGC1-α in muscle contributes to the conversion of type IIb fibers into type II and type I fibers, which are rich in mitochondria [49]. PGC1-α interacts with PPAR-γ (peroxisome proliferator-activated receptor-γ) and ERRs (estrogen-related receptors) and is regulated by numerous post-translational mechanisms [50]. These mechanisms include methylation, acetylation, and phosphorylation by upstream kinases. Different in vitro studies indicate that PGC1-α harbors two sites susceptible of being phosphorylated by AMPK, specifically Thr177 and Ser538. In addition, AMPK can indirectly regulate PGC1-α through phosphorylation of additional targets, such as HDAC5, SIRT1 and p38 MAPK. Moreover, AMPK promotes the activity of TFEB (transcription factor EB), which activates the gene encoding PGC1-α, *PPARGC1A*, as well as different genes involved in autophagy [51–54].

AMPK-dependent mitochondrial biogenesis has been specifically studied in skeletal muscle in response to exercise. Exercise activates AMPK in myocytes, leading to mitochondrial biogenesis upregulation [55]. Several studies have shown that overexpression of a constitutively active AMPK γ3-subunit induces mitochondrial biogenesis in mice [56]. In addition, AMPK activation improves muscle regeneration and protects muscle from age-related pathologies, in part by increasing autophagic activity [57].

4. AMPK: Regulation of Autophagy

4.1. Autophagy Regulation

Autophagy is an essential catabolic pathway conserved in all known nucleated cells [7,58]. Although autophagic degradation is constitutively active at basal levels, the main physiological autophagy inducer is nutrient deprivation and/or energy scarcity. This intrinsic characteristic has remained unaltered in organisms ranging from yeast to humans, which makes the involvement of AMPK in the regulation of this process logical.

An autophagy pathway starts with the formation of double-membrane vesicles called autophagosomes. These autophagosomes sequester cytoplasmic cargo (both through specific and non-specific mechanisms) and move along the cellular microtubule network until they eventually fuse with lysosomes [59]. Autophagosome-lysosome fusion allows for the degradation of autophagosome cargo and the autophagosomal inner membrane. Once degradation has occurred, the resulting biomolecules, such as amino acids, lipids, or nucleotides are recycled back to the cytoplasm and will be reused by the cell to synthesize new biomolecules [60]. From a molecular perspective, the autophagy pathway requires the involvement of a group of evolutionarily conserved genes/proteins called ATG (AuTophaGy-related) proteins [61]. These proteins were originally described in yeast and are required for autophagosome formation, maturation, transport, or degradation, being involved in the different steps of the autophagic pathway in a hierarchical and temporally-coordinated fashion (Figure 3) [61,62].

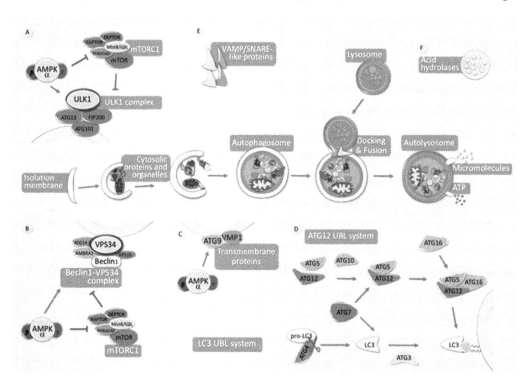

Figure 3. Scheme of the main steps for autophagy and their regulation by AMPK. An autophagy pathway starts with the formation of the isolation membrane, also known as phagophore. The autophagy implicates the coordinated temporal and spatial activation of numerous molecular components. (**A**): The ULK1-FIP200-ATG13-ATG101 complex is responsible for initiating the autophagic process. The activity of this protein complex is antagonistically regulated by mTORC1 (inhibitory phosphorylation) and by AMPK, which both activates the ULK1 complex as well as inhibits the activity of the mTORC1 complex. (**B**): The ClassIII PI3K complex formed by VPS34, Beclin1, ATG14, AMBRA1, and other subunits creates a membrane domain enriched in PtsIns3P, which drives the nucleation of ATG (AuTophaGy-related) proteins in the phagophore, either directly or indirectly. AMPK is able to increase the pro-autophagic function of this complex and to enhance its formation, whereas mTORC1 activity negatively regulates its function. (**C**): Two different transmembrane proteins, the vacuole membrane protein 1 (VMP1) and ATG9 participate in the recruitment of membranes to the phagophore. AMPK is able to phosphorylate ATG9, which increases its recruitment towards autophagosome formation sites. (**D**): Two ubiquitin-like (UBL) protein conjugation systems (ATG12- and LC3- UBLs) involving the participation of ATG4 cysteine proteinases (which activate LC3 by cleaving its carboxyl terminus), the E1-like enzyme ATG7 (common to both conjugation systems), and the E2-like enzymes ATG10 (ATG12 system), and ATG3 (LC3 system). In coordination, the activity of both systems is required to conjugate LC3 (and other members of this protein family homologous to yeast ATG8) to a phosphatidyl-ethanolamine lipid at the nascent pre-autophagosomal membrane. (**E**): Upon completion, fully-formed autophagosomes move along the microtubule network, eventually fusing with a lysosome, thus acquiring hydrolytic activity, and thus becoming autolysosomes. Several SNARE-like proteins (i.e., Syntaxin17 and VAMP8, among others) are required for efficient fusion between lysosomes and autophagosomes. Once content and inner membrane are degraded by acidic hydrolases, the resultant molecules (amino acids, nucleotides, lipids, etc.) are recycled back to the cytoplasm by membrane permeases.

In mammalian cells, autophagosome biogenesis requires the combined activity of two protein complexes, namely the Class III PI3-Kinase protein complex and ULK1/2 (unc-51-like kinase1/2)-containing complexes, which are recruited to autophagosome forming sites during autophagy initiation [63–65]. Mammalian ULK proteins are a family of serine/threonine kinases which are the orthologues of yeast ATG1 and whose activity is essential for the recruitment of autophagy-relevant proteins involved in autophagosome biogenesis [66,67]. There are four members of the ULK family in mammalian cells (ULK1-4). ULK1 is the main ATG1 functional orthologue in mammalian cells [68], although ULK2 has been shown to compensate for ULK1 loss. By contrast, ULK3 and ULK4 seem to have evolved to perform biological functions unrelated to autophagy [68]. ULK1 is part of a protein complex containing ATG13, FIP200, also known as RB1CC1 (RB1-inducible coiled-coil protein 1), and ATG101. As we will discuss later, the autophagy-promoting activity of this protein complex can be modulated through specific phosphorylation of its subunits.

The activity of ULK1 is not enough to promote efficient autophagosome biogenesis. In fact, the activity of the Class III PI3-Kinase protein complex, which contains the catalytic subunit VPS34 (Vacuolar protein sorting 34) and other variable regulatory subunits, is also required for autophagosome formation. This protein complex, which acts as a lipid kinase, generates PIP3P-enriched membrane domains at the site of autophagosome formation, which are required to recruit essential factors for autophagosome formation. Apart from the autophagy-relevant Class III PI3K complex, which is formed by VPS34, VPS15, Beclin1 (Coiled-coil, myosin-like BCL2 interacting protein), ATG14, and AMBRA1 (Activating molecule in Beclin1-regulated autophagy protein 1), VPS34 can form part of other Class III PI3K complexes. These alternative VPS34 protein complexes are comprised of different subunits and regulate vesicular trafficking in processes such as endocytosis or Golgi-mediated protein secretion [69,70]. The relative abundance of these different VPS43-containing complexes is variable and will be regulated according to cellular needs. Thus, in conditions of autophagy induction, most VPS34-containing complexes will consist of autophagy-relevant subunits. Thus, autophagy regulation at this level is achieved both by direct modulation of VPS34 kinase activity and by increasing the formation of autophagy-relevant VPS34-containing complexes.

In addition to the activity of these protein complexes and other ATG proteins, which are specifically involved in autophagy execution, multiple signaling cascades are able to regulate autophagic activity. The relative importance of the different regulatory inputs for autophagy execution is, in many cases, cell type-specific and tissue-dependent. However, there are some major autophagy regulatory circuits which have remained conserved through evolution and are present in most tissues/cell types from most multicellular organisms, including mammals, such as AMPK or mTORC1 [59,71].

4.2. AMPK Antagonizes mTORC1 to Regulate ULK Complex Activity

In mammalian cells, although many different pathways or signaling events may influence autophagy, the two main regulators for autophagic degradation are mTOR (mechanistic/mammalian Target Of Rapamycin) and AMPK kinases [72,73]. mTOR can be found forming two protein complexes, mTORC1 and mTORC2. Although mTORC2 involvement in autophagy dynamics is not totally neglectable, its contribution to autophagy regulation is marginal. By contrast, mTORC1 can be considered as the main autophagy suppressor in mammalian cells. mTORC1 is normally active in situations of high energy levels, high amino acid cellular content, or growth factors stimulation, all of which have a negative impact in autophagic degradation [74]. By contrast, AMPK activation positively regulates autophagic activity, as one may expect due to its pro-catabolic functions. Due to their antagonistic roles, mTORC1 and AMPK activities are molecularly connected, inhibition of mTORC1 activity being one of the main mechanisms by which AMPK increases autophagic degradation and vice versa.

When energy/growth factors or amino acids are abundant, mTORC1 represses autophagy through inhibitory phosphorylation of ATG13, which reduces the activity of the ULK1 complex, thus decreasing the rate of autophagosome formation [75,76]. In the same sense, ULK1 itself is a direct target of mTORC1, so mTORC1 can inhibit the autophagic process by acting both on ULK1 and ATG13 [75]. AMPK plays an opposite role to mTORC1 regarding ULK1 complex activity, thus positively regulating the first steps of autophagosome formation in response to a variety of pro-autophagic stimuli [77,78]. In fact, AMPK increases ULK1 activity by directly phosphorylating Ser467, Ser555, Thr574, and Ser637, which increases the recruitment of autophagy-relevant proteins (ATG proteins) to the membrane domains in which autophagosome formation takes place [79].

In addition, AMPK negatively regulates mTORC1 activity, which blocks its inhibitory effect on ULK1 by two complementary actions [74]. First, AMPK activates TSC2 (Tuberous sclerosis complex 2) by phosphorylating Thr1227 and Ser1345 residues, thus favoring the assembly of TSC1/TSC2 heterodimer, which negatively impacts mTORC1 activity [80]. Second, AMPK can inhibit mTORC1 by direct phosphorylation of RAPTOR (regulatory-associated protein of mTOR) Ser722 and Ser792 residues [81]. In addition, AMPK is able to promote autophagy by acting differentially at different levels of autophagy regulation, through specific phosphorylation in components of autophagy-initiating protein complexes. Again, the activities of AMPK and mTORC1 are antagonistic in relation to autophagy regulation, and they act together to couple autophagy regulation with multiple signaling pathways, with the ULK1 complex being one of the main checkpoints for the regulation of autophagy initiation (Table 1).

Table 1. Regulation of autophagy relevant proteins by AMPK. H, human; M, mouse; R, rat.

Protein	Phosphorylation Site(s)	Stage of Autophagy	Autophagy Function	Ref.
ATG9	Ser761(H, M, R)	Autophagosome elongation	Participates in the recruitment of lipids to the isolation membrane	[82]
BECN1	Ser91(M, R) Ser94(M, R)	Autophagosome biogenesis	Part of the III PI3KC3 complex	[83]
mTOR (RAPTOR)	Ser722(H, M) Ser792(H, M)	Regulation of Autophagy	Negative regulator of Autophagy	[81]
mTOR	Thr2446(H)	Regulation of Autophagy	Negative regulator of Autophagy	[84]
PAQR3	Thr32(H, M)	Autophagosome biogenesis	Facilitates the formation of pro-autophagic PI3KC3 III complex	[85]
RACK1	Thr50(H, M, R)	Autophagosome biogenesis	Promoting the assembly of the III PI3KC3 complex	[86]
TSC2	Ser1387(H, M, R) Thr1271(H, R)	Regulation of Autophagy	Negative regulator of Mtor	[80,87]
ULK1	Ser555(M, R) Ser467(H, M, R) Thr574(M, R) Ser637(M, R)	Autophagy Initiation	Part of the ULK1-complex/early steps of autophagosome biogenesis	[79]
VPS34	Thr163(H, M, R) Ser165(H, M, R)	Autophagosome biogenesis	Part of the III PI3KC3 complex	[83]

4.3. AMPK Regulates Class III PI3K Complex Activity

Apart from its role in regulating ULK1 activity, AMPK is also involved in the regulation of Class III PI3K complex activity. ULK1 itself exerts its pro-autophagic activity by phosphorylating several components of this complex, including Beclin1, AMBRA1, or the catalytic subunit VPS34 [65]. Interestingly, mTORC1 also inhibits autophagosome biogenesis through phosphorylation of ATG14L, an essential component of the pro-autophagic VPS34 complex [64].

AMPK regulation of autophagy also operates through phosphorylation in different subunits of the different Class III PI3K complexes, including VPS34 itself, which modifies their affinity for other components of the complex. Thus, AMPK regulates the relative abundance of the different

VPS34-containing complexes, thus connecting the activity of processes involving vesicle trafficking to cellular energy status. In this regard, different biochemical studies have shown how AMPK regulates the composition of the Class III PI3K complex. For example, AMPK phosphorylation of Beclin1 at Thr388 increases Beclin1 binding to VPS34 and ATG14, which promotes higher autophagy activity upon glucose withdrawal than the wild-type control [83,88]. Similarly, AMPK phosphorylation of mouse Beclin1 at Ser-91 and Ser-94 increases the rate of autophagosome formation under nutrient stress conditions [83]. Apart from its activity towards components of the different Class III PI3K complexes, AMPK can also influence their composition by phosphorylating other proteins, which are relevant for the formation/stability of VPS34-containing complexes. Thus, AMPK-mediated phosphorylation of Thr32 on PAQR3 (progestin and adipo-Q receptors member 3), an ATG14L/VPS34 scaffolding protein, or that of Thr50 on the VPS34 associated protein RACK1 (Receptor for activated C kinase 1) has also been shown to enhance stability and pro-autophagic activity of Class III PI3K complexes [85,86].

In parallel to its activating phosphorylation in diverse components of the pro-autophagic VPS34 complexes, AMPK inhibits VPS34 complexes that do not contain pro-autophagic factors and are thus involved in different cellular vesicle trafficking processes, by direct phosphorylation of VPS34 on Thr163 and Ser165 [83]. Hence, in autophagy-promoting conditions, AMPK activation both enhances the activity of pro-autophagic VPS34 complexes and inhibits the formation of other different Class III PI3K complexes involved in autophagy-independent processes.

4.4. Additional AMPK Regulation of Autophagy

Apart from its direct activity towards autophagy-initiating complexes, AMPK can also influence autophagic activity by specific phosphorylation of ATG9, a transmembrane protein involved in autophagosome biogenesis by supplying vesicles which contribute to autophagosome elongation. In fact, AMPK-mediated phosphorylation in Ser761 of ATG9 increases recruitment of ATG9A (and ATG9-containing vesicles) to LC3-positive autophagosomes, thus enhancing autophagosome biogenesis [82].

Additionally, AMPK is able to influence autophagic activity in a transcriptional fashion (Table 2). In fact, under stress situations, AMPK directly phosphorylates the FOXO3 (Forkhead box O3) transcription factor, which regulates genes implicated in autophagy execution [89]. This activity antagonizes that of the mTOR, which on the other hand phosphorylates other members of the FOX (Forkhead box) family, such as FOXK2 (Forkhead box protein K2) and FOXK1 (Forkhead box protein K1), which compete with FOXO3 to repress genes implicated in autophagy [90]. A similar situation in which AMPK and mTOR activities antagonize each other can be found in relation with TFEB/TFE transcription factors, which control the expression of a variety of genes involved in lysosomal biogenesis and autophagy [91].

In situations of high energy, mTOR phosphorylates these factors of transcription and inhibits their function. By contrast, recent reports have shown that TFEB/TFE nuclear translocation is highly reduced either in cells deficient for AMPK or treated with AMPK inhibitors [92,93]. Consistently, it has been recently reported that AMPK activity is required for efficient dissociation of the transcriptional repressor BRD4 (Bromodomain-containing protein 4) from autophagy gene promoters in response to starvation [94].

Apart from its ability to regulate autophagy-relevant transcription factors through direct phosphorylation, AMPK also regulates different transcriptional regulators, such as EP300 [95] or Class IIa HDACs [47], which are involved in metabolism, autophagy and lysosomal functions (Table 2).

Table 2. Transcriptional regulation of autophagy through AMPK phosphorylation. H, human; M, mouse; R, rat.

Transcription Factor	Phosphorylation Site(s)	Target Gene (s)	Ref.
CHOP	Ser30(H, M, R)	*ATG5, MAP1LC3B*	[96]
FOXO3	Thr179(H) Ser399(H) Ser413(H) Ser555(H) Ser588(H) Ser626(H)	*ATG4B, GABARAPL1, ATG12, ATG14, GLUL, MAP1LC3, BECN1, PIK3CA, PIK3C3, ULK1, BNIP3, FBXO32*	[89]
HSF1	Ser121(H, M, R)	*ATG7*	[97]
Nrf2	Ser558(H, M)	*SQSTM1*	[98]
p53	Ser15(H, R)	*AEN, DRAM1, BAX, IGFBP3, BBC3, C12orf5, PRKAB1, PRKAB2, CDKN2A, SESN1, SESN2, DAPK1, BCL2, MCL1*	[99]
p73	Ser426(H)	*ATG5, DRAM1, ATG7, UVRAG*	[100]

4.5. Selective Degradation of Mitochondria by Autophagy

Apart from its general role in the regulation of bulk autophagic degradation, AMPK specifically participates in the regulation of mitophagy, the selective elimination of defective mitochondria through autophagy [101]. Mitochondria are organized in a dynamic network that changes its morphology through the combined actions of fission and fusion. Thus, mitochondria can be found in different distributions ranging from a single closed network to large numbers of small fragments. Recent studies have shown that an increase in mitochondrial fission is required in order to facilitate mitophagy [102]. This renders mitochondria susceptible to being engulfed by pre-autophagosomal isolation membranes, thus allowing mitophagy to take place. Consistently, mitochondrial stressors, such as electron transport chain poisons, or other stressors that damage mitochondria (and thus would increase mitophagy) have been shown to increase mitochondrial fission [103]. AMPK activation by energy imbalance and also a variety of other cellular and mitochondrial stressors promotes mitochondrial fission, thus coupling mitochondrial dynamics with mitophagic degradation [104]. This effect mainly relies on the ability of AMPK to phosphorylate and to activate the MFF (mitochondrial fission factor). MFF is a mitochondrial outer-membrane protein, which recruits cytoplasmic DRP1 (dynamin 1 like protein) to the mitochondrial outer membrane [105]. DRP recruitment to the mitochondrial outer-membrane increases mitochondrial fission, enabling the resulting fragmented mitochondria to undergo mitophagy [106].

In this context, AMPK phosphorylation of ULK1 on Ser555 has been shown to be critical for the development of exercise-induced mitophagy [107]. Thus, through its combined actions on key factors for autophagy regulation (by acting on ULK1 and other major autophagy regulators) and mitochondrial network dynamics (by specifically activating MFF), AMPK is able to coordinate autophagosome formation and mitochondrial size in order to enable efficient autophagic degradation of mitochondria.

5. Conclusions and Future Perspectives

Autophagy regulation has become increasingly complex in high eukaryotes, in which multiple signaling cascades are connected to autophagy key factors. However, AMPK probably remains as the major molecular autophagy inducer, counteracting the activity of mTORC1, which has also evolutionarily remained as the main molecular autophagy inhibitor. In parallel to its function in autophagy regulation, AMPK's role of adapting cellular metabolism to energetic availability has also been conserved in a diversity of organisms, from yeast to mammals. Thus, it is not surprising that substantial efforts have been made to identify new pharmacological AMPK activators. Recently, a variety of compounds able to increase AMPK activity, such as AICAR, Compound-13, PT-1, A769662 or benzimidazole have been identified [108]. Many of these drugs have shown great potential as research tools to modulate AMPK activity, and some of them have been successfully tested in animal models. However, and despite the substantial advances in this field, metformin (clinically developed in the late 1950s) is still the only AMPK-activating drug widely used in human patients. Interestingly, therapeutic AMPK modulation by metformin has shown promising results in the context of diverse metabolic conditions such as type II diabetes, fatty liver diseases, Alzheimer's, and in diverse types of cancers [109]. In fact, the beneficial effects of metformin are sometimes beyond the scope of its a priori potential. The fact that AMPK plays a pivotal role in autophagy regulation, together with the wide variety of processes for which autophagic activity is beneficial, suggests a potential mechanistic involvement of autophagy for some of the positive effects of AMPK activation. Future studies aimed at dissecting the precise molecular mechanisms by which AMPK exerts its wide variety of beneficial effects for human health will shed more light into these questions.

Abbreviation

AID	autoinhibitory domain
AMPK	AMP-activated protein kinase
CBM	carbohydrate-binding module
CBS	cystathionine β-synthase repeats
ChREBP	carbohydrate-responsive element-binding protein
CREB	cAMP response element-binding protein
CTD	C-terminus domain
DEPTOR	DEP domain-containing mTOR-interacting protein
FOXO	forkhead box protein O
HDAC	histone deacetylase
HMGCR	HMG-CoA reductase
HNF4α	hepatocyte nuclear factor 4α
MCL1	myeloid cell leukaemia sequence 1
mLST8	mammalian lethal with SEC13 protein
NTD	N-terminus domain
PGC1α	peroxisome proliferator-activated receptor-γ co-activator 1α
PLD1	phospholipase D1
PRAS40	40 kDa Pro-rich AKT substrate
RAPTOR	regulatory-associated protein of mTOR
RIM	regulatory-subunit-interacting motif
SREBP1	sterol regulatory element-binding protein 1
ST-loop	serine/threonine enriched loop

References

1. Carlson, M.; Osmond, B.; Botstein, D. Mutants of yeast defective in sucrose utilization. *Genetics* **1981**, *98*, 25–40. [PubMed]

2. Hedbacker, K.; Carlson, M. SNF1/AMPK pathways in yeast. *Front. Biosci.* **2008**, *13*, 2408–2420. [CrossRef] [PubMed]

3. Hardie, D.G. The AMP-activated protein kinase pathway—New players upstream and downstream. *J. Cell Sci.* **2004**, *117*, 5479–5487. [CrossRef] [PubMed]

4. Herzig, S.; Shaw, R.J. AMPK: Guardian of metabolism and mitochondrial homeostasis. *Nat. Rev. Mol. Cell Biol.* **2018**, *19*, 121–135. [CrossRef] [PubMed]

5. Ke, R.; Xu, Q.; Li, C.; Luo, L.; Huang, D. Mechanisms of AMPK in the maintenance of atp balance during energy metabolism. *Cell Biol. Int.* **2018**, *42*, 384–392. [CrossRef] [PubMed]

6.   Kroemer, G.; Marino, G.; Levine, B. Autophagy and the integrated stress response. *Mol. Cell* **2010**, *40*, 280–293. [CrossRef] [PubMed]

7.   Marino, G.; Lopez-Otin, C. Autophagy: Molecular mechanisms, physiological functions and relevance in human pathology. *Cell. Mol. Life Sci.* **2004**, *61*, 1439–1454. [CrossRef] [PubMed]

8.   Kaushik, S.; Cuervo, A.M. Chaperone-mediated autophagy: A unique way to enter the lysosome world. *Trends Cell Biol.* **2012**, *22*, 407–417. [CrossRef] [PubMed]

9.   Sahu, R.; Kaushik, S.; Clement, C.C.; Cannizzo, E.S.; Scharf, B.; Follenzi, A.; Potolicchio, I.; Nieves, E.; Cuervo, A.M.; Santambrogio, L. Microautophagy of cytosolic proteins by late endosomes. *Dev. Cell* **2011**, *20*, 131–139. [CrossRef] [PubMed]

10.   Levine, B.; Kroemer, G. Autophagy in the pathogenesis of disease. *Cell* **2008**, *132*, 27–42. [CrossRef] [PubMed]

11.   Levine, B.; Mizushima, N.; Virgin, H.W. Autophagy in immunity and inflammation. *Nature* **2011**, *469*, 323–335. [CrossRef] [PubMed]

12.   Hardie, D.G. AMPK: A key regulator of energy balance in the single cell and the whole organism. *Int. J. Obes. (Lond.)* **2008**, *32* (Suppl. 4), S7–S12. [CrossRef]

13.   Willows, R.; Navaratnam, N.; Lima, A.; Read, J.; Carling, D. Effect of different gamma-subunit isoforms on the regulation of AMPK. *Biochem. J.* **2017**, *474*, 1741–1754. [CrossRef] [PubMed]

14.   Ross, F.A.; Jensen, T.E.; Hardie, D.G. Differential regulation by AMP and ADP of AMPK complexes containing different gamma subunit isoforms. *Biochem. J.* **2016**, *473*, 189–199. [CrossRef] [PubMed]

15.   Stein, S.C.; Woods, A.; Jones, N.A.; Davison, M.D.; Carling, D. The regulation of AMP-activated protein kinase by phosphorylation. *Biochem. J.* **2000**, *345 Pt 3*, 437–443. [CrossRef]

16.   Hardie, D.G. AMPK: Positive and negative regulation, and its role in whole-body energy homeostasis. *Curr. Opin. Cell Biol.* **2015**, *33*, 1–7. [CrossRef] [PubMed]

17.   Fogarty, S.; Hawley, S.A.; Green, K.A.; Saner, N.; Mustard, K.J.; Hardie, D.G. Calmodulin-dependent protein kinase kinase-β activates AMPK without forming a stable complex: Synergistic effects of Ca^{2+} and AMP. *Biochem. J.* **2010**, *426*, 109–118. [CrossRef] [PubMed]

18.   Xie, M.; Zhang, D.; Dyck, J.R.; Li, Y.; Zhang, H.; Morishima, M.; Mann, D.L.; Taffet, G.E.; Baldini, A.; Khoury, D.S.; et al. A pivotal role for endogenous TGF-β-activated kinase-1 in the LKB1/AMP-activated protein kinase energy-sensor pathway. *Proc. Natl. Acad. Sci. USA* **2006**, *103*, 17378–17383. [CrossRef] [PubMed]

19.   Herrero-Martin, G.; Hoyer-Hansen, M.; Garcia-Garcia, C.; Fumarola, C.; Farkas, T.; Lopez-Rivas, A.; Jaattela, M. Tak1 activates AMPK-dependent cytoprotective autophagy in trail-treated epithelial cells. *EMBO J.* **2009**, *28*, 677–685. [CrossRef] [PubMed]

20.   Xiao, B.; Heath, R.; Saiu, P.; Leiper, F.C.; Leone, P.; Jing, C.; Walker, P.A.; Haire, L.; Eccleston, J.F.; Davis, C.T.; et al. Structural basis for AMP binding to mammalian AMP-activated protein kinase. *Nature* **2007**, *449*, 496–500. [CrossRef] [PubMed]

21.   Oakhill, J.S.; Steel, R.; Chen, Z.P.; Scott, J.W.; Ling, N.; Tam, S.; Kemp, B.E. AMPK is a direct adenylate charge-regulated protein kinase. *Science* **2011**, *332*, 1433–1435. [CrossRef] [PubMed]

22.   Hawley, S.A.; Boudeau, J.; Reid, J.L.; Mustard, K.J.; Udd, L.; Makela, T.P.; Alessi, D.R.; Hardie, D.G. Complexes between the LKB1 tumor suppressor, STRADα/β and MO25α/β are upstream kinases in the AMP-activated protein kinase cascade. *J. Biol.* **2003**, *2*, 28. [CrossRef] [PubMed]

23.   Lin, S.C.; Hardie, D.G. AMPK: Sensing glucose as well as cellular energy status. *Cell Metab.* **2018**, *27*, 299–313. [CrossRef] [PubMed]

24.   Gowans, G.J.; Hawley, S.A.; Ross, F.A.; Hardie, D.G. AMP is a true physiological regulator of AMP-activated protein kinase by both allosteric activation and enhancing net phosphorylation. *Cell Metab.* **2013**, *18*, 556–566. [CrossRef] [PubMed]

25.   Mihaylova, M.M.; Shaw, R.J. The AMPK signalling pathway coordinates cell growth, autophagy and metabolism. *Nat. Cell Biol.* **2011**, *13*, 1016–1023. [CrossRef] [PubMed]

26.   Garcia, D.; Shaw, R.J. AMPK: Mechanisms of cellular energy sensing and restoration of metabolic balance. *Mol. Cell* **2017**, *66*, 789–800. [CrossRef] [PubMed]

27.   Fullerton, M.D.; Galic, S.; Marcinko, K.; Sikkema, S.; Pulinilkunnil, T.; Chen, Z.P.; O'Neill, H.M.; Ford, R.J.; Palanivel, R.; O'Brien, M.; et al. Single phosphorylation sites in Acc1 and Acc2 regulate lipid homeostasis and the insulin-sensitizing effects of metformin. *Nat. Med.* **2013**, *19*, 1649–1654. [CrossRef] [PubMed]

28. Willows, R.; Sanders, M.J.; Xiao, B.; Patel, B.R.; Martin, S.R.; Read, J.; Wilson, J.R.; Hubbard, J.; Gamblin, S.J.; Carling, D. Phosphorylation of AMPK by upstream kinases is required for activity in mammalian cells. *Biochem. J.* **2017**, *474*, 3059–3073. [CrossRef] [PubMed]

29. Ahmadian, M.; Abbott, M.J.; Tang, T.; Hudak, C.S.; Kim, Y.; Bruss, M.; Hellerstein, M.K.; Lee, H.Y.; Samuel, V.T.; Shulman, G.I.; et al. Desnutrin/ATGL is regulated by AMPK and is required for a brown adipose phenotype. *Cell Metab.* **2011**, *13*, 739–748. [CrossRef] [PubMed]

30. Kerner, J.; Hoppel, C. Fatty acid import into mitochondria. *Biochim. Biophys. Acta* **2000**, *1486*, 1–17. [CrossRef]

31. Saggerson, D. Malonyl-coa, a key signaling molecule in mammalian cells. *Annu. Rev. Nutr.* **2008**, *28*, 253–272. [CrossRef] [PubMed]

32. Fantino, M. Role of lipids in the control of food intake. *Curr. Opin. Clin. Nutr. Metab. Care* **2011**, *14*, 138–144. [CrossRef] [PubMed]

33. Li, Y.; Xu, S.; Mihaylova, M.M.; Zheng, B.; Hou, X.; Jiang, B.; Park, O.; Luo, Z.; Lefai, E.; Shyy, J.Y.; et al. AMPK phosphorylates and inhibits SREBP activity to attenuate hepatic steatosis and atherosclerosis in diet-induced insulin-resistant mice. *Cell Metab.* **2011**, *13*, 376–388. [CrossRef] [PubMed]

34. Sato, S.; Jung, H.; Nakagawa, T.; Pawlosky, R.; Takeshima, T.; Lee, W.R.; Sakiyama, H.; Laxman, S.; Wynn, R.M.; Tu, B.P.; et al. Metabolite regulation of nuclear localization of carbohydrate-response element-binding protein (ChREBP): Role of AMP as an allosteric inhibitor. *J. Biol. Chem.* **2016**, *291*, 10515–10527. [CrossRef] [PubMed]

35. Sato, Y.; Tsuyama, T.; Sato, C.; Karim, M.F.; Yoshizawa, T.; Inoue, M.; Yamagata, K. Hypoxia reduces HNF4α/MODY1 protein expression in pancreatic β-cells by activating AMP-activated protein kinase. *J. Biol. Chem.* **2017**, *292*, 8716–8728. [CrossRef] [PubMed]

36. Elhanati, S.; Kanfi, Y.; Varvak, A.; Roichman, A.; Carmel-Gross, I.; Barth, S.; Gibor, G.; Cohen, H.Y. Multiple regulatory layers of SREBP1/2 by SIRT6. *Cell Rep.* **2013**, *4*, 905–912. [CrossRef] [PubMed]

37. Hardie, D.G. AMPK: A target for drugs and natural products with effects on both diabetes and cancer. *Diabetes* **2013**, *62*, 2164–2172. [CrossRef] [PubMed]

38. Wu, N.; Zheng, B.; Shaywitz, A.; Dagon, Y.; Tower, C.; Bellinger, G.; Shen, C.H.; Wen, J.; Asara, J.; McGraw, T.E.; et al. AMPK-dependent degradation of TXNIP upon energy stress leads to enhanced glucose uptake via GLUT1. *Mol. Cell* **2013**, *49*, 1167–1175. [CrossRef] [PubMed]

39. Chavez, J.A.; Roach, W.G.; Keller, S.R.; Lane, W.S.; Lienhard, G.E. Inhibition of glut4 translocation by tbc1d1, a rab gtpase-activating protein abundant in skeletal muscle, is partially relieved by AMP-activated protein kinase activation. *J. Biol. Chem.* **2008**, *283*, 9187–9195. [CrossRef] [PubMed]

40. Domenech, E.; Maestre, C.; Esteban-Martinez, L.; Partida, D.; Pascual, R.; Fernandez-Miranda, G.; Seco, E.; Campos-Olivas, R.; Perez, M.; Megias, D.; et al. AMPK and PFKFB3 mediate glycolysis and survival in response to mitophagy during mitotic arrest. *Nat. Cell Biol.* **2015**, *17*, 1304–1316. [CrossRef] [PubMed]

41. Hardie, D.G. Ampk—Sensing energy while talking to other signaling pathways. *Cell Metab.* **2014**, *20*, 939–952. [CrossRef] [PubMed]

42. Bultot, L.; Guigas, B.; Von Wilamowitz-Moellendorff, A.; Maisin, L.; Vertommen, D.; Hussain, N.; Beullens, M.; Guinovart, J.J.; Foretz, M.; Viollet, B.; et al. AMP-activated protein kinase phosphorylates and inactivates liver glycogen synthase. *Biochem. J.* **2012**, *443*, 193–203. [CrossRef] [PubMed]

43. Hingst, J.R.; Bruhn, L.; Hansen, M.B.; Rosschou, M.F.; Birk, J.B.; Fentz, J.; Foretz, M.; Viollet, B.; Sakamoto, K.; Faergeman, N.J.; et al. Exercise-induced molecular mechanisms promoting glycogen supercompensation in human skeletal muscle. *Mol. Metab.* **2018**, *16*, 24–34. [CrossRef] [PubMed]

44. Janzen, N.R.; Whitfield, J.; Hoffman, N.J. Interactive roles for AMPK and glycogen from cellular energy sensing to exercise metabolism. *Int. J. Mol. Sci.* **2018**, *19*, 3344. [CrossRef] [PubMed]

45. Lee, J.M.; Seo, W.Y.; Song, K.H.; Chanda, D.; Kim, Y.D.; Kim, D.K.; Lee, M.W.; Ryu, D.; Kim, Y.H.; Noh, J.R.; et al. AMPK-dependent repression of hepatic gluconeogenesis via disruption of CREB/CRTC2 complex by orphan nuclear receptor small heterodimer partner. *J. Biol. Chem.* **2010**, *285*, 32182–32191. [CrossRef] [PubMed]

46. Di Giorgio, E.; Brancolini, C. Regulation of class iia hdac activities: It is not only matter of subcellular localization. *Epigenomics* **2016**, *8*, 251–269. [CrossRef] [PubMed]

47. Mihaylova, M.M.; Vasquez, D.S.; Ravnskjaer, K.; Denechaud, P.D.; Yu, R.T.; Alvarez, J.G.; Downes, M.; Evans, R.M.; Montminy, M.; Shaw, R.J. Class iia histone deacetylases are hormone-activated regulators of FOXO and mammalian glucose homeostasis. *Cell* **2011**, *145*, 607–621. [CrossRef] [PubMed]

48. Kiriyama, Y.; Nochi, H. Intra- and intercellular quality control mechanisms of mitochondria. *Cells* **2017**, *7*, 1. [CrossRef] [PubMed]

49. Spiegelman, B.M. Transcriptional control of mitochondrial energy metabolism through the PGC1 coactivators. *Novartis Found. Symp.* **2007**, *287*, 60–63; Discussion 63–69. [PubMed]

50. Eichner, L.J.; Giguere, V. Estrogen related receptors (errs): A new dawn in transcriptional control of mitochondrial gene networks. *Mitochondrion* **2011**, *11*, 544–552. [CrossRef] [PubMed]

51. Puigserver, P.; Rhee, J.; Lin, J.; Wu, Z.; Yoon, J.C.; Zhang, C.Y.; Krauss, S.; Mootha, V.K.; Lowell, B.B.; Spiegelman, B.M. Cytokine stimulation of energy expenditure through p38 map kinase activation of ppargamma coactivator-1. *Mol. Cell* **2001**, *8*, 971–982. [CrossRef]

52. Li, X.; Monks, B.; Ge, Q.; Birnbaum, M.J. Akt/pkb regulates hepatic metabolism by directly inhibiting PGC-1α transcription coactivator. *Nature* **2007**, *447*, 1012–1016. [CrossRef] [PubMed]

53. Teyssier, C.; Ma, H.; Emter, R.; Kralli, A.; Stallcup, M.R. Activation of nuclear receptor coactivator PGC-1α by arginine methylation. *Genes Dev.* **2005**, *19*, 1466–1473. [CrossRef] [PubMed]

54. Rodgers, J.T.; Lerin, C.; Haas, W.; Gygi, S.P.; Spiegelman, B.M.; Puigserver, P. Nutrient control of glucose homeostasis through a complex of PGC-1α and sirt1. *Nature* **2005**, *434*, 113–118. [CrossRef] [PubMed]

55. Kim, Y.; Triolo, M.; Hood, D.A. Impact of aging and exercise on mitochondrial quality control in skeletal muscle. *Oxid. Med. Cell. Longev.* **2017**, *2017*, 3165396. [CrossRef] [PubMed]

56. Pinter, K.; Grignani, R.T.; Watkins, H.; Redwood, C. Localisation of AMPK gamma subunits in cardiac and skeletal muscles. *J. Muscle Res. Cell Motil.* **2013**, *34*, 369–378. [CrossRef] [PubMed]

57. Burkewitz, K.; Zhang, Y.; Mair, W.B. AMPK at the nexus of energetics and aging. *Cell Metab.* **2014**, *20*, 10–25. [CrossRef] [PubMed]

58. Yin, Z.; Pascual, C.; Klionsky, D.J. Autophagy: Machinery and regulation. *Microb. Cell* **2016**, *3*, 588–596. [CrossRef] [PubMed]

59. Yang, Z.; Klionsky, D.J. Mammalian autophagy: Core molecular machinery and signaling regulation. *Curr. Opin. Cell Biol.* **2010**, *22*, 124–131. [CrossRef] [PubMed]

60. Singh, R.; Cuervo, A.M. Autophagy in the cellular energetic balance. *Cell Metab.* **2011**, *13*, 495–504. [CrossRef] [PubMed]

61. Itakura, E.; Mizushima, N. Characterization of autophagosome formation site by a hierarchical analysis of mammalian Atg proteins. *Autophagy* **2010**, *6*, 764–776. [CrossRef] [PubMed]

62. Koyama-Honda, I.; Itakura, E.; Fujiwara, T.K.; Mizushima, N. Temporal analysis of recruitment of mammalian Atg proteins to the autophagosome formation site. *Autophagy* **2013**, *9*, 1491–1499. [CrossRef] [PubMed]

63. Kaizuka, T.; Mizushima, N. Atg13 is essential for autophagy and cardiac development in mice. *Mol. Cell. Biol.* **2016**, *36*, 585–595. [CrossRef] [PubMed]

64. Park, J.M.; Jung, C.H.; Seo, M.; Otto, N.M.; Grunwald, D.; Kim, K.H.; Moriarity, B.; Kim, Y.M.; Starker, C.; Nho, R.S.; et al. The ULK1 complex mediates mtorc1 signaling to the autophagy initiation machinery via binding and phosphorylating ATG14. *Autophagy* **2016**, *12*, 547–564. [CrossRef] [PubMed]

65. Russell, R.C.; Tian, Y.; Yuan, H.; Park, H.W.; Chang, Y.Y.; Kim, J.; Kim, H.; Neufeld, T.P.; Dillin, A.; Guan, K.L. ULK1 induces autophagy by phosphorylating beclin-1 and activating VPS34 lipid kinase. *Nat. Cell Biol.* **2013**, *15*, 741–750. [CrossRef] [PubMed]

66. Hara, T.; Takamura, A.; Kishi, C.; Iemura, S.; Natsume, T.; Guan, J.L.; Mizushima, N. Fip200, a ulk-interacting protein, is required for autophagosome formation in mammalian cells. *J. Cell Biol.* **2008**, *181*, 497–510. [CrossRef] [PubMed]

67. Chan, E.Y.; Kir, S.; Tooze, S.A. Sirna screening of the kinome identifies ULK1 as a multidomain modulator of autophagy. *J. Biol. Chem.* **2007**, *282*, 25464–25474. [CrossRef] [PubMed]

68. Mizushima, N. The role of the Atg1/ULK1 complex in autophagy regulation. *Curr. Opin. Cell Biol.* **2010**, *22*, 132–139. [CrossRef] [PubMed]

69. Stjepanovic, G.; Baskaran, S.; Lin, M.G.; Hurley, J.H. Vps34 kinase domain dynamics regulate the autophagic pi 3-kinase complex. *Mol. Cell* **2017**, *67*, 528–534. [CrossRef] [PubMed]

70. Stjepanovic, G.; Baskaran, S.; Lin, M.G.; Hurley, J.H. Unveiling the role of vps34 kinase domain dynamics in regulation of the autophagic pi3k complex. *Mol. Cell Oncol.* **2017**, *4*, e1367873. [CrossRef] [PubMed]

71. Russell, R.C.; Yuan, H.X.; Guan, K.L. Autophagy regulation by nutrient signaling. *Cell Res.* **2014**, *24*, 42–57. [CrossRef] [PubMed]

72. Inoki, K.; Kim, J.; Guan, K.L. AMPK and mtor in cellular energy homeostasis and drug targets. *Annu. Rev. Pharmacol. Toxicol.* **2012**, *52*, 381–400. [CrossRef] [PubMed]

73. Kim, J.; Kundu, M.; Viollet, B.; Guan, K.L. AMPK and mtor regulate autophagy through direct phosphorylation of ULK1. *Nat. Cell Biol.* **2011**, *13*, 132–141. [CrossRef] [PubMed]

74. Hardie, D.G. Cell biology. Why starving cells eat themselves. *Science* **2011**, *331*, 410–411. [CrossRef] [PubMed]

75. Puente, C.; Hendrickson, R.C.; Jiang, X. Nutrient-regulated phosphorylation of ATG13 inhibits starvation-induced autophagy. *J. Biol. Chem.* **2016**, *291*, 6026–6035. [CrossRef] [PubMed]

76. Kamada, Y.; Yoshino, K.; Kondo, C.; Kawamata, T.; Oshiro, N.; Yonezawa, K.; Ohsumi, Y. Tor directly controls the Atg1 kinase complex to regulate autophagy. *Mol. Cell. Biol.* **2010**, *30*, 1049–1058. [CrossRef] [PubMed]

77. Dite, T.A.; Ling, N.X.Y.; Scott, J.W.; Hoque, A.; Galic, S.; Parker, B.L.; Ngoei, K.R.W.; Langendorf, C.G.; O'Brien, M.T.; Kundu, M.; et al. The autophagy initiator ulk1 sensitizes AMPK to allosteric drugs. *Nat. Commun.* **2017**, *8*, 571. [CrossRef] [PubMed]

78. Alers, S.; Loffler, A.S.; Wesselborg, S.; Stork, B. Role of AMPK-mtor-ULK1/2 in the regulation of autophagy: Cross talk, shortcuts, and feedbacks. *Mol. Cell. Biol.* **2012**, *32*, 2–11. [CrossRef] [PubMed]

79. Egan, D.F.; Shackelford, D.B.; Mihaylova, M.M.; Gelino, S.; Kohnz, R.A.; Mair, W.; Vasquez, D.S.; Joshi, A.; Gwinn, D.M.; Taylor, R.; et al. Phosphorylation of ULK1 (hATG1) by AMP-activated protein kinase connects energy sensing to mitophagy. *Science* **2011**, *331*, 456–461. [CrossRef] [PubMed]

80. Inoki, K.; Zhu, T.; Guan, K.L. Tsc2 mediates cellular energy response to control cell growth and survival. *Cell* **2003**, *115*, 577–590. [CrossRef]

81. Gwinn, D.M.; Shackelford, D.B.; Egan, D.F.; Mihaylova, M.M.; Mery, A.; Vasquez, D.S.; Turk, B.E.; Shaw, R.J. AMPK phosphorylation of raptor mediates a metabolic checkpoint. *Mol. Cell* **2008**, *30*, 214–226. [CrossRef] [PubMed]

82. Weerasekara, V.K.; Panek, D.J.; Broadbent, D.G.; Mortenson, J.B.; Mathis, A.D.; Logan, G.N.; Prince, J.T.; Thomson, D.M.; Thompson, J.W.; Andersen, J.L. Metabolic-stress-induced rearrangement of the 14-3-3ζ interactome promotes autophagy via a ULK1- and AMPK-regulated 14-3-3ζ interaction with phosphorylated Atg9. *Mol. Cell. Biol.* **2014**, *34*, 4379–4388. [CrossRef] [PubMed]

83. Kim, J.; Kim, Y.C.; Fang, C.; Russell, R.C.; Kim, J.H.; Fan, W.; Liu, R.; Zhong, Q.; Guan, K.L. Differential regulation of distinct vps34 complexes by AMPK in nutrient stress and autophagy. *Cell* **2013**, *152*, 290–303. [CrossRef] [PubMed]

84. Cheng, S.W.; Fryer, L.G.; Carling, D.; Shepherd, P.R. Thr2446 is a novel mammalian target of rapamycin (mtor) phosphorylation site regulated by nutrient status. *J. Biol. Chem.* **2004**, *279*, 15719–15722. [CrossRef] [PubMed]

85. Xu, D.Q.; Wang, Z.; Wang, C.Y.; Zhang, D.Y.; Wan, H.D.; Zhao, Z.L.; Gu, J.; Zhang, Y.X.; Li, Z.G.; Man, K.Y.; et al. PAQR3 controls autophagy by integrating AMPK signaling to enhance ATG14L-associated PI3K activity. *EMBO J.* **2016**, *35*, 496–514. [CrossRef] [PubMed]

86. Zhao, Y.; Wang, Q.; Qiu, G.; Zhou, S.; Jing, Z.; Wang, J.; Wang, W.; Cao, J.; Han, K.; Cheng, Q.; et al. RACK1 promotes autophagy by enhancing the Atg14l-beclin 1-Vps34-Vps15 complex formation upon phosphorylation by AMPK. *Cell Rep.* **2015**, *13*, 1407–1417. [CrossRef] [PubMed]

87. Huang, J.; Manning, B.D. The tsc1-tsc2 complex: A molecular switchboard controlling cell growth. *Biochem. J.* **2008**, *412*, 179–190. [CrossRef] [PubMed]

88. Kim, J.; Guan, K.L. AMPK connects energy stress to pik3c3/vps34 regulation. *Autophagy* **2013**, *9*, 1110–1111. [CrossRef] [PubMed]

89. Greer, E.L.; Oskoui, P.R.; Banko, M.R.; Maniar, J.M.; Gygi, M.P.; Gygi, S.P.; Brunet, A. The energy sensor AMP-activated protein kinase directly regulates the mammalian foxo3 transcription factor. *J. Biol. Chem.* **2007**, *282*, 30107–30119. [CrossRef] [PubMed]

90. Bowman, C.J.; Ayer, D.E.; Dynlacht, B.D. Foxk proteins repress the initiation of starvation-induced atrophy and autophagy programs. *Nat. Cell Biol.* **2014**, *16*, 1202–1214. [CrossRef] [PubMed]

91. Lapierre, L.R.; Kumsta, C.; Sandri, M.; Ballabio, A.; Hansen, M. Transcriptional and epigenetic regulation of autophagy in aging. *Autophagy* **2015**, *11*, 867–880. [CrossRef] [PubMed]

92. Young, N.P.; Kamireddy, A.; Van Nostrand, J.L.; Eichner, L.J.; Shokhirev, M.N.; Dayn, Y.; Shaw, R.J. AMPK governs lineage specification through tfeb-dependent regulation of lysosomes. *Genes Dev.* **2016**, *30*, 535–552. [CrossRef] [PubMed]

93.   Kim, S.H.; Kim, G.; Han, D.H.; Lee, M.; Kim, I.; Kim, B.; Kim, K.H.; Song, Y.M.; Yoo, J.E.; Wang, H.J.; et al. Ezetimibe ameliorates steatohepatitis via AMP activated protein kinase-tfeb-mediated activation of autophagy and nlrp3 inflammasome inhibition. *Autophagy* **2017**, *13*, 1767–1781. [CrossRef] [PubMed]

94.   Sakamaki, J.I.; Wilkinson, S.; Hahn, M.; Tasdemir, N.; O'Prey, J.; Clark, W.; Hedley, A.; Nixon, C.; Long, J.S.; New, M.; et al. Bromodomain protein brd4 is a transcriptional repressor of autophagy and lysosomal function. *Mol. Cell* **2017**, *66*, 517–532. [CrossRef] [PubMed]

95.   Lin, Y.Y.; Kiihl, S.; Suhail, Y.; Liu, S.Y.; Chou, Y.H.; Kuang, Z.; Lu, J.Y.; Khor, C.N.; Lin, C.L.; Bader, J.S.; et al. Functional dissection of lysine deacetylases reveals that hdac1 and p300 regulate AMPK. *Nature* **2012**, *482*, 251–255. [CrossRef] [PubMed]

96.   Dai, X.; Ding, Y.; Liu, Z.; Zhang, W.; Zou, M.H. Phosphorylation of chop (c/ebp homologous protein) by the AMP-activated protein kinase Alpha 1 in macrophages promotes chop degradation and reduces injury-induced neointimal disruption in vivo. *Circ. Res.* **2016**, *119*, 1089–1100. [CrossRef] [PubMed]

97.   Dai, S.; Tang, Z.; Cao, J.; Zhou, W.; Li, H.; Sampson, S.; Dai, C. Suppression of the hsf1-mediated proteotoxic stress response by the metabolic stress sensor AMPK. *EMBO J.* **2015**, *34*, 275–293. [CrossRef] [PubMed]

98.   Joo, M.S.; Kim, W.D.; Lee, K.Y.; Kim, J.H.; Koo, J.H.; Kim, S.G. AMPK facilitates nuclear accumulation of Nrf2 by phosphorylating at serine 550. *Mol. Cell. Biol.* **2016**, *36*, 1931–1942. [CrossRef] [PubMed]

99.   Jones, R.G.; Plas, D.R.; Kubek, S.; Buzzai, M.; Mu, J.; Xu, Y.; Birnbaum, M.J.; Thompson, C.B. Amp-activated protein kinase induces a p53-dependent metabolic checkpoint. *Mol. Cell* **2005**, *18*, 283–293. [CrossRef] [PubMed]

100. Adamovich, Y.; Adler, J.; Meltser, V.; Reuven, N.; Shaul, Y. AMPK couples p73 with p53 in cell fate decision. *Cell Death Differ.* **2014**, *21*, 1451–1459. [CrossRef] [PubMed]

101. Zhang, C.S.; Lin, S.C. AMPK promotes autophagy by facilitating mitochondrial fission. *Cell Metab.* **2016**, *23*, 399–401. [CrossRef] [PubMed]

102. Shirihai, O.S.; Song, M.; Dorn, G.W. How mitochondrial dynamism orchestrates mitophagy. *Circ. Res.* **2015**, *116*, 1835–1849. [CrossRef] [PubMed]

103. Youle, R.J.; van der Bliek, A.M. Mitochondrial fission, fusion, and stress. *Science* **2012**, *337*, 1062–1065. [CrossRef] [PubMed]

104. Toyama, E.Q.; Herzig, S.; Courchet, J.; Lewis, T.L., Jr.; Loson, O.C.; Hellberg, K.; Young, N.P.; Chen, H.; Polleux, F.; Chan, D.C.; et al. Metabolism. AMP-activated protein kinase mediates mitochondrial fission in response to energy stress. *Science* **2016**, *351*, 275–281. [CrossRef] [PubMed]

105. Otera, H.; Wang, C.; Cleland, M.M.; Setoguchi, K.; Yokota, S.; Youle, R.J.; Mihara, K. Mff is an essential factor for mitochondrial recruitment of Drp1 during mitochondrial fission in mammalian cells. *J. Cell Biol.* **2010**, *191*, 1141–1158. [CrossRef] [PubMed]

106. Wang, C.; Youle, R. Cell biology: Form follows function for mitochondria. *Nature* **2016**, *530*, 288–289. [CrossRef] [PubMed]

107. Laker, R.C.; Drake, J.C.; Wilson, R.J.; Lira, V.A.; Lewellen, B.M.; Ryall, K.A.; Fisher, C.C.; Zhang, M.; Saucerman, J.J.; Goodyear, L.J.; et al. AMPK phosphorylation of ulk1 is required for targeting of mitochondria to lysosomes in exercise-induced mitophagy. *Nat. Commun.* **2017**, *8*, 548. [CrossRef] [PubMed]

108. Kim, J.; Yang, G.; Kim, Y.; Kim, J.; Ha, J. AMPK activators: Mechanisms of action and physiological activities. *Exp. Mol. Med.* **2016**, *48*, e224. [CrossRef] [PubMed]

109. Schulten, H.J. Pleiotropic effects of metformin on cancer. *Int. J. Mol. Sci.* **2018**, *19*, 2850. [CrossRef] [PubMed]

7

Hypothalamic AMPK as a Mediator of Hormonal Regulation of Energy Balance

Baile Wang [1,2] and Kenneth King-Yip Cheng [3,*]

[1] State Key Laboratory of Pharmaceutical Biotechnology, The University of Hong Kong,
Hong Kong, China; blwong@connect.hku.hk
[2] Department of Medicine, The University of Hong Kong, Hong Kong, China
[3] Department of Health Technology and Informatics, The Hong Kong Polytechnic University,
Hong Kong, China
* Correspondence: kenneth.ky.cheng@polyu.edu.hk

Abstract: As a cellular energy sensor and regulator, adenosine monophosphate (AMP)-activated protein kinase (AMPK) plays a pivotal role in the regulation of energy homeostasis in both the central nervous system (CNS) and peripheral organs. Activation of hypothalamic AMPK maintains energy balance by inducing appetite to increase food intake and diminishing adaptive thermogenesis in adipose tissues to reduce energy expenditure in response to food deprivation. Numerous metabolic hormones, such as leptin, adiponectin, ghrelin and insulin, exert their energy regulatory effects through hypothalamic AMPK via integration with the neural circuits. Although activation of AMPK in peripheral tissues is able to promote fatty acid oxidation and insulin sensitivity, its chronic activation in the hypothalamus causes obesity by inducing hyperphagia in both humans and rodents. In this review, we discuss the role of hypothalamic AMPK in mediating hormonal regulation of feeding and adaptive thermogenesis, and summarize the diverse underlying mechanisms by which central AMPK maintains energy homeostasis.

Keywords: hypothalamus; adenosine monophosphate-activated protein kinase; adipose tissue; food intake; adaptive thermogenesis; beiging

1. The Hypothalamus and Energy Balance

The hypothalamus, a central integrator of the central nervous system (CNS), plays a critical role in the homeostatic regulation of appetite and energy expenditure by integrating hormonal, neuronal, and environmental signals [1]. It senses peripheral and central nutrients availability to modulate food intake and energy metabolism. Dysfunction of this highly-regulated system leads to energy imbalance, which initiates the development and progression of obesity and its related metabolic complications. There are different areas in the hypothalamus, which are believed to exert diverse functions in energy balance. Early in 1940, Hetherington and Ranson found that electrolytic lesions in the lateral hypothalamic area (LHA) cause inhibition of food intake, identifying LHA as a "feeding center" in the brain [2]. Subsequent studies showed that electrical stimulation of the LHA increases food intake [3], whereas lesions in the ventromedial hypothalamus (VMH) lead to the similar appetite-inducing effect [4]. Follow-up works demonstrated that not only lesions in the LHA and the VMH, but also disruption of other hypothalamic nuclei, including the arcuate nucleus (ARC), the dorsomedial hypothalamus (DMH), and the paraventricular nucleus (PVN), results in energy imbalance and obesity [5–8]. Among all of these regions, the ARC is critical in regulating feeding behavior and energy metabolism. The ARC is located near the median eminence (ME), which has abundant fenestrated capillaries that can lead to a 'penetrable' blood–brain barrier (BBB). The distinguished feature of the ME facilitates ARC neurons to sense hormonal and nutritional signals from the periphery [9], which is the reason why the ARC serves as the integration center of central and peripheral neural inputs.

The recent development of advanced techniques, including electrophysiology, optogenetics, and chemogenetics, enable us to identify distinct neuronal populations in the hypothalamus in rodents. The two best-studied neuronal populations in the ARC, which have opposite effects in appetite regulation, are the orexigenic Neuropeptide Y (NPY) and Agouti-related protein (AgRP) co-expressing neurons and the anorexigenic pro-opiomelanocortin (POMC) neurons. NPY/AgRP neurons are activated under a fasting condition, which drives hunger to promote food intake [10], whereas POMC neurons release alpha-melanocyte-stimulating hormone (α-MSH) to send satiety signals [11]. These two neuronal populations project to many second-order neurons in the PVN, VMH, DMH, and LHA [12–14]. The activity of these neurons is regulated by numerous neurotransmitters and/or hormones. For instance, the neurotransmitter serotonin exerts its anorexigenic effects by stimulating POMC neurons and suppressing NPY/AgRP neurons [15,16].

2. AMPK, an Energy Sensor and Regulator

AMP-activated protein kinase (AMPK), an evolutionarily conserved serine/threonine protein kinase, is a nutrient sensor that senses the ratio of AMP: adenosine triphosphate (ATP) or adenosine diphosphate (ADP): ATP to maintain energy balance in both peripheral tissues and the CNS. The heterotrimeric complex AMPK consists of three subunits, i.e., α catalytic subunit and β and γ regulatory subunits. Each subunit has several isoforms (α1, α2, β1, β2, γ1, γ2, γ3), suggesting 12 possibilities of heterotrimer combinations [17]. Some of these isoforms are tissue-specific and exert different functions under different physiological conditions [18,19]. For instance, heterotrimers containing the α1 isoform mainly exist in adipose tissues and the liver, whereas those containing α2 are predominantly expressed in skeletal muscles, the heart, and the brain [20,21]. The activity of AMPK can be regulated by both allosteric activation and phosphorylation at threonine 172 (Thr172) in the α-subunit. Specifically, allosteric activation is triggered by the increased intracellular AMP:ATP (or ADP:ATP) ratio, which facilitates the binding of AMP and/or ADP to the γ-subunit [22], while phosphorylation of AMPK is regulated by several upstream kinases, including liver kinase B1 (LKB1) [23,24], calcium-/calmodulin-dependent kinase kinase β (CaMKKβ) [25–27], TGFβ-activated kinase 1 (TAK1) [28,29], and the phosphatases including Mg^{2+}-/Mn^{2+}-dependent protein phosphatase 1E (PPM1E) [30], protein phosphatase 2A (PP2A), and protein phosphatase 2C (PP2C) [31]. The activated AMPK then shuts down ATP consumption and converts to ATP-producing pathways to stimulate carbohydrate and lipid metabolism by enhancing mitochondrial functions [21]. On the other hand, phosphorylation of AMPK at serine 485 (Ser485) in the α1 subunit or at serine 491 (Ser491) in the α2 subunit by protein kinase A (PKA) [32,33], autophosphorylation [32,34], or other protein kinases (such as Akt (also known as protein kinase B) [35,36] or the 70-kDa ribosomal protein S6 kinase (p70S6K) [37]) inhibits AMPK activity. The reduced AMPK activity in peripheral tissues, including liver, skeletal muscle, and adipose tissue causes glucose intolerance and lower exercise capacity, resulting in type 2 diabetes and obesity [21]. On the contrary, activation of AMPK by metabolic hormones, such as adiponectin and leptin, or a pharmacological compound, such as metformin, promotes insulin sensitivity and fatty acid oxidation in the peripheral tissues. Therefore, AMPK has been proposed as a promising drug target for obesity and type 2 diabetes [38,39].

3. Hypothalamic AMPK in the Regulation of Energy Balance

Apart from its crucial role in peripheral tissues, AMPK also plays a pivotal role in theCNS, especially in the hypothalamus. Activity of AMPK in the hypothalamus is induced by fasting but inhibited by feeding, hypothermia, and leptin, whereas high-fat diet (HFD) feeding blunts the leptin action and increases AMPK activity in the hypothalamus [40–43]. Specifically, AMPK activity is increased in AgRP-expressing neurons under fasting condition [44]. Hypothalamic AMPK modulates the functions of different neuronal populations (such as POMC and NPY/AgRP neurons), thereby controlling appetite and energy consumption to maintain energy homeostasis [45,46]. In addition, hypothalamic AMPK has been shown to control dietary selection, first- and second-phase insulin secretion, lipid metabolism, and hepatic gluconeogenesis, all of which are crucial for energy balance at the whole-body level [47–50]. Early studies revealed that pharmacological or adenovirus-mediated activation of AMPK in the medial hypothalamus significantly promotes food intake as a result of increased transcriptional levels of *NPY* and *AgRP* [51,52]. On the contrary, inhibition of AMPK by adenovirus-mediated overexpression of the dominant negative form of AMPK inhibits food intake [53]. Genetic-specific deletion of AMPKα2 in POMC neurons reduces energy expenditure and hence increases adiposity in mice, whereas deletion of this energy sensor in AgRP neurons prevents age-dependent obesity by promoting the anorexigenic effect of melanocortin [54]. Oh TS et al. recently demonstrated that AMPK regulates *NPY* and *POMC* transcription via autophagy in response to glucose deprivation in the mouse hypothalamic cell line [55]. A knockin mouse model with an activating mutation of AMPKγ2 (R302Q) gradually develops obesity due to elevated excitability of AgRP neurons and its associated hyperphagia [56]. Indeed, humans carrying this activating mutation have higher adiposity and dysregulated glucose balance [56]. Another protein-altering variant in AMPKγ1 has been recently shown to be associated with body mass index (BMI), which is identified by exome-targeted genotyping array [57]. Lentivirus-mediated overexpression of the constitutive active form of AMPK in corticotropin-releasing hormone (CRH) positive neurons in PVN leads to a food preference to a high carbohydrate diet over a HFD and obesity in mice [48]. In addition, AMPK activates the p21-activated kinase (PAK) signaling pathway in AgRP neurons, thereby mediating fasting-induced excitatory synaptic plasticity, neuronal activation, and feeding [44]. Apart from its direct action on POMC, NPY, and AgRP neurons, AMPK activity is also crucial to maintain excitatory synaptic input to AgRP neurons upon food deprivation [58]. In the following sections, we will discuss the key hormonal factors that positively or negatively regulate hypothalamic AMPK activity to control appetite and the underlying neuronal regulation.

4. Key Hormonal Factors That Regulate Food Intake via Hypothalamic AMPK

4.1. Leptin

Adipose tissue is an active and dynamic endocrine organ that secretes an array of hormones, bioactive peptides, and metabolites (collectively called adipokines), which control systemic energy, lipid and glucose homeostasis [59]. Leptin is the first identified adipokine that plays an indispensable role in controlling food intake and energy expenditure by mediating the crosstalk between adipose tissues and the hypothalamus [60]. The leptin receptor is abundantly expressed in POMC and NPY neurons in different regions of the hypothalamus [61–63]. Mutations in the *ob* gene (which encodes leptin) or the *leptin receptor* gene lead to severe obesity in humans and rodents mainly due to hyperphagia [64–66]. Leptin stimulates AMPK activation in skeletal muscle but reduces AMPK activity in the hypothalamus [38]. Noticeably, the inhibitory effect of leptin on AMPK activity is independent of the classic leptin signal transducer and activator of transcription 3 (STAT3) pathway [52]. The reduction of AMPK activity by leptin leads to an altered expression of neuropeptides, including NPY, AgRP, and α-MSH, in the ARC and the PVN [51,52]. Leptin selectively depolarizes POMC neurons and stimulates β-endorphin and α-MSH to inhibit AMPK activity [58,67,68]. AMPK also coordinates with other signaling networks, including mammalian target of rapamycin complex 1 (mTORC1) and phosphatidylinositol 3 kinase (PI3K), to fine-tune the hypothalamic actions of leptin [37,52,69]. For instance, PI3K-Akt-mTOR-p70S6K has been shown to phosphorylate AMPK at Ser485 and Ser491 in the hypothalamus upon leptin stimulation, which in turn reduces AMPK activity, leading to the inhibitory effect on food intake [37]. A more recent study also reports that leptin activates mTORC1 to repress AMPK activity via the PI3K-Akt axis [70]. Furthermore, the well-established downstream targets of AMPK in the peripheral tissues, such as acetyl-CoA carboxylase (ACC) and peroxisome proliferator-activated receptor gamma coactivator 1-alpha (PGC-1α), have been shown to mediate the hypothalamic function of AMPK [71,72]. Inhibition of AMPK by leptin increases the intracellular level of malonyl-CoA in the ARC and palmitoyl-CoA in the PVN through ACC [71]. Pharmacological blocking of the increase of these fatty acids attenuates leptin-induced suppression of food intake. A subsequent study demonstrated that inactivation of ACC by knocking in Serine 79 and Serine 212 with alanine in ACC impairs appetite in response to both fasting and cold in mice [73]. Genetic deletion of PGC-1α in AgRP neurons but not POMC neurons blunts the anorexigenic effect of leptin [72]. At the molecular level, knockdown of PGC-1α significantly reduces the mRNA level of *AgRP* in an AgRP-immortalized cell line under starvation but not fed state [72].

4.2. Adiponectin

Adiponectin is the most abundant adipokine that exerts multiple beneficial effects on the cardiometabolic system mainly via its insulin-sensitizing and anti-inflammatory properties [74,75]. In contrast to the increased level of leptin, the circulating level of adiponectin is reduced in humans with obesity and diabetes [76,77]. Adiponectin promotes glucose uptake and fatty acid oxidation in the skeletal muscle, suppresses glucose production in the liver, and induces vasorelaxation in the blood vessels [74,75]. These metabolic and vascular actions of adiponectin are largely mediated by AMPK [74,75]. Apart from its endocrine actions in the peripheral tissues, adiponectin also regulates feeding and energy expenditure via the hypothalamus [78]. Adiponectin can be detected in the cerebrospinal fluid (CSF) of mice after intravenous injection of recombinant full-length adiponectin, which promotes adaptive thermogenesis in brown adipose tissue (BAT) via the sympathetic nervous system (SNS)-uncoupling protein 1 (UCP1) axis [79]. Subsequent studies demonstrate that adiponectin is also detectable in human CSF, despite some studies having argued that adiponectin cannot pass through the blood–brain barrier [80–83]. In stark contrast to its abundant expression in circulation, only a trace amount of the trimeric and low-molecular-mass hexameric form (~0.1% of serum concentration), but no high-molecular-weight form, of adiponectin can be detected in CSF [80]. Importantly, the key signaling molecules (including the adiponectin receptors AdipoR1 and AdipoR2, the adaptor proteins containing an NH$_2$-terminal Bin/Amphiphiphysin/Rvs domain, a central pleckstrin homology domain, and a COOH-terminal phosphotyrosine binding domain (APPL)1 and APPL2) mediating adiponectin actions in peripheral tissues can also be detected in different regions of the hypothalamus [84–88]. With regard to feeding regulation, two early studies demonstrated opposite effects of adiponectin on food intake via distinct mechanisms in the hypothalamus [89,90]. The first study by Kubota et al. demonstrated that intravenous injection of full-length adiponectin increases AMPK activity in the hypothalamus, which in turn promotes food intake and decreases energy expenditure under a refeeding condition [89]. These adiponectin actions are abolished by siRNA-mediated knockdown expression of AdipoR1 or adenovirus-mediated overexpression of dominant negative AMPK. Genetic abrogation of adiponectin has a similar effect on hypothalamic AMPK activity and appetite. On the contrary, the study by Coope A et al. showed that intracerebroventricular (i.c.v.) injection of adiponectin reduces food intake via AdipoR1 in a fasted state [90]. Such change is accompanied by activations of insulin (increased phosphorylation of insulin receptor substrate 1[IRS1], Akt, and forkhead box protein O1 (FOXO1)) and leptin (STAT3 phosphorylation) signaling as well as an increase of AdipoR1-APPL1 interaction. Consistent with Coope A et al.'s study, a recent study demonstrated that i.c.v. injection of adiponectin decreases body weight as a consequence of reduced food consumption and increased adaptive thermogenesis in the BAT, and such effect of adiponectin is diminished in rats with a nutritional imbalance during their neonatal period [91]. On the other hand, the peroxisome proliferator-activated receptor (PPAR)γ agonist pioglitazone, a well-established insulin-sensitizing drug, boosts food intake and reduces energy expenditure by inducing adiponectin production in adipocytes, which in turn increases and decreases mRNA expression of *NPY* and *POMC* in the hypothalamus, respectively, via the AdipoR1-AMPK-dependent pathway [92]. Surprisingly, patch-clamp electrophysiology experiments reveal that adiponectin specifically depolarizes POMC neurons and inhibits NPY neurons in a

PI3K-dependent and AMPK-independent manner [93]. The discrepancy of adiponectin actions on food intake and hypothalamic neuronal activity may be due to the different nutritional states and concentrations of glucose used in the experiments. Indeed, two recent studies from Yada's research group show that adiponectin exerts opposite effects on feeding and POMC neuron activity under low and high glucose concentration, despite the fact that the suppressive effect of adiponectin on NPY neurons is independent of glucose [94,95]. In addition to its direct action on hypothalamic AMPK, several studies report that adiponectin is able to modulate the actions of insulin and leptin on the hypothalamus, thereby controlling energy homeostasis [89,90,93,96].

Recently, Okada-Iwabu et al. discovered an orally active synthetic small molecule of adiponectin receptor agonist (namely AdipoRon) [97]. Treatment with AdipoRon not only improves metabolic health but also prolongs the lifespan in obese and diabetic mouse models [97]. Intraperitoneal injection of AdipoRon attenuates corticosterone-induced body weight gain, depression, and neuroinflammation in mice, indicating that AdipoRon can penetrate and target the CNS [98]. Since AdipoRon has been proposed for the treatment of type 2 diabetes, it is, therefore, interesting to investigate whether AdipoRon has any effect on hypothalamic function, as adiponectin, in the regulation of feeding and energy expenditure.

4.3. Ghrelin

The stomach-derived hormone ghrelin, released during fasting, is the first circulating factor that has been reported to stimulate appetite in humans [99]. The orexigenic action of ghrelin is mediated by NPY and AgRP peptides [100]. Central or peripheral administration of ghrelin upregulates hypothalamic AMPK activity in both the ARC and the VMH in rats via growth hormone secretagogue receptor [51,101–103]. AMPK activation by ghrelin can be controlled at the transcriptional level by the transcriptional factor ALL1-fused gene from chromosome 4 (AF4), its upstream kinase CaMKKβ, the Sirtuin 1 (SIRT1)-p53 pathway, or glucose availability [104–107]. Inhibition of AMPK activity abolishes the orexigenic action of ghrelin [101,108–110]. On the contrary, knockin of an activating mutation in AMPKγ2 potentiates the orexigenic action of ghrelin under a refeeding condition [56]. There are multiple downstream targets of hypothalamic AMPK to mediate the orexigenic effect of ghrelin. First, ghrelin increases the release of intracellular Ca^{2+} to activate the CaMKKβ pathway, and, thus, facilitates AMPK phosphorylation in NPY neurons in the ARC [58,111,112]. Second, ghrelin activates AMPK and increases cytosolic Ca^{2+} in NPY neurons in the ARC [112]. Third, ghrelin triggers a hypothalamic mitochondrial function via uncoupling protein 2 (UCP2), which antagonizes the reactive oxidative species (ROS) production, allowing AMPK-mediated fatty acid oxidation for the support of synaptic plasticity and neuronal activation of NPY neurons [108]. Fourth, López et al. report that ghrelin inhibits fatty acid synthesis via the AMPK-ACC-dependent pathway, leading to reduced production of malonyl-CoA (the product of ACC), which in turn promotes carnitine palmitoyltransferase I (CPT1) activity in the mitochondria [101]. A subsequent study indicated that the regulation of ghrelin on fatty acid metabolism only occurs in the VMH but not the ARC [102]. Lastly, like leptin, ghrelin modulates the activity of presynaptic neurons that activate NPY neurons in AMPK-dependent and positive feedback loop manners [58].

4.4. Insulin

Insulin is exclusively produced by pancreatic β cells, and secreted in response to different nutrient stimuli, including glucose, fatty acid, and amino acids, after a meal. Apart from its glucose lowering and lipogenic actions, insulin also acts as an anorexigenic hormone. Insulin-deficient animals are hyperphagia, whereas their voracious appetite could be rectified by central administration of insulin [113,114]. Brain insulin resistance, a status in which neurons fail to respond to a physiological concentration of insulin, causes a dysregulation of energy homeostasis and cognitive functions [115,116]. The role of central insulin signaling in maintaining energy balance could be verified, at least in part, using neuron-specific insulin receptor knockout (NIRKO) mice. NIRKO mice have an elevated plasma insulin level, increased food consumption, and are susceptible to diet-induced obesity without alterations in brain development or neuronal survival [117]. In addition, i.c.v. injection of insulin [118–120] or insulin analogues [121] reduces both body weight and food intake, while intrahypothalamic infusion of an anti-insulin antibody results in opposite effects [122]. Insulin exerts a broad suppressive effect on AMPKα2 activity in different regions of the hypothalamus, including the PVN, the ARC and the LHA [52]. Indeed, the suppressive action of insulin on AMPKα2 activity is comparable to that of leptin [52]. Streptozotocin (STZ)-induced β cell loss and subsequent insulin deficiency lead to activation of AMPK and increase expression of NPY in the hypothalamus, resulting in hyperphagia in rats [53]. Insulin treatment reverses STZ-induced AMPK activation in the hypothalamus. Pharmacological or molecular inhibition of AMPK in the hypothalamus reverses STZ-induced hyperphagia [53]. Moreover, the inhibitory effect of insulin on AMPK activity and food intake can be further potentiated by i.c.v. injection of the amino acid taurine [123], and the effect of insulin on AMPK activation depends on the extracellular glucose concentration [124]. On the other hand, the anorexigenic action and the inhibitory effect of insulin on hypothalamic AMPK are largely abolished by cold exposure [125]. Apart from its direct action, Han et al. found that hypoglycemia triggered by insulin increases AMPKα2 activity in the hypothalamus [126]. This phenomenon was remarkable in hypothalamic ARC, VMH, and PVN [126]. Interestingly, insulin has been shown to inhibit AMPK activity by inducing phosphorylation of AMPK at Ser485 and Ser491 in skeletal muscle, ischemic heart, and hepatoma HepG2 cells in an Akt-dependent manner; however, whether insulin exerts a similar effect on hypothalamic AMPK phosphorylations in response to feeding remains elusive [35,127].

4.5. Glucagon-Like Peptide-1 (GLP-1)

GLP-1 is not only an incretin hormone secreted by intestinal L cells [128], but also a neuropeptide produced by preproglucagon neurons in the nucleus of the solitary tract (NTS) in the brainstem, which projects to hypothalamic nuclei to regulate appetite [129,130]. Hypothalamic GLP-1 level is reduced under fasting condition, while central administration of GLP-1 inhibits food intake in fasted rats [131,132]. This anorectic effect of GLP-1 is mediated by its inhibitory effect on fasting-induced hypothalamic AMPK activation [132,133]. HFD or central administration of fructose has been shown to inhibit the anorectic action of GLP-1 [134,135]. In addition, expression of the *proglucagon* gene (which encodes GLP-1) in the brain is regulated by transcription factor 7 like 2 (TCF7L2), which is associated with the risk of diabetes [136]. Transgenic overexpression of the dominant negative form of TCF7L2 driven by the proglucagon promoter represses the expression of GLP-1 in the brain, leading to a defective repression of AMPK activity in response to feeding. The defect can be reversed by treatment with the cyclic adenosine monophosphate (cAMP)-promoting agent forskolin, indicating that GLP-1 mediates its anorectic effect via the AMPK-PKA-cAMP axis [136]. Of note, this signaling axis also mediates the anorectic effect of the GLP-1 receptor in the hindbrain [137]. Similar to GLP-1, targeted injection of the GLP-1 receptor agonist liraglutide or exendin-4 into the VMH inhibits food intake in humans and rodents [138,139]. The anorexic effect of exendin-4 can be reversed by pre-injection with the AMPK activator AICAR in the VMH [139]. Further analysis reveals that mTOR but not ACC acts as a downstream mediator of AMPK for the hypophagic effect of exendin-4 [139].

5. The Role of Hypothalamic AMPK in the Regulation of Energy Expenditure

Total energy expenditure consists of basal metabolism, physical activity, and adaptive thermogenesis. Among these three components, adaptive thermogenesis in response to cold temperature or dietary intake is predominantly controlled by the hypothalamus. Adaptive thermogenesis is mainly mediated by BAT, which dissipates heat via the mitochondrial protein UCP1 in brown adipocytes. In the past few years, great advances have been made to broaden our knowledge on the inducible thermogenic adipocytes (beige adipocytes) in subcutaneous white adipose tissue (sWAT). Under certain circumstances (such as cold exposure, β-adrenergic stimulation, intermittent fasting, or exercise), beige adipocytes could be induced within WAT, especially in sWAT [140–144]. This process is called white fat beiging or browning, and is largely regulated by the crosstalk between the hypothalamus, the SNS, and adipose tissues. Beiging of WAT not only enhances energy expenditure, but also improves glucose metabolism, insulin sensitivity, and hyperlipidemia to ameliorate obesity and its related cardiometabolic complications [145–148]. The activation of BAT and beiging of WAT is, at least in part, controlled by the hypothalamus-SNS axis [149]. As the interscapular brown adipocytes only exist in human infants [150], and the gene expression profiles of the inducible UCP-1 positive cells in human adults share high similarities with mouse beige adipocytes rather than classical brown adipocytes [151,152], it is possible that induction of beige adipocytes in humans could be a potential therapeutic target for the prevention of obesity and its related metabolic syndromes.

An early study showed that whole-body depletion of AMPKα2 leads to elevated sympathetic activity and increased catecholamine secretion [153], suggesting the potential role of AMPK in beiging via the SNS. Indeed, emerging evidence suggests that numerous hormonal factors regulate adipose tissue beiging and adaptive thermogenesis via inhibition of hypothalamic AMPK activity, which will be further discussed in the following sections.

5.1. Leptin

As mentioned above, leptin is known to inhibit AMPK activity in the hypothalamus, which is accompanied by enhanced whole-body energy expenditure [52,60]. Mice with deletion of protein tyrosine phosphatase 1B (PTP1B), an inhibitor of leptin signaling, have diminished activation of hypothalamic AMPKα2, accompanied by an upregulation of UCP1 expression and mitochondrial density in BAT [154]. Central administration of leptin increases sympathetic outflow to adipose tissues via AMPKα2 [155]. Further studies demonstrated that sensitizing the leptin signaling in POMC neurons by deletion of PTP1B increases energy expenditure and promotes the conversion of WAT into BAT [156,157], although whether the inhibition of AMPK contributes to these changes is unknown. In addition, sympathetic denervation abolishes the potentiating effect of PTP1B deletion on WAT beiging [157]. These data collectively indicate that leptin regulates adaptive thermogenesis in adipose tissues via the SNS.

5.2. Thyroid Hormones

Thyroid hormones, including triiodothyronine (T3) and thyroxine (T4), have been found to raise energy expenditure via their peripheral actions in BAT or central action in the hypothalamus [158,159]. Stereotaxic injection of T3 (an active form of thyroid hormone) into the VMH (where AMPK and thyroid hormone receptors are highly co-expressed) stimulates SNS activity and BAT thermogenesis by inactivating hypothalamic AMPK via the thyroid hormone receptors [158]. Subsequent studies indicate that administration of T3 in the VMH but not the ARC is able to induce beiging in sWAT and thermogenesis via inhibition of hypothalamic AMPK activity [160,161]. Inactivation of the lipogenic pathway in the VMH attenuates the central action of T3 on BAT thermogenesis [158]. Genetic deletion of AMPKα1 in steroidogenic factor 1 (SF1) neurons in the VMH mimics the central effect of T3 on BAT metabolism [162]. Ablation of UCP1 completely abolishes the thermogenic action of central T3 administration [161]. At the molecular level, T3 relives endoplasmic reticulum (ER) stress and ceramide level in the VMH via an AMPK-dependent pathway, which has been shown to promote beiging in WAT and reduce obesity [162]. Substitution therapy with Levothroxin, a synthetic form of T4, promotes the basal metabolic rate and BAT activity in human subjects with a condition of hypothyroid state [163], but it remains unknown whether this is mediated by central or peripheral action of AMPK.

5.3. BMP8B

Bone morphogenetic protein 8B (BMP8B), a member of transforming growth factor β, acts both centrally and peripherally to increase BAT thermogenesis in female rodents [164]. mRNA of *BMP8B* can be detected in the brain and its receptors, including ALK4, ALK5, and ALK7, which are expressed in the VMH and the LHA. BMP8B-deficient mice display impaired thermogenesis and reduced AMPK activity in the VMH [164,165]. Acute i.c.v. injection of BMP8B in the VMH rather than the LHA enhances sympathetic outflow to BAT but not to the kidney, which can be abolished by expression of the constitutively active form of AMPK or potentiated by expression of the dominant negative form of AMPK in the VMH, respectively [164,165]. In addition, central administration of BMP8B exerts its thermogenic effect in BAT via upregulating orexin, a key modulator of BAT thermogenesis, in the LHA via glutamatergic signaling [165].

5.4. GLP-1

Activation of GLP-1 receptors in the hypothalamus not only inhibits appetite but also regulates BAT function. Stimulation of GLP-1 receptors by their agonist liraglutide in the VMH triggers both BAT thermogenesis and WAT beiging in mice, which are mediated by AMPK inhibition [138]. Central administration of the GLP-1 receptor agonist increases sympathetic outflow to BAT, leading to increased ability of glucose and lipid clearance and thermogenesis in BAT [166]. Although injection of liraglutide in DMH has no effect on BAT thermogenesis, injection of native GLP-1 in DMH increases the core body temperature and thermogenic program in BAT [138,167]. However, whether the thermogenic actions of native GLP-1 and exendin-4 are also mediated by AMPK signaling remains unclear, which warrants further investigation. With regard to human studies, the effect of GLP-1 receptor agonists and GLP-1 on energy expenditure remains inconclusive [168].

5.5. Estradiol

Estrogens are known to play a key role in the regulation of energy balance. The central action of estradiol on BAT thermogenesis has been recently identified and linked to the AMPK pathway [169,170]. Estradiol binds to its receptor (estrogen receptor α) in the VMH to diminish AMPK activity and enhances BAT thermogenesis without affecting feeding behavior [169]. Similar to the action of thyroid hormone, estradiol is able to relieve endoplasmic reticulum (ER) stress and reduce ceramide synthesis in the VMH, which in turn promotes BAT thermogenesis.

6. Conclusions and Future Perspectives

The diverse mechanisms driven by the hormonal factors convey on the hypothalamic AMPK signaling axis, supporting a critical role of AMPK in controlling feeding behavior and energy expenditure to maintain whole-body energy homeostasis (Table 1). AMPK in the VMH is crucial for BAT thermogenesis and beiging of sWAT, whereas AMPK in the ARC regulates food intake. Inhibition of AMPK activity by estradiol and thyroid hormone protects the hypothalamus from lipotoxicity and ER stress, which are the central pathogenic pathways that contribute to insulin and leptin resistance in obesity [171]. A recent study pinpointed that AMPK in SF1 neurons in the VMH regulates BAT thermogenesis via the SNS [172]; however, whether other hypothalamic neuronal population(s) mediates the inhibitory effect of AMPK activation on BAT functions remains unclear.

Considering the vital roles of hypothalamic AMPK, drugs that specifically target central AMPK are worth developing to prevent obesity and its related metabolic syndromes. As the regulatory effects of AMPK are differential in the periphery and centrally, the best therapeutic strategy is to specifically target hypothalamic AMPK without altering its functions in peripheral tissues. In this respect, the use of nanoparticles or exosomes [173], optogenetic neuromodulations [58], or chimeric proteins (targeting peptides associated with effective molecules or steroid hormones) [174,175], drawing from the implementations in other diseases, might be innovative strategies to achieve specific modulation of hypothalamic AMPK activity. However, despite the high specificity, we cannot exclude the possibility that these strategies may also affect other neuronal populations near the target hypothalamic region, which would result in limited efficacy and undesired side effects [176]. Another important issue is how to address the long-term influence of the altered hypothalamic AMPK activity. As AMPK is a canonical regulator of glucose and lipid metabolism, whether the sustained inhibition of hypothalamic AMPK may lead to lipotoxicity or other deleterious effects in neurons still needs further investigation. Taken together, great endeavors are required to advance our understanding of neuronal and hormonal regulation of hypothalamic AMPK, and AMPK in the hypothalamus will be a fascinating therapeutic target if we can address all of the above concerns properly.

Table 1. Actions of hormonal factors on hypothalamic AMPK activity, food intake, and energy expenditure.

Hormonal Factors	Hypothalamic AMPK Activity	Food Intake	Energy Expenditure
Adiponectin	↑	↑↓	↑↓
Ghrelin	↑	↑	-
Leptin	↓	↓	↑
Insulin	↓	↓	↑
GLP-1 and its analogues	↓	↓	↑
Thyroid hormones	↓	-	↑
BMP8B	↓	-	↑
Estradiol	↓	-	↑

References

1. Dietrich, M.O.; Horvath, T.L. Hypothalamic control of energy balance: Insights into the role of synaptic plasticity. *Trends Neurosci.* **2013**, *36*, 65–73. [CrossRef] [PubMed]
2. Hetherington, A.W.; Ranson, S.W. Hypothalamic lesions and adiposity in the rat. *Anat. Rec.* **1940**, *78*, 149–172. [CrossRef]
3. Anand, B.K.; Brobeck, J.R. Hypothalamic control of food intake in rats and cats. *Yale J. Biol. Med.* **1951**, *24*, 123–140. [PubMed]
4. Brobeck, J.R.; Tepperman, J.; Long, C. Experimental hypothalamic hyperphagia in the albino rat. *Yale J. Biol. Med.* **1943**, *15*, 831–853. [PubMed]
5. Bernardis, L.L. Disruption of diurnal feeding and weight gain cycles in weanling rats by ventromedial and dorsomedial hypothalamic lesions. *Physiol. Behav.* **1973**, *10*, 855–861. [CrossRef]
6. Leibowitz, S.F.; Hammer, N.J.; Chang, K. Hypothalamic paraventricular nucleus lesions produce overeating and obesity in the rat. *Physiol. Behav.* **1981**, *27*, 1031–1040. [CrossRef]
7. Fukushima, M.; Tokunaga, K.; Lupien, J.; Kemnitz, J.; Bray, G. Dynamic and static phases of obesity following lesions in PVN and VMH. *Am. J. Physiol. Regul. Integr. Comp. Physiol.* **1987**, *253*, R523–R529. [CrossRef] [PubMed]
8. Choi, S.; Dallman, M.F. Hypothalamic Obesity: Multiple Routes Mediated by Loss of Function in Medial Cell Groups 1. *Endocrinology* **1999**, *140*, 4081–4088. [CrossRef] [PubMed]
9. Rodriguez, E.M.; Blazquez, J.L.; Guerra, M. The design of barriers in the hypothalamus allows the median eminence and the arcuate nucleus to enjoy private milieus: The former opens to the portal blood and the latter to the cerebrospinal fluid. *Peptides* **2010**, *31*, 757–776. [CrossRef] [PubMed]
10. Mizuno, T.M.; Makimura, H.; Silverstein, J.; Roberts, J.L.; Lopingco, T.; Mobbs, C.V. Fasting regulates hypothalamic neuropeptide Y., agouti-related peptide, and proopiomelanocortin in diabetic mice independent of changes in leptin or insulin. *Endocrinology* **1999**, *140*, 4551–4557. [CrossRef] [PubMed]
11. Biebermann, H.; Kühnen, P.; Kleinau, G.; Krude, H. *Appetite Control*; The Neuroendocrine Circuitry Controlled by POMC, MSH, and AGRP; Springer: Berlin, Germany, 2012; pp. 47–75.
12. Bagnol, D.; Lu, X.Y.; Kaelin, C.B.; Day, H.E.; Ollmann, M.; Gantz, I.; Akil, H.; Barsh, G.S.; Watson, S.J. Anatomy of an endogenous antagonist: Relationship between Agouti-related protein and proopiomelanocortin in brain. *J. Neurosci.* **1999**, *19*, RC26. [CrossRef] [PubMed]

13. Kleinridders, A.; Konner, A.C.; Bruning, J.C. CNS-targets in control of energy and glucose homeostasis. *Curr. Opin. Pharmacol.* **2009**, *9*, 794–804. [CrossRef] [PubMed]

14. Waterson, M.J.; Horvath, T.L. Neuronal Regulation of Energy Homeostasis: Beyond the Hypothalamus and Feeding. *Cell Metab.* **2015**, *22*, 962–970. [CrossRef] [PubMed]

15. Sohn, J.W.; Xu, Y.; Jones, J.E.; Wickman, K.; Williams, K.W.; Elmquist, J.K. Serotonin 2C receptor activates a distinct population of arcuate pro-opiomelanocortin neurons via TRPC channels. *Neuron* **2011**, *71*, 488–497. [CrossRef] [PubMed]

16. Heisler, L.K.; Jobst, E.E.; Sutton, G.M.; Zhou, L.; Borok, E.; Thornton-Jones, Z.; Liu, H.Y.; Zigman, J.M.; Balthasar, N.; Kishi, T.; et al. Serotonin reciprocally regulates melanocortin neurons to modulate food intake. *Neuron* **2006**, *51*, 239–249. [CrossRef] [PubMed]

17. Hardie, D.G. AMPK–sensing energy while talking to other signaling pathways. *Cell Metab.* **2014**, *20*, 939–952. [CrossRef] [PubMed]

18. Dasgupta, B.; Chhipa, R.R. Evolving Lessons on the Complex Role of AMPK in Normal Physiology and Cancer. *Trends Pharmacol. Sci.* **2016**, *37*, 192–206. [CrossRef] [PubMed]

19. Ross, F.A.; Jensen, T.E.; Hardie, D.G. Differential regulation by AMP and ADP of AMPK complexes containing different gamma subunit isoforms. *Biochem. J.* **2016**, *473*, 189–199. [CrossRef] [PubMed]

20. O'Neill, H.M. AMPK and Exercise: Glucose Uptake and Insulin Sensitivity. *Diabetes Metab. J.* **2013**, *37*, 1–21. [CrossRef] [PubMed]

21. Steinberg, G.R.; Kemp, B.E. AMPK in Health and Disease. *Physiol. Rev.* **2009**, *89*, 1025–1078. [CrossRef] [PubMed]

22. Hardie, D.G.; Ross, F.A.; Hawley, S.A. AMPK: A nutrient and energy sensor that maintains energy homeostasis. *Nat. Rev. Mol. Cell Biol.* **2012**, *13*, 251–262. [CrossRef] [PubMed]

23. Woods, A.; Johnstone, S.R.; Dickerson, K.; Leiper, F.C.; Fryer, L.G.; Neumann, D.; Schlattner, U.; Wallimann, T.; Carlson, M.; Carling, D. LKB1 is the upstream kinase in the AMP-activated protein kinase cascade. *Curr. Biol.* **2003**, *13*, 2004–2008. [CrossRef] [PubMed]

24. Shaw, R.J.; Kosmatka, M.; Bardeesy, N.; Hurley, R.L.; Witters, L.A.; DePinho, R.A.; Cantley, L.C. The tumor suppressor LKB1 kinase directly activates AMP-activated kinase and regulates apoptosis in response to energy stress. *Proc. Natl. Acad. Sci USA* **2004**, *101*, 3329–3335. [CrossRef] [PubMed]

25. Woods, A.; Dickerson, K.; Heath, R.; Hong, S.P.; Momcilovic, M.; Johnstone, S.R.; Carlson, M.; Carling, D. Ca^{2+}/calmodulin-dependent protein kinase kinase-beta acts upstream of AMP-activated protein kinase in mammalian cells. *Cell Metab.* **2005**, *2*, 21–33. [CrossRef] [PubMed]

26. Hawley, S.A.; Pan, D.A.; Mustard, K.J.; Ross, L.; Bain, J.; Edelman, A.M.; Frenguelli, B.G.; Hardie, D.G. Calmodulin-dependent protein kinase kinase-beta is an alternative upstream kinase for AMP-activated protein kinase. *Cell Metab.* **2005**, *2*, 9–19. [CrossRef] [PubMed]

27. Hurley, R.L.; Anderson, K.A.; Franzone, J.M.; Kemp, B.E.; Means, A.R.; Witters, L.A. The Ca^{2+}/calmodulin-dependent protein kinase kinases are AMP-activated protein kinase kinases. *J. Biol. Chem.* **2005**, *280*, 29060–29066. [CrossRef] [PubMed]

28. Momcilovic, M.; Hong, S.-P.; Carlson, M. Mammalian TAK1 activates Snf1 protein kinase in yeast and phosphorylates AMP-activated protein kinase in vitro. *J. Biol. Chem.* **2006**, *281*, 25336–25343. [CrossRef] [PubMed]

29. Xie, M.; Zhang, D.; Dyck, J.R.; Li, Y.; Zhang, H.; Morishima, M.; Mann, D.L.; Taffet, G.E.; Baldini, A.; Khoury, D.S. A pivotal role for endogenous TGF-β-activated kinase-1 in the LKB1/AMP-activated protein kinase energy-sensor pathway. *Proc. Natl. Acad. Sci. USA* **2006**, *103*, 17378–17383. [CrossRef] [PubMed]

30. Voss, M.; Paterson, J.; Kelsall, I.R.; Martin-Granados, C.; Hastie, C.J.; Peggie, M.W.; Cohen, P.T. Ppm1E is an in cellulo AMP-activated protein kinase phosphatase. *Cell Signal.* **2011**, *23*, 114–124. [CrossRef] [PubMed]

31. Davies, S.P.; Helps, N.R.; Cohen, P.T.; Hardie, D.G. 5′-AMP inhibits dephosphorylation, as well as promoting phosphorylation, of the AMP-activated protein kinase. Studies using bacterially expressed human protein phosphatase-2C alpha and native bovine protein phosphatase-2AC. *FEBS Lett.* **1995**, *377*, 421–425. [CrossRef] [PubMed]

32. Hurley, R.L.; Barre, L.K.; Wood, S.D.; Anderson, K.A.; Kemp, B.E.; Means, A.R.; Witters, L.A. Regulation of AMP-activated protein kinase by multisite phosphorylation in response to agents that elevate cellular cAMP. *J. Biol. Chem.* **2006**, *281*, 36662–36672. [CrossRef] [PubMed]

33. Pulinilkunnil, T.; He, H.; Kong, D.; Asakura, K.; Peroni, O.D.; Lee, A.; Kahn, B.B. Adrenergic regulation of AMP-activated protein kinase in brown adipose tissue in vivo. *J. Biol. Chem.* **2011**, *286*, 8798–8809. [CrossRef] [PubMed]

34. Hawley, S.A.; Ross, F.A.; Gowans, G.J.; Tibarewal, P.; Leslie, N.R.; Hardie, D.G. Phosphorylation by Akt within the ST loop of AMPK-alpha1 down-regulates its activation in tumour cells. *Biochem. J.* **2014**, *459*, 275–287. [CrossRef] [PubMed]

35. Horman, S.; Vertommen, D.; Heath, R.; Neumann, D.; Mouton, V.; Woods, A.; Schlattner, U.; Wallimann, T.; Carling, D.; Hue, L.; et al. Insulin antagonizes ischemia-induced Thr172 phosphorylation of AMP-activated protein kinase alpha-subunits in heart via hierarchical phosphorylation of Ser485/491. *J. Biol. Chem.* **2006**, *281*, 5335–5340. [CrossRef] [PubMed]

36. Ning, J.; Xi, G.; Clemmons, D.R. Suppression of AMPK activation via S485 phosphorylation by IGF-I during hyperglycemia is mediated by AKT activation in vascular smooth muscle cells. *Endocrinology* **2011**, *152*, 3143–3154. [CrossRef] [PubMed]

37. Dagon, Y.; Hur, E.; Zheng, B.; Wellenstein, K.; Cantley, L.C.; Kahn, B.B. p70S6 kinase phosphorylates AMPK on serine 491 to mediate leptin's effect on food intake. *Cell Metab.* **2012**, *16*, 104–112. [CrossRef] [PubMed]

38. Minokoshi, Y.; Kim, Y.B.; Peroni, O.D.; Fryer, L.G.; Muller, C.; Carling, D.; Kahn, B.B. Leptin stimulates fatty-acid oxidation by activating AMP-activated protein kinase. *Nature* **2002**, *415*, 339–343. [CrossRef] [PubMed]

39. Yamauchi, T.; Kamon, J.; Minokoshi, Y.; Ito, Y.; Waki, H.; Uchida, S.; Yamashita, S.; Noda, M.; Kita, S.; Ueki, K.; et al. Adiponectin stimulates glucose utilization and fatty-acid oxidation by activating AMP-activated protein kinase. *Nat. Med.* **2002**, *8*, 1288–1295. [CrossRef] [PubMed]

40. Martin, T.L.; Alquier, T.; Asakura, K.; Furukawa, N.; Preitner, F.; Kahn, B.B. Diet-induced obesity alters AMP kinase activity in hypothalamus and skeletal muscle. *J. Biol. Chem.* **2006**, *281*, 18933–18941. [CrossRef] [PubMed]

41. Cavaliere, G.; Viggiano, E.; Trinchese, G.; De Filippo, C.; Messina, A.; Monda, V.; Valenzano, A.; Cincione, R.I.; Zammit, C.; Cimmino, F.; et al. Long Feeding High-Fat Diet Induces Hypothalamic Oxidative Stress and Inflammation, and Prolonged Hypothalamic AMPK Activation in Rat Animal Model. *Front. Physiol.* **2018**, *9*, 818. [CrossRef] [PubMed]

42. McCrimmon, R.J.; Fan, X.; Ding, Y.; Zhu, W.; Jacob, R.J.; Sherwin, R.S. Potential role for AMP-activated protein kinase in hypoglycemia sensing in the ventromedial hypothalamus. *Diabetes* **2004**, *53*, 1953–1958. [CrossRef] [PubMed]

43. Cao, C.; Gao, T.; Cheng, M.; Xi, F.; Zhao, C.; Yu, W. Mild hypothermia ameliorates muscle wasting in septic rats associated with hypothalamic AMPK-induced autophagy and neuropeptides. *Biochem. Biophys. Res. Commun.* **2017**, *490*, 882–888. [CrossRef] [PubMed]

44. Kong, D.; Dagon, Y.; Campbell, J.N.; Guo, Y.; Yang, Z.; Yi, X.; Aryal, P.; Wellenstein, K.; Kahn, B.B.; Sabatini, B.L.; et al. A Postsynaptic AMPK → p21-Activated Kinase Pathway Drives Fasting-Induced Synaptic Plasticity in AgRP Neurons. *Neuron* **2016**, *91*, 25–33. [CrossRef] [PubMed]

45. Huynh, M.K.; Kinyua, A.W.; Yang, D.J.; Kim, K.W. Hypothalamic AMPK as a Regulator of Energy Homeostasis. *Neural Plast.* **2016**, *2016*, 2754078. [CrossRef] [PubMed]

46. Lopez, M.; Nogueiras, R.; Tena-Sempere, M.; Dieguez, C. Hypothalamic AMPK: A canonical regulator of whole-body energy balance. *Nat. Rev. Endocrinol.* **2016**, *12*, 421–432. [CrossRef] [PubMed]

47. Yang, C.S.; Lam, C.K.; Chari, M.; Cheung, G.W.; Kokorovic, A.; Gao, S.; Leclerc, I.; Rutter, G.A.; Lam, T.K. Hypothalamic AMP-activated protein kinase regulates glucose production. *Diabetes* **2010**, *59*, 2435–2443. [CrossRef] [PubMed]

48. Okamoto, S.; Sato, T.; Tateyama, M.; Kageyama, H.; Maejima, Y.; Nakata, M.; Hirako, S.; Matsuo, T.; Kyaw, S.; Shiuchi, T.; et al. Activation of AMPK-Regulated CRH Neurons in the PVH is Sufficient and Necessary to Induce Dietary Preference for Carbohydrate over Fat. *Cell Rep.* **2018**, *22*, 706–721. [CrossRef] [PubMed]

49. Kume, S.; Kondo, M.; Maeda, S.; Nishio, Y.; Yanagimachi, T.; Fujita, Y.; Haneda, M.; Kondo, K.; Sekine, A.; Araki, S.I.; et al. Hypothalamic AMP-Activated Protein Kinase Regulates Biphasic Insulin Secretion from Pancreatic beta Cells during Fasting and in Type 2 Diabetes. *eBioMedicine* **2016**, *13*, 168–180. [CrossRef] [PubMed]

50. Park, S.; Kim, D.S.; Kang, S.; Shin, B.K. Chronic activation of central AMPK attenuates glucose-stimulated insulin secretion and exacerbates hepatic insulin resistance in diabetic rats. *Brain Res. Bull.* **2014**, *108*, 18–26. [CrossRef] [PubMed]

51. Andersson, U.; Filipsson, K.; Abbott, C.R.; Woods, A.; Smith, K.; Bloom, S.R.; Carling, D.; Small, C.J. AMP-activated protein kinase plays a role in the control of food intake. *J. Biol. Chem.* **2004**, *279*, 12005–12008. [CrossRef] [PubMed]

52. Minokoshi, Y.; Alquier, T.; Furukawa, N.; Kim, Y.B.; Lee, A.; Xue, B.; Mu, J.; Foufelle, F.; Ferre, P.; Birnbaum, M.J.; et al. AMP-kinase regulates food intake by responding to hormonal and nutrient signals in the hypothalamus. *Nature* **2004**, *428*, 569–574. [CrossRef] [PubMed]

53. Namkoong, C.; Kim, M.S.; Jang, P.G.; Han, S.M.; Park, H.S.; Koh, E.H.; Lee, W.J.; Kim, J.Y.; Park, I.S.; Park, J.Y. Enhanced hypothalamic AMP-activated protein kinase activity contributes to hyperphagia in diabetic rats. *Diabetes* **2005**, *54*, 63–68. [CrossRef] [PubMed]

54. Claret, M.; Smith, M.A.; Batterham, R.L.; Selman, C.; Choudhury, A.I.; Fryer, L.G.; Clements, M.; Al-Qassab, H.; Heffron, H.; Xu, A.W.; et al. AMPK is essential for energy homeostasis regulation and glucose sensing by POMC and AgRP neurons. *J. Clin. Investig.* **2007**, *117*, 2325–2336. [CrossRef] [PubMed]

55. Oh, T.S.; Cho, H.; Cho, J.H.; Yu, S.W.; Kim, E.K. Hypothalamic AMPK-induced autophagy increases food intake by regulating NPY and POMC expression. *Autophagy* **2016**, *12*, 2009–2025. [CrossRef] [PubMed]

56. Yavari, A.; Stocker, C.J.; Ghaffari, S.; Wargent, E.T.; Steeples, V.; Czibik, G.; Pinter, K.; Bellahcene, M.; Woods, A.; Martinez de Morentin, P.B.; et al. Chronic Activation of gamma2 AMPK Induces Obesity and Reduces beta Cell Function. *Cell Metab.* **2016**, *23*, 821–836. [CrossRef] [PubMed]

57. Turcot, V.; Lu, Y.; Highland, H.M.; Schurmann, C.; Justice, A.E.; Fine, R.S.; Bradfield, J.P.; Esko, T.; Giri, A.; Graff, M.; et al. Protein-altering variants associated with body mass index implicate pathways that control energy intake and expenditure in obesity. *Nat. Genet.* **2018**, *50*, 26–41. [CrossRef] [PubMed]

58. Yang, Y.; Atasoy, D.; Su, H.H.; Sternson, S.M. Hunger states switch a flip-flop memory circuit via a synaptic AMPK-dependent positive feedback loop. *Cell* **2011**, *146*, 992–1003. [CrossRef] [PubMed]

59. Kwok, K.H.; Lam, K.S.; Xu, A. Heterogeneity of white adipose tissue: Molecular basis and clinical implications. *Exp. Mol. Med.* **2016**, *48*, e215. [CrossRef] [PubMed]

60. Friedman, J.M.; Halaas, J.L. Leptin and the regulation of body weight in mammals. *Nature* **1998**, *395*, 763–770. [CrossRef] [PubMed]

61. Cheung, C.C.; Clifton, D.K.; Steiner, R.A. Proopiomelanocortin neurons are direct targets for leptin in the hypothalamus. *Endocrinology* **1997**, *138*, 4489–4492. [CrossRef] [PubMed]

62. Hakansson, M.L.; Brown, H.; Ghilardi, N.; Skoda, R.C.; Meister, B. Leptin receptor immunoreactivity in chemically defined target neurons of the hypothalamus. *J. Neurosci.* **1998**, *18*, 559–572. [CrossRef] [PubMed]

63. Mercer, J.G.; Hoggard, N.; Williams, L.M.; Lawrence, C.B.; Hannah, L.T.; Morgan, P.J.; Trayhurn, P. Coexpression of leptin receptor and prepropeptide Y mRNA in arcuate nucleus of mouse hypothalamus. *J. Neuroendocrinol.* **1996**, *8*, 733–735. [CrossRef] [PubMed]

64. Pelleymounter, M.A.; Cullen, M.J.; Baker, M.B.; Hecht, R. Effects of the obese gene product on body weight regulation in ob/ob mice. *Science* **1995**, *269*, 540–543. [CrossRef] [PubMed]

65. Chen, H.; Charlat, O.; Tartaglia, L.A.; Woolf, E.A.; Weng, X.; Ellis, S.J.; Lakey, N.D.; Culpepper, J.; Moore, K.J.; Breitbart, R.E.; et al. Evidence that the diabetes gene encodes the leptin receptor: Identification of a mutation in the leptin receptor gene in db/db mice. *Cell* **1996**, *84*, 491–495. [CrossRef]

66. Clement, K.; Vaisse, C.; Lahlou, N.; Cabrol, S.; Pelloux, V.; Cassuto, D.; Gourmelen, M.; Dina, C.; Chambaz, J.; Lacorte, J.M.; et al. A mutation in the human leptin receptor gene causes obesity and pituitary dysfunction. *Nature* **1998**, *392*, 398–401. [CrossRef] [PubMed]

67. Cowley, M.A.; Smart, J.L.; Rubinstein, M.; Cerdan, M.G.; Diano, S.; Horvath, T.L.; Cone, R.D.; Low, M.J. Leptin activates anorexigenic POMC neurons through a neural network in the arcuate nucleus. *Nature* **2001**, *411*, 480–484. [CrossRef] [PubMed]

68. Poleni, P.E.; Akieda-Asai, S.; Koda, S.; Sakurai, M.; Bae, C.R.; Senba, K.; Cha, Y.S.; Furuya, M.; Date, Y. Possible involvement of melanocortin-4-receptor and AMP-activated protein kinase in the interaction of glucagon-like peptide-1 and leptin on feeding in rats. *Biochem. Biophys. Res. Commun.* **2012**, *420*, 36–41. [CrossRef] [PubMed]

69. Cota, D.; Matter, E.K.; Woods, S.C.; Seeley, R.J. The role of hypothalamic mammalian target of rapamycin complex 1 signaling in diet-induced obesity. *J. Neurosci.* **2008**, *28*, 7202–7208. [CrossRef] [PubMed]
70. Watterson, K.R.; Bestow, D.; Gallagher, J.; Hamilton, D.L.; Ashford, F.B.; Meakin, P.J.; Ashford, M.L. Anorexigenic and orexigenic hormone modulation of mammalian target of rapamycin complex 1 activity and the regulation of hypothalamic agouti-related protein mRNA expression. *Neurosignals* **2013**, *21*, 28–41. [CrossRef] [PubMed]
71. Gao, S.; Kinzig, K.P.; Aja, S.; Scott, K.A.; Keung, W.; Kelly, S.; Strynadka, K.; Chohnan, S.; Smith, W.W.; Tamashiro, K.L.; et al. Leptin activates hypothalamic acetyl-CoA carboxylase to inhibit food intake. *Proc. Natl. Acad. Sci. USA* **2007**, *104*, 17358–17363. [CrossRef] [PubMed]
72. Gill, J.F.; Delezie, J.; Santos, G.; Handschin, C. PGC-1alpha expression in murine AgRP neurons regulates food intake and energy balance. *Mol. Metab.* **2016**, *5*, 580–588. [CrossRef] [PubMed]
73. Galic, S.; Loh, K.; Murray-Segal, L.; Steinberg, G.R.; Andrews, Z.B.; Kemp, B.E. AMPK signaling to acetyl-CoA carboxylase is required for fasting- and cold-induced appetite but not thermogenesis. *eLife* **2018**, *7*. [CrossRef] [PubMed]
74. Wang, Z.V.; Scherer, P.E. Adiponectin, the past two decades. *J. Mol. Cell Biol.* **2016**, *8*, 93–100. [CrossRef] [PubMed]
75. Cheng, K.K.; Lam, K.S.; Wang, B.; Xu, A. Signaling mechanisms underlying the insulin-sensitizing effects of adiponectin. *Best Pract. Res. Clin. Endocrinol. Metab.* **2014**, *28*, 3–13. [CrossRef] [PubMed]
76. Hotta, K.; Funahashi, T.; Arita, Y.; Takahashi, M.; Matsuda, M.; Okamoto, Y.; Iwahashi, H.; Kuriyama, H.; Ouchi, N.; Maeda, K.; et al. Plasma concentrations of a novel, adipose-specific protein, adiponectin, in type 2 diabetic patients. *Arterioscler. Thromb. Vasc. Biol.* **2000**, *20*, 1595–1599. [CrossRef] [PubMed]
77. Weyer, C.; Funahashi, T.; Tanaka, S.; Hotta, K.; Matsuzawa, Y.; Pratley, R.E.; Tataranni, P.A. Hypoadiponectinemia in obesity and type 2 diabetes: Close association with insulin resistance and hyperinsulinemia. *J. Clin. Endocrinol. Metab.* **2001**, *86*, 1930–1935. [CrossRef] [PubMed]
78. Thundyil, J.; Pavlovski, D.; Sobey, C.G.; Arumugam, T.V. Adiponectin receptor signalling in the brain. *Br. J. Pharmacol.* **2012**, *165*, 313–327. [CrossRef] [PubMed]
79. Qi, Y.; Takahashi, N.; Hileman, S.M.; Patel, H.R.; Berg, A.H.; Pajvani, U.B.; Scherer, P.E.; Ahima, R.S. Adiponectin acts in the brain to decrease body weight. *Nat. Med.* **2004**, *10*, 524–529. [CrossRef] [PubMed]
80. Kusminski, C.M.; McTernan, P.G.; Schraw, T.; Kos, K.; O'Hare, J.P.; Ahima, R.; Kumar, S.; Scherer, P.E. Adiponectin complexes in human cerebrospinal fluid: Distinct complex distribution from serum. *Diabetologia* **2007**, *50*, 634–642. [CrossRef] [PubMed]
81. Neumeier, M.; Weigert, J.; Buettner, R.; Wanninger, J.; Schaffler, A.; Muller, A.M.; Killian, S.; Sauerbruch, S.; Schlachetzki, F.; Steinbrecher, A.; et al. Detection of adiponectin in cerebrospinal fluid in humans. *Am. J. Physiol. Endocrinol. Metab.* **2007**, *293*, E965–E969. [CrossRef] [PubMed]
82. Spranger, J.; Verma, S.; Gohring, I.; Bobbert, T.; Seifert, J.; Sindler, A.L.; Pfeiffer, A.; Hileman, S.M.; Tschop, M.; Banks, W.A. Adiponectin does not cross the blood-brain barrier but modifies cytokine expression of brain endothelial cells. *Diabetes* **2006**, *55*, 141–147. [CrossRef] [PubMed]
83. Pan, W.; Tu, H.; Kastin, A.J. Differential BBB interactions of three ingestive peptides: Obestatin, ghrelin, and adiponectin. *Peptides* **2006**, *27*, 911–916. [CrossRef] [PubMed]
84. Klein, I.; Sanchez-Alavez, M.; Tabarean, I.; Schaefer, J.; Holmberg, K.H.; Klaus, J.; Xia, F.; Marcondes, M.C.; Dubins, J.S.; Morrison, B.; et al. AdipoR1 and 2 are expressed on warm sensitive neurons of the hypothalamic preoptic area and contribute to central hyperthermic effects of adiponectin. *Brain Res.* **2011**, *1423*, 1–9. [CrossRef] [PubMed]
85. Guillod-Maximin, E.; Roy, A.F.; Vacher, C.; Aubourg, A.; Bailleux, V.; Lorsignol, A.; Pénicaud, L.; Parquet, M.; Taouis, M. Adiponectin receptors are expressed in hypothalamus and colocalized with proopiomelanocortin and neuropeptide Y in rodent arcuate neurons. *J. Endocrinol.* **2009**, *200*, 93–105. [CrossRef] [PubMed]
86. Benomar, Y.; Amine, H.; Crepin, D.; Al Rifai, S.; Riffault, L.; Gertler, A.; Taouis, M. Central Resistin/TLR4 Impairs Adiponectin Signaling, Contributing to Insulin and FGF21 Resistance. *Diabetes* **2016**, *65*, 913–926. [CrossRef] [PubMed]
87. Wang, B.; Li, A.; Li, X.; Ho, P.W.; Wu, D.; Wang, X.; Liu, Z.; Wu, K.K.; Yau, S.S.; Xu, A.; et al. Activation of hypothalamic RIP-Cre neurons promotes beiging of WAT via sympathetic nervous system. *EMBO Rep.* **2018**, *19*. [CrossRef] [PubMed]

88. Cheng, K.K.; Lam, K.S.; Wang, Y.; Huang, Y.; Carling, D.; Wu, D.; Wong, C.; Xu, A. Adiponectin-induced endothelial nitric oxide synthase activation and nitric oxide production are mediated by APPL1 in endothelial cells. *Diabetes* **2007**, *56*, 1387–1394. [CrossRef] [PubMed]

89. Kubota, N.; Yano, W.; Kubota, T.; Yamauchi, T.; Itoh, S.; Kumagai, H.; Kozono, H.; Takamoto, I.; Okamoto, S.; Shiuchi, T.; et al. Adiponectin stimulates AMP-activated protein kinase in the hypothalamus and increases food intake. *Cell Metab.* **2007**, *6*, 55–68. [CrossRef] [PubMed]

90. Coope, A.; Milanski, M.; Araujo, E.P.; Tambascia, M.; Saad, M.J.; Geloneze, B.; Velloso, L.A. AdipoR1 mediates the anorexigenic and insulin/leptin-like actions of adiponectin in the hypothalamus. *FEBS Lett.* **2008**, *582*, 1471–1476. [CrossRef] [PubMed]

91. Halah, M.P.; Marangon, P.B.; Antunes-Rodrigues, J.; Elias, L.L.K. Neonatal nutritional programming impairs adiponectin effects on energy homeostasis in adult life of male rats. *Am. J. Physiol. Endocrinol. Metab.* **2018**, *315*, E29–E37. [CrossRef] [PubMed]

92. Quaresma, P.G.; Reencober, N.; Zanotto, T.M.; Santos, A.C.; Weissmann, L.; de Matos, A.H.; Lopes-Cendes, I.; Folli, F.; Saad, M.J.; Prada, P.O. Pioglitazone treatment increases food intake and decreases energy expenditure partially via hypothalamic adiponectin/adipoR1/AMPK pathway. *Int. J. Obes.* **2016**, *40*, 138–146. [CrossRef] [PubMed]

93. Sun, J.; Gao, Y.; Yao, T.; Huang, Y.; He, Z.; Kong, X.; Yu, K.J.; Wang, R.T.; Guo, H.; Yan, J.; et al. Adiponectin potentiates the acute effects of leptin in arcuate Pomc neurons. *Mol. Metab.* **2016**, *5*, 882–891. [CrossRef] [PubMed]

94. Suyama, S.; Maekawa, F.; Maejima, Y.; Kubota, N.; Kadowaki, T.; Yada, T. Glucose level determines excitatory or inhibitory effects of adiponectin on arcuate POMC neuron activity and feeding. *Sci. Rep.* **2016**, *6*, 30796. [CrossRef] [PubMed]

95. Suyama, S.; Lei, W.; Kubota, N.; Kadowaki, T.; Yada, T. Adiponectin at physiological level glucose-independently enhances inhibitory postsynaptic current onto NPY neurons in the hypothalamic arcuate nucleus. *Neuropeptides* **2017**, *65*, 1–9. [CrossRef] [PubMed]

96. Wang, C.; Mao, X.; Wang, L.; Liu, M.; Wetzel, M.D.; Guan, K.L.; Dong, L.Q.; Liu, F. Adiponectin sensitizes insulin signaling by reducing p70 S6 kinase-mediated serine phosphorylation of IRS-1. *J. Biol. Chem.* **2007**, *282*, 7991–7996. [CrossRef] [PubMed]

97. Okada-Iwabu, M.; Yamauchi, T.; Iwabu, M.; Honma, T.; Hamagami, K.; Matsuda, K.; Yamaguchi, M.; Tanabe, H.; Kimura-Someya, T.; Shirouzu, M.; et al. A small-molecule AdipoR agonist for type 2 diabetes and short life in obesity. *Nature* **2013**, *503*, 493–499. [CrossRef] [PubMed]

98. Nicolas, S.; Debayle, D.; Bechade, C.; Maroteaux, L.; Gay, A.S.; Bayer, P.; Heurteaux, C.; Guyon, A.; Chabry, J. Adiporon, an adiponectin receptor agonist acts as an antidepressant and metabolic regulator in a mouse model of depression. *Transl. Psychiatry* **2018**, *8*, 159. [CrossRef] [PubMed]

99. Wren, A.; Seal, L.; Cohen, M.; Brynes, A.; Frost, G.; Murphy, K.; Dhillo, W.; Ghatei, M.; Bloom, S. Ghrelin enhances appetite and increases food intake in humans. *J. Clin. Endocrinol. Metab.* **2001**, *86*, 5992–5995. [CrossRef] [PubMed]

100. Chen, H.Y.; Trumbauer, M.E.; Chen, A.S.; Weingarth, D.T.; Adams, J.R.; Frazier, E.G.; Shen, Z.; Marsh, D.J.; Feighner, S.D.; Guan, X.M.; et al. Orexigenic action of peripheral ghrelin is mediated by neuropeptide Y and agouti-related protein. *Endocrinology* **2004**, *145*, 2607–2612. [CrossRef] [PubMed]

101. López, M.; Lage, R.; Saha, A.K.; Pérez-Tilve, D.; Vázquez, M.J.; Varela, L.; Sangiao-Alvarellos, S.; Tovar, S.; Raghay, K.; Rodríguez-Cuenca, S. Hypothalamic fatty acid metabolism mediates the orexigenic action of ghrelin. *Cell Metab.* **2008**, *7*, 389–399. [CrossRef] [PubMed]

102. Gao, S.; Casals, N.; Keung, W.; Moran, T.H.; Lopaschuk, G.D. Differential effects of central ghrelin on fatty acid metabolism in hypothalamic ventral medial and arcuate nuclei. *Physiol. Behav.* **2013**, *118*, 165–170. [CrossRef] [PubMed]

103. Lim, C.T.; Kola, B.; Feltrin, D.; Perez-Tilve, D.; Tschop, M.H.; Grossman, A.B.; Korbonits, M. Ghrelin and cannabinoids require the ghrelin receptor to affect cellular energy metabolism. *Mol. Cell. Endocrinol.* **2013**, *365*, 303–308. [CrossRef] [PubMed]

104. Anderson, K.A.; Ribar, T.J.; Lin, F.; Noeldner, P.K.; Green, M.F.; Muehlbauer, M.J.; Witters, L.A.; Kemp, B.E.; Means, A.R. Hypothalamic CaMKK2 contributes to the regulation of energy balance. *Cell Metab.* **2008**, *7*, 377–388. [CrossRef] [PubMed]

105. Komori, T.; Doi, A.; Nosaka, T.; Furuta, H.; Akamizu, T.; Kitamura, T.; Senba, E.; Morikawa, Y. Regulation of AMP-activated protein kinase signaling by AFF4 protein, member of AF4 (ALL1-fused gene from chromosome 4) family of transcription factors, in hypothalamic neurons. *J. Biol. Chem.* **2012**, *287*, 19985–19996. [CrossRef] [PubMed]

106. Velásquez, D.A.; Martínez, G.; Romero, A.; Vázquez, M.J.; Boit, K.D.; Dopeso-Reyes, I.G.; López, M.; Vidal, A.; Nogueiras, R.; Diéguez, C. The central sirtuin1/p53 pathway is essential for the orexigenic action of ghrelin. *Diabetes* **2011**, DB_100802.

107. Lockie, S.H.; Stark, R.; Mequinion, M.; Ch'ng, S.; Kong, D.; Spanswick, D.C.; Lawrence, A.J.; Andrews, Z.B. Glucose Availability Predicts the Feeding Response to Ghrelin in Male Mice, an Effect Dependent on AMPK in AgRP Neurons. *Endocrinology* **2018**, *159*, 3605–3614. [CrossRef] [PubMed]

108. Andrews, Z.B.; Liu, Z.W.; Walllingford, N.; Erion, D.M.; Borok, E.; Friedman, J.M.; Tschöp, M.H.; Shanabrough, M.; Cline, G.; Shulman, G.I. UCP2 mediates ghrelin's action on NPY/AgRP neurons by lowering free radicals. *Nature* **2008**, *454*, 846–851. [CrossRef] [PubMed]

109. Kola, B.; Hubina, E.; Tucci, S.A.; Kirkham, T.C.; Garcia, E.A.; Mitchell, S.E.; Williams, L.M.; Hawley, S.A.; Hardie, D.G.; Grossman, A.B. Cannabinoids and ghrelin have both central and peripheral metabolic and cardiac effects via AMP-activated protein kinase. *J. Biol. Chem.* **2005**, *280*, 25196–25201. [CrossRef] [PubMed]

110. Wren, A.; Small, C.; Ward, H.; Murphy, K.; Dakin, C.; Taheri, S.; Kennedy, A.; Roberts, G.; Morgan, D.; Ghatei, M. The novel hypothalamic peptide ghrelin stimulates food intake and growth hormone secretion. *Endocrinology* **2000**, *141*, 4325–4328. [CrossRef] [PubMed]

111. Andrews, Z.B. Central mechanisms involved in the orexigenic actions of ghrelin. *Peptides* **2011**, *32*, 2248–2255. [CrossRef] [PubMed]

112. Kohno, D.; Sone, H.; Minokoshi, Y.; Yada, T. Ghrelin raises $[Ca^{2+}]$ i via AMPK in hypothalamic arcuate nucleus NPY neurons. *Biochem. Biophys. Res. Commun.* **2008**, *366*, 388–392. [CrossRef] [PubMed]

113. Sipols, A.J.; Baskin, D.G.; Schwartz, M.W. Effect of intracerebroventricular insulin infusion on diabetic hyperphagia and hypothalamic neuropeptide gene expression. *Diabetes* **1995**, *44*, 147–151. [CrossRef] [PubMed]

114. Schwartz, M.W.; Figlewicz, D.P.; Baskin, D.G.; Woods, S.C.; Porte, D., Jr. Insulin in the brain: A hormonal regulator of energy balance. *Endocr. Rev.* **1992**, *13*, 387–414. [PubMed]

115. Arnold, S.E.; Arvanitakis, Z.; Macauley-Rambach, S.L.; Koenig, A.M.; Wang, H.Y.; Ahima, R.S.; Craft, S.; Gandy, S.; Buettner, C.; Stoeckel, L.E.; et al. Brain insulin resistance in type 2 diabetes and Alzheimer disease: Concepts and conundrums. *Nat. Rev. Neurol.* **2018**, *14*, 168–181. [CrossRef] [PubMed]

116. Cetinkalp, S.; Simsir, I.Y.; Ertek, S. Insulin resistance in brain and possible therapeutic approaches. *Curr. Vasc. Pharmacol.* **2014**, *12*, 553–564. [CrossRef] [PubMed]

117. Bruning, J.C.; Gautam, D.; Burks, D.J.; Gillette, J.; Schubert, M.; Orban, P.C.; Klein, R.; Krone, W.; Muller-Wieland, D.; Kahn, C.R. Role of brain insulin receptor in control of body weight and reproduction. *Science* **2000**, *289*, 2122–2125. [CrossRef] [PubMed]

118. Woods, S.C.; Lotter, E.C.; McKay, L.D.; Porte, D., Jr. Chronic intracerebroventricular infusion of insulin reduces food intake and body weight of baboons. *Nature* **1979**, *282*, 503–505. [CrossRef] [PubMed]

119. Woods, S.C.; Seeley, R.J.; Porte, D., Jr.; Schwartz, M.W. Signals that regulate food intake and energy homeostasis. *Science* **1998**, *280*, 1378–1383. [CrossRef] [PubMed]

120. Richardson, R.D.; Ramsay, D.S.; Lernmark, A.; Scheurink, A.J.; Baskin, D.G.; Woods, S.C. Weight loss in rats following intraventricular transplants of pancreatic islets. *Am. J. Physiol.* **1994**, *266*, R59–R64. [CrossRef] [PubMed]

121. Air, E.L.; Strowski, M.Z.; Benoit, S.C.; Conarello, S.L.; Salituro, G.M.; Guan, X.M.; Liu, K.; Woods, S.C.; Zhang, B.B. Small molecule insulin mimetics reduce food intake and body weight and prevent development of obesity. *Nat. Med.* **2002**, *8*, 179–183. [CrossRef] [PubMed]

122. McGowan, M.K.; Andrews, K.M.; Grossman, S.P. Chronic intrahypothalamic infusions of insulin or insulin antibodies alter body weight and food intake in the rat. *Physiol. Behav.* **1992**, *51*, 753–766. [CrossRef]

123. Solon, C.S.; Franci, D.; Ignacio-Souza, L.M.; Romanatto, T.; Roman, E.A.; Arruda, A.P.; Morari, J.; Torsoni, A.S.; Carneiro, E.M.; Velloso, L.A. Taurine enhances the anorexigenic effects of insulin in the hypothalamus of rats. *Amino Acids* **2012**, *42*, 2403–2410. [CrossRef] [PubMed]

124. Cai, F.; Gyulkhandanyan, A.V.; Wheeler, M.B.; Belsham, D.D. Glucose regulates AMP-activated protein kinase activity and gene expression in clonal, hypothalamic neurons expressing proopiomelanocortin: Additive effects of leptin or insulin. *J. Endocrinol.* **2007**, *192*, 605–614. [CrossRef] [PubMed]

125. Roman, E.A.; Cesquini, M.; Stoppa, G.R.; Carvalheira, J.B.; Torsoni, M.A.; Velloso, L.A. Activation of AMPK in rat hypothalamus participates in cold-induced resistance to nutrient-dependent anorexigenic signals. *J. Physiol.* **2005**, *568*, 993–1001. [CrossRef] [PubMed]

126. Han, S.M.; Namkoong, C.; Jang, P.G.; Park, I.S.; Hong, S.W.; Katakami, H.; Chun, S.; Kim, S.W.; Park, J.Y.; Lee, K.U.; et al. Hypothalamic AMP-activated protein kinase mediates counter-regulatory responses to hypoglycaemia in rats. *Diabetologia* **2005**, *48*, 2170–2178. [CrossRef] [PubMed]

127. Valentine, R.J.; Coughlan, K.A.; Ruderman, N.B.; Saha, A.K. Insulin inhibits AMPK activity and phosphorylates AMPK Ser(4)(8)(5)/(4)(9)(1) through Akt in hepatocytes, myotubes and incubated rat skeletal muscle. *Arch. Biochem. Biophys.* **2014**, *562*, 62–69. [CrossRef] [PubMed]

128. Lim, G.E.; Brubaker, P.L. Glucagon-like peptide 1 secretion by the L-cell: The view from within. *Diabetes* **2006**, *55*, S70–S77. [CrossRef]

129. Goldstone, A.P.; Morgan, I.; Mercer, J.G.; Morgan, D.G.; Moar, K.M.; Ghatei, M.A.; Bloom, S.R. Effect of leptin on hypothalamic GLP-1 peptide and brain-stem pre-proglucagon mRNA. *Biochem. Biophys. Res. Commun.* **2000**, *269*, 331–335. [CrossRef] [PubMed]

130. Trapp, S.; Richards, J.E. The gut hormone glucagon-like peptide-1 produced in brain: Is this physiologically relevant? *Curr. Opin. Pharmacol.* **2013**, *13*, 964–969. [CrossRef] [PubMed]

131. Turton, M.D.; O'Shea, D.; Gunn, I.; Beak, S.A.; Edwards, C.M.; Meeran, K.; Choi, S.J.; Taylor, G.M.; Heath, M.M.; Lambert, P.D.; et al. A role for glucagon-like peptide-1 in the central regulation of feeding. *Nature* **1996**, *379*, 69–72. [CrossRef] [PubMed]

132. Seo, S.; Ju, S.; Chung, H.; Lee, D.; Park, S. Acute effects of glucagon-like peptide-1 on hypothalamic neuropeptide and AMP activated kinase expression in fasted rats. *Endocr. J.* **2008**, *55*, 867–874. [CrossRef] [PubMed]

133. Hurtado-Carneiro, V.; Sanz, C.; Roncero, I.; Vazquez, P.; Blazquez, E.; Alvarez, E. Glucagon-like peptide 1 (GLP-1) can reverse AMP-activated protein kinase (AMPK) and S6 kinase (P70S6K) activities induced by fluctuations in glucose levels in hypothalamic areas involved in feeding behaviour. *Mol. Neurobiol.* **2012**, *45*, 348–361. [CrossRef] [PubMed]

134. Williams, D.L.; Hyvarinen, N.; Lilly, N.; Kay, K.; Dossat, A.; Parise, E.; Torregrossa, A.M. Maintenance on a high-fat diet impairs the anorexic response to glucagon-like-peptide-1 receptor activation. *Physiol. Behav.* **2011**, *103*, 557–564. [CrossRef] [PubMed]

135. Burmeister, M.A.; Ayala, J.; Drucker, D.J.; Ayala, J.E. Central glucagon-like peptide 1 receptor-induced anorexia requires glucose metabolism-mediated suppression of AMPK and is impaired by central fructose. *Am. J. Physiol. Endocrinol. Metab.* **2013**, *304*, E677–E685. [CrossRef] [PubMed]

136. Shao, W.; Wang, D.; Chiang, Y.T.; Ip, W.; Zhu, L.; Xu, F.; Columbus, J.; Belsham, D.D.; Irwin, D.M.; Zhang, H.; et al. The Wnt signaling pathway effector TCF7L2 controls gut and brain proglucagon gene expression and glucose homeostasis. *Diabetes* **2013**, *62*, 789–800. [CrossRef] [PubMed]

137. Hayes, M.R.; Leichner, T.M.; Zhao, S.; Lee, G.S.; Chowansky, A.; Zimmer, D.; De Jonghe, B.C.; Kanoski, S.E.; Grill, H.J.; Bence, K.K. Intracellular signals mediating the food intake-suppressive effects of hindbrain glucagon-like peptide-1 receptor activation. *Cell Metab.* **2011**, *13*, 320–330. [CrossRef] [PubMed]

138. Beiroa, D.; Imbernon, M.; Gallego, R.; Senra, A.; Herranz, D.; Villarroya, F.; Serrano, M.; Ferno, J.; Salvador, J.; Escalada, J.; et al. GLP-1 agonism stimulates brown adipose tissue thermogenesis and browning through hypothalamic AMPK. *Diabetes* **2014**, *63*, 3346–3358. [CrossRef] [PubMed]

139. Burmeister, M.A.; Brown, J.D.; Ayala, J.E.; Stoffers, D.A.; Sandoval, D.A.; Seeley, R.J.; Ayala, J.E. The glucagon-like peptide-1 receptor in the ventromedial hypothalamus reduces short-term food intake in male mice by regulating nutrient sensor activity. *Am. J. Physiol. Endocrinol. Metab.* **2017**, *313*, E651–E662. [CrossRef] [PubMed]

140. Barbatelli, G.; Murano, I.; Madsen, L.; Hao, Q.; Jimenez, M.; Kristiansen, K.; Giacobino, J.; De Matteis, R.; Cinti, S. The emergence of cold-induced brown adipocytes in mouse white fat depots is determined predominantly by white to brown adipocyte transdifferentiation. *Am. J. Physiol. Endocrinol. Metab.* **2010**, *298*, E1244–E1253. [CrossRef] [PubMed]

141. Rosen, E.D.; Spiegelman, B.M. What we talk about when we talk about fat. *Cell* **2014**, *156*, 20–44. [CrossRef] [PubMed]

142. Li, G.; Xie, C.; Lu, S.; Nichols, R.G.; Tian, Y.; Li, L.; Patel, D.; Ma, Y.; Brocker, C.N.; Yan, T.; et al. Intermittent Fasting Promotes White Adipose Browning and Decreases Obesity by Shaping the Gut Microbiota. *Cell Metab.* **2017**, *26*, 672–685. [CrossRef] [PubMed]

143. Stanford, K.I.; Middelbeek, R.J.; Goodyear, L.J. Exercise Effects on White Adipose Tissue: Beiging and Metabolic Adaptations. *Diabetes* **2015**, *64*, 2361–2368. [CrossRef] [PubMed]

144. Bartelt, A.; Bruns, O.T.; Reimer, R.; Hohenberg, H.; Ittrich, H.; Peldschus, K.; Kaul, M.G.; Tromsdorf, U.I.; Weller, H.; Waurisch, C.; et al. Brown adipose tissue activity controls triglyceride clearance. *Nat. Med.* **2011**, *17*, 200–205. [CrossRef] [PubMed]

145. Seale, P.; Conroe, H.M.; Estall, J.; Kajimura, S.; Frontini, A.; Ishibashi, J.; Cohen, P.; Cinti, S.; Spiegelman, B.M. Prdm16 determines the thermogenic program of subcutaneous white adipose tissue in mice. *J. Clin. Investig.* **2011**, *121*, 96–105. [CrossRef] [PubMed]

146. Liu, W.; Bi, P.; Shan, T.; Yang, X.; Yin, H.; Wang, Y.-X.; Liu, N.; Rudnicki, M.A.; Kuang, S. miR-133a regulates adipocyte browning in vivo. *PLoS Genet.* **2013**, *9*, e1003626. [CrossRef] [PubMed]

147. Castillo-Quan, J.I. From white to brown fat through the PGC-1α-dependent myokine irisin: Implications for diabetes and obesity. *Dis. Model. Mech.* **2012**, *5*, 293–295. [CrossRef] [PubMed]

148. Bi, P.; Shan, T.; Liu, W.; Yue, F.; Yang, X.; Liang, X.R.; Wang, J.; Li, J.; Carlesso, N.; Liu, X.; et al. Inhibition of Notch signaling promotes browning of white adipose tissue and ameliorates obesity. *Nat. Med.* **2014**, *20*, 911–918. [CrossRef] [PubMed]

149. Contreras, C.; Nogueiras, R.; Dieguez, C.; Medina-Gomez, G.; Lopez, M. Hypothalamus and thermogenesis: Heating the BAT, browning the WAT. *Mol. Cell Endocrinol.* **2016**, *438*, 107–115. [CrossRef] [PubMed]

150. Lidell, M.E.; Betz, M.J.; Dahlqvist Leinhard, O.; Heglind, M.; Elander, L.; Slawik, M.; Mussack, T.; Nilsson, D.; Romu, T.; Nuutila, P.; et al. Evidence for two types of brown adipose tissue in humans. *Nat. Med.* **2013**, *19*, 631–634. [CrossRef] [PubMed]

151. Wu, J.; Bostrom, P.; Sparks, L.M.; Ye, L.; Choi, J.H.; Giang, A.H.; Khandekar, M.; Virtanen, K.A.; Nuutila, P.; Schaart, G.; et al. Beige adipocytes are a distinct type of thermogenic fat cell in mouse and human. *Cell* **2012**, *150*, 366–376. [CrossRef] [PubMed]

152. Harms, M.; Seale, P. Brown and beige fat: Development, function and therapeutic potential. *Nat. Med.* **2013**, *19*, 1252–1263. [CrossRef] [PubMed]

153. Viollet, B.; Andreelli, F.; Jorgensen, S.B.; Perrin, C.; Geloen, A.; Flamez, D.; Mu, J.; Lenzner, C.; Baud, O.; Bennoun, M.; et al. The AMP-activated protein kinase alpha2 catalytic subunit controls whole-body insulin sensitivity. *J. Clin. Investig.* **2003**, *111*, 91–98. [CrossRef] [PubMed]

154. Xue, B.; Pulinilkunnil, T.; Murano, I.; Bence, K.K.; He, H.; Minokoshi, Y.; Asakura, K.; Lee, A.; Haj, F.; Furukawa, N.; et al. Neuronal protein tyrosine phosphatase 1B deficiency results in inhibition of hypothalamic AMPK and isoform-specific activation of AMPK in peripheral tissues. *Mol. Cell. Biol.* **2009**, *29*, 4563–4573. [CrossRef] [PubMed]

155. Tanida, M.; Yamamoto, N.; Shibamoto, T.; Rahmouni, K. Involvement of hypothalamic AMP-activated protein kinase in leptin-induced sympathetic nerve activation. *PLoS ONE* **2013**, *8*, e56660. [CrossRef] [PubMed]

156. Banno, R.; Zimmer, D.; De Jonghe, B.C.; Atienza, M.; Rak, K.; Yang, W.; Bence, K.K. PTP1B and SHP2 in POMC neurons reciprocally regulate energy balance in mice. *J. Clin. Investig.* **2010**, *120*, 720–734. [CrossRef] [PubMed]

157. Dodd, G.T.; Decherf, S.; Loh, K.; Simonds, S.E.; Wiede, F.; Balland, E.; Merry, T.L.; Munzberg, H.; Zhang, Z.Y.; Kahn, B.B.; et al. Leptin and insulin act on POMC neurons to promote the browning of white fat. *Cell* **2015**, *160*, 88–104. [CrossRef] [PubMed]

158. Lopez, M.; Varela, L.; Vazquez, M.J.; Rodriguez-Cuenca, S.; Gonzalez, C.R.; Velagapudi, V.R.; Morgan, D.A.; Schoenmakers, E.; Agassandian, K.; Lage, R.; et al. Hypothalamic AMPK and fatty acid metabolism mediate thyroid regulation of energy balance. *Nat. Med.* **2010**, *16*, 1001–1008. [CrossRef] [PubMed]

159. Sjogren, M.; Alkemade, A.; Mittag, J.; Nordstrom, K.; Katz, A.; Rozell, B.; Westerblad, H.; Arner, A.; Vennstrom, B. Hypermetabolism in mice caused by the central action of an unliganded thyroid hormone receptor alpha1. *EMBO J.* **2007**, *26*, 4535–4545. [CrossRef] [PubMed]

160. Martinez-Sanchez, N.; Moreno-Navarrete, J.M.; Contreras, C.; Rial-Pensado, E.; Ferno, J.; Nogueiras, R.; Dieguez, C.; Fernandez-Real, J.M.; Lopez, M. Thyroid hormones induce browning of white fat. *J. Endocrinol.* **2017**, *232*, 351–362. [CrossRef] [PubMed]

161. Alvarez-Crespo, M.; Csikasz, R.I.; Martinez-Sanchez, N.; Dieguez, C.; Cannon, B.; Nedergaard, J.; Lopez, M. Essential role of UCP1 modulating the central effects of thyroid hormones on energy balance. *Mol. Metab.* **2016**, *5*, 271–282. [CrossRef] [PubMed]

162. Martinez-Sanchez, N.; Seoane-Collazo, P.; Contreras, C.; Varela, L.; Villarroya, J.; Rial-Pensado, E.; Buque, X.; Aurrekoetxea, I.; Delgado, T.C.; Vazquez-Martinez, R.; et al. Hypothalamic AMPK-ER Stress-JNK1 Axis Mediates the Central Actions of Thyroid Hormones on Energy Balance. *Cell Metab.* **2017**, *26*, 212–229. [CrossRef] [PubMed]

163. Broeders, E.P.; Vijgen, G.H.; Havekes, B.; Bouvy, N.D.; Mottaghy, F.M.; Kars, M.; Schaper, N.C.; Schrauwen, P.; Brans, B.; van Marken Lichtenbelt, W.D. Thyroid Hormone Activates Brown Adipose Tissue and Increases Non-Shivering Thermogenesis–A Cohort Study in a Group of Thyroid Carcinoma Patients. *PLoS ONE* **2016**, *11*, e0145049. [CrossRef] [PubMed]

164. Whittle, A.J.; Carobbio, S.; Martins, L.; Slawik, M.; Hondares, E.; Vazquez, M.J.; Morgan, D.; Csikasz, R.I.; Gallego, R.; Rodriguez-Cuenca, S.; et al. BMP8B increases brown adipose tissue thermogenesis through both central and peripheral actions. *Cell* **2012**, *149*, 871–885. [CrossRef] [PubMed]

165. Martins, L.; Seoane-Collazo, P.; Contreras, C.; Gonzalez-Garcia, I.; Martinez-Sanchez, N.; Gonzalez, F.; Zalvide, J.; Gallego, R.; Dieguez, C.; Nogueiras, R.; et al. A Functional Link between AMPK and Orexin Mediates the Effect of BMP8B on Energy Balance. *Cell Rep.* **2016**, *16*, 2231–2242. [CrossRef] [PubMed]

166. Kooijman, S.; Wang, Y.; Parlevliet, E.T.; Boon, M.R.; Edelschaap, D.; Snaterse, G.; Pijl, H.; Romijn, J.A.; Rensen, P.C. Central GLP-1 receptor signalling accelerates plasma clearance of triacylglycerol and glucose by activating brown adipose tissue in mice. *Diabetologia* **2015**, *58*, 2637–2646. [CrossRef] [PubMed]

167. Lee, S.J.; Sanchez-Watts, G.; Krieger, J.P.; Pignalosa, A.; Norell, P.N.; Cortella, A.; Pettersen, K.G.; Vrdoljak, D.; Hayes, M.R.; Kanoski, S.E.; et al. Loss of dorsomedial hypothalamic GLP-1 signaling reduces BAT thermogenesis and increases adiposity. *Mol. Metab.* **2018**, *11*, 33–46. [CrossRef] [PubMed]

168. Maciel, M.G.; Beserra, B.T.S.; Oliveira, F.C.B.; Ribeiro, C.M.; Coelho, M.S.; Neves, F.A.R.; Amato, A.A. The effect of glucagon-like peptide 1 and glucagon-like peptide 1 receptor agonists on energy expenditure: A systematic review and meta-analysis. *Diabetes Res. Clin. Pract.* **2018**, *142*, 222–235. [CrossRef] [PubMed]

169. Martinez de Morentin, P.B.; Gonzalez-Garcia, I.; Martins, L.; Lage, R.; Fernandez-Mallo, D.; Martinez-Sanchez, N.; Ruiz-Pino, F.; Liu, J.; Morgan, D.A.; Pinilla, L.; et al. Estradiol regulates brown adipose tissue thermogenesis via hypothalamic AMPK. *Cell Metab.* **2014**, *20*, 41–53. [CrossRef] [PubMed]

170. Gonzalez-Garcia, I.; Contreras, C.; Estevez-Salguero, A.; Ruiz-Pino, F.; Colsh, B.; Pensado, I.; Linares-Pose, L.; Rial-Pensado, E.; Martinez de Morentin, P.B.; Ferno, J.; et al. Estradiol Regulates Energy Balance by Ameliorating Hypothalamic Ceramide-Induced ER Stress. *Cell Rep.* **2018**, *25*, 413–423. [CrossRef] [PubMed]

171. Konner, A.C.; Bruning, J.C. Selective insulin and leptin resistance in metabolic disorders. *Cell Metab.* **2012**, *16*, 144–152. [CrossRef] [PubMed]

172. Seoane-Collazo, P.; Roa, J.; Rial-Pensado, E.; Linares-Pose, L.; Beiroa, D.; Ruiz-Pino, F.; Lopez-Gonzalez, T.; Morgan, D.A.; Pardavila, J.A.; Sanchez-Tapia, M.J.; et al. SF1-Specific AMPKα1 deletion protects against diet-induced obesity. *Diabetes* **2018**, *67*, 2213–2226. [CrossRef] [PubMed]

173. Milbank, E.; Martinez, M.C.; Andriantsitohaina, R. Extracellular vesicles: Pharmacological modulators of the peripheral and central signals governing obesity. *Pharmacol. Ther.* **2016**, *157*, 65–83. [CrossRef] [PubMed]

174. Finan, B.; Yang, B.; Ottaway, N.; Stemmer, K.; Muller, T.D.; Yi, C.X.; Habegger, K.; Schriever, S.C.; Garcia-Caceres, C.; Kabra, D.G.; et al. Targeted estrogen delivery reverses the metabolic syndrome. *Nat. Med.* **2012**, *18*, 1847–1856. [CrossRef] [PubMed]

175. Finan, B.; Clemmensen, C.; Zhu, Z.; Stemmer, K.; Gauthier, K.; Muller, L.; De Angelis, M.; Moreth, K.; Neff, F.; Perez-Tilve, D.; et al. Chemical hybridization of glucagon and thyroid hormone optimizes therapeutic impact for metabolic disease. *Cell* **2016**, *167*, 843–857. [CrossRef] [PubMed]

176. Shughrue, P.J.; Lane, M.V.; Merchenthaler, I. Comparative distribution of estrogen receptor-alpha and -beta mRNA in the rat central nervous system. *J. Comp. Neurol.* **1997**, *388*, 507–525. [CrossRef]

8

Structure and Physiological Regulation of AMPK

Yan Yan [1,2], X. Edward Zhou [1], H. Eric Xu [1,2] and Karsten Melcher [1,*]

[1] Center for Cancer and Cell Biology, Van Andel Research Institute, 333 Bostwick Ave. N.E., Grand Rapids, MI 49503, USA; yan.yan@vai.org (Y.Y.); edward.zhou@vai.org (X.E.Z.); eric.xu@vai.org (H.E.X.)

[2] VARI/SIMM Center, Center for Structure and Function of Drug Targets, CAS-Key Laboratory of Receptor Research, Shanghai Institute of Materia Medica, Chinese Academy of Sciences, Shanghai 201203, China

* Correspondence: Karsten.melcher@vai.org

Abstract: Adenosine monophosphate (AMP)-activated protein kinase (AMPK) is a heterotrimeric $\alpha\beta\gamma$ complex that functions as a central regulator of energy homeostasis. Energy stress manifests as a drop in the ratio of adenosine triphosphate (ATP) to AMP/ADP, which activates AMPK's kinase activity, allowing it to upregulate ATP-generating catabolic pathways and to reduce energy-consuming catabolic pathways and cellular programs. AMPK senses the cellular energy state by competitive binding of the three adenine nucleotides AMP, ADP, and ATP to three sites in its γ subunit, each, which in turn modulates the activity of AMPK's kinase domain in its α subunit. Our current understanding of adenine nucleotide binding and the mechanisms by which differential adenine nucleotide occupancies activate or inhibit AMPK activity has been largely informed by crystal structures of AMPK in different activity states. Here we provide an overview of AMPK structures, and how these structures, in combination with biochemical, biophysical, and mutational analyses provide insights into the mechanisms of adenine nucleotide binding and AMPK activity modulation.

Keywords: energy metabolism; AMPK; activation loop; AID; α-linker; β-linker; CBS; LKB1; CaMKK2; αRIM

---

1. AMPK Is a Master Regulator of Energy Homeostasis That Is Dysregulated in Disease

AMPK is the primary energy sensor and regulator of energy homeostasis in eukaryotes. It is activated by energy stress in response to increased ATP consumption (e.g., exercise, cell proliferation, anabolism) or decreased ATP production (e.g., low glucose levels, oxidative stress, hypoxia), which are sensed as low ratios of ATP to AMP and ADP. Upon activation, AMPK phosphorylates downstream targets to directly or indirectly modulate the activities of rate-limiting metabolic enzymes, transcription and translation factors, proliferation and growth pathways, and epigenetic regulators. Collectively, this increases oxidative phosphorylation, autophagy, and uptake and metabolism of glucose and fatty acids, and decreases the synthesis of fatty acids, cholesterol, proteins, and ribosomal RNAs (rRNAs), as well as decreasing cell growth and proliferation [1–6]. Due to its central roles in metabolism, AMPK is dysregulated in diabetes, obesity, cardiometabolic disease, and cancer, and it is a promising pharmacological target [1,2,5,7–10], especially for the treatment of type 2 diabetes [11–13].

2. AMPK Consists of a Stable Core Attached to Moveable Domains

AMPK is a heterotrimeric $\alpha\beta\gamma$ protein kinase. In mammals, it is encoded by two alternative α subunits (α1 and α2), two alternative β subunits (β1 and β2), and three alternative γ subunits (γ1, γ2, and γ3) that can form up to 12 different $\alpha\beta\gamma$ isoforms [14]. The α subunits contain a canonical Ser/Thr kinase domain (KD), an autoinhibitory domain (AID), an adenine nucleotide sensor segment termed an α-linker, and a β subunit-interacting C-terminal domain (α-CTD), the latter of which contains the ST loop, which harbors proposed phosphorylation sites for AKT [15], PKA [16], and GSK [17].

The β subunits are composed of a myristoylated, unstructured N-terminus, a glycogen-binding carbohydrate-binding module (CBM), a scaffolding C-terminal domain (β-CTD) that interacts with both the γ subunit, and the α-CTD, and the extended β-linker loop that connects the CBM with the β-CTD (Figure 1A,B). The three alternative γ subunits consist of N-termini of different lengths and unknown function, followed by a conserved adenine nucleotide-binding domain that contains four cystathione β-synthetase (CBS) AMP/ADP/ATP binding sites (Figure 1). CBS1, 3, and 4 are functional, whereas in CBS2, the ribose-binding Asp residue is replaced by an Arg, and no nucleotide binding has been observed for CBS2 in heterotrimer structures.

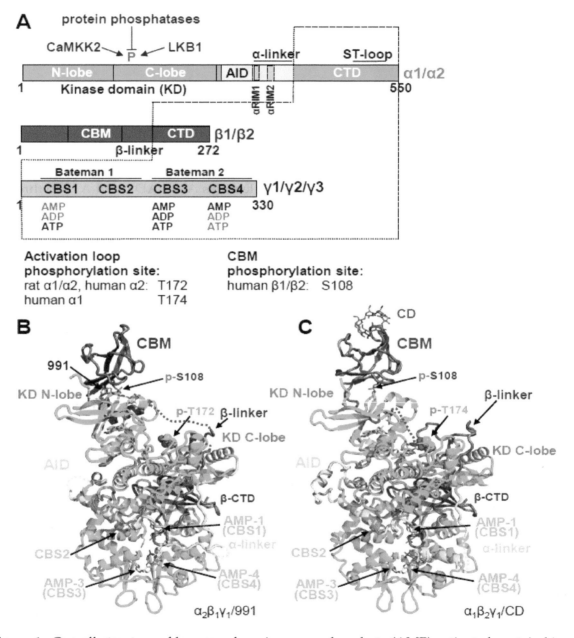

Figure 1. Overall structure of human adenosine monophosphate (AMP)-activated protein kinase (AMPK). (**A**). Domain structure and AMPK isoforms. Activation loop and carbohydrate-binding module (CBM) phosphorylation sites of different isoforms are indicated below the domain map (**B,C**). Crystal structures of phosphorylated, AMP-bound AMPK $\alpha_2\beta_1\gamma_1$/991 ((**B**); PDB: 4CFE) and $\alpha_1\beta_2\gamma_1$/cyclodextrin (CD) ((**C**); PDB: 4RER).

AMPK is a highly dynamic complex with a stable core formed by the γ subunit and the α- and β-CTDs, in which the β-CTD is sandwiched between the α and γ subunits (Figure 1A, core highlighted by dotted lines). Attached to the core are moveable domains whose position is determined by ligand binding and posttranslational modifications. As such, the holo-complex cannot be crystallized in the absence of multiple stabilizing ligands and/or protein engineering. Consequently, the first structures of AMPK consisted of isolated domains, e.g., the KD [18–21], the CBM bound to the glycogen mimic cyclodextrin [22], the yeast and mammalian nucleotide-bound scaffolding cores [23–26], the AID [27], and the yeast KD–AID complex [21] (Figure 2).

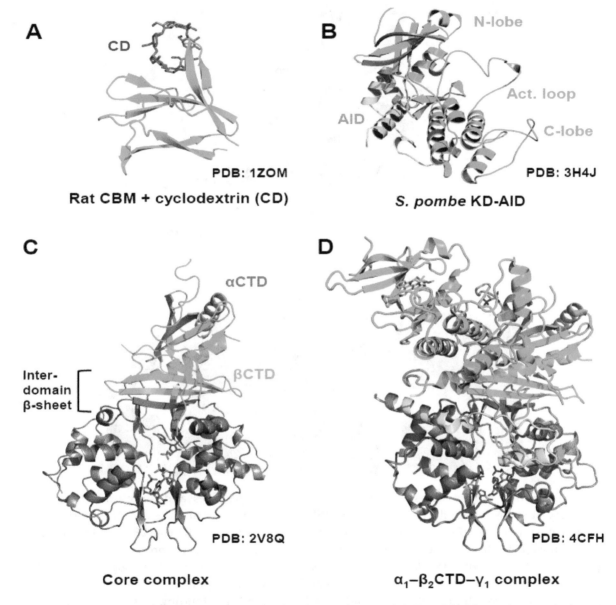

Figure 2. Structure of AMPK domains and subcomplexes. (**A**) Rat CBM bound to cyclodextrin; (**B**) Fission yeast kinase domain–autoinhibitory domain (KD-AID) complex; (**C**) AMP-bound, phosphorylated mammalian AMPK core complex (rat α_1-human β_2-rat γ_1); (**D**) AMP-bound, phosphorylated rat α_1—human β_2CTD—rat γ_1 complex.

Activation Loop Phosphorylation Orchestrates the Catalytic Center for Phosphoryl Transfer

Kinase domains have a highly conserved structure consisting of a smaller N-terminal lobe (N-lobe), composed of a β-sheet and the αB and αC helices, and a larger α-helical C-terminal lobe (C-lobe; see Figures 1B and 2B). The cleft between the lobes is the binding site for substrate peptides and Mg^{2+}–ATP. The two lobes are separated by a flexible hinge at the back that allows them to move towards each other to cycle through substrate-accessible open and catalytically-competent closed conformations as part of the kinase catalytic cycle. Key regulatory elements of the KD are: (i) the activation loop at the entrance of the catalytic cleft; (ii) the αC helix in the N-lobe, which positions the ATP-binding lysine (K47 in human α1) and the Mg^{2+}-binding DFG (Asp-Phe-Gly) loop; and (iii) the peptide substrate-binding catalytic loop in the C-lobe (Figure 3) [28–30].

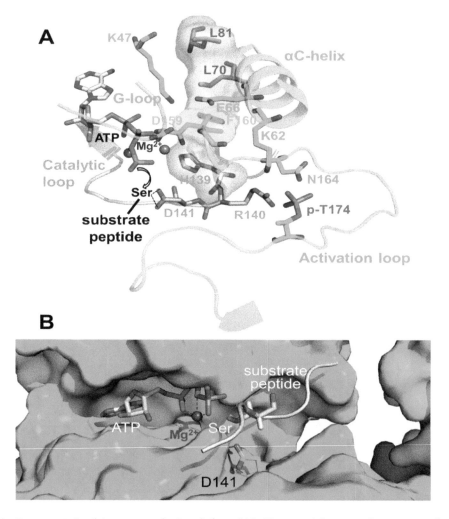

Figure 3. Active protein kinase catalytic cleft. (**A**) Key residues and structural elements of phosphorylated AMP-bound $\alpha_1\beta_2\gamma_1$ AMPK (4RER). Active kinase structures are characterized by a precisely positioned set of motifs for substrate- and adenosine triphosphate (ATP)-binding, in which four residues (L70, L81, H139, F160; shown in stick plus translucent surface presentation) are stacked against each other to form a regulatory spine. In this conformation, the activation loop p-T174 (p-T172 in human α_2) positions R140 and D141 from the catalytic loop for peptide substrate binding, and K62 from the αC-helix for aligning the ATP-binding K47 and the Mg^{2+}-binding DFG loop. The AMPK active protein kinase cleft resembles the canonical protein kinase A (PKA) site. To better visualize the active structure, we modeled the serine residue of a substrate peptide and the co-substrate ATP from the structure of PKA (PDB: 1ATP) in the catalytic cleft. Spheres: Mg^{2+} ions. (**B**) Surface presentation of the AMPK catalytic cleft (4RER) overlaid with a stick model of the aligned substrate peptide and ATP from the structure of substrate-bound CDK2 (PDB: 1QMZ). The Ser hydroxyl-positioning AMPK D141 is shown in green stick representation.

AMPK belongs to the RD (Arg-Asp) kinases, in many of which phosphorylation stabilizes the activation loop through a charge interaction between the negatively charged activation loop phosphate, and the positively charged residues from the αC helix (K62 in AMPK α1), the activation loop (N164), and the catalytic loop (R140). This conformation in turn stabilizes the αC helix and positions the arginine (R) and adjacent aspartate (D) of the catalytic loop for substrate binding (Figure 3). The hallmark of active protein kinases is therefore a precisely positioned set of motifs for substrate- and ATP-binding, in which four residues from the catalytic loop (H139), the Mg^{2+}-binding DFG loop (F160), the αC helix (L70) and the αC-αD loop (L81) are stacked against each other [28–30], as found in structures of active AMPK (Figure 3).

3. AMPK Is Activated Both by Direct Allosteric Activation and by Increasing Net Activation Loop Phosphorylation

AMPK activity is regulated at three different levels: at the level of (i) activation loop phosphorylation by upstream kinases, (ii) protection against activation loop dephosphorylation by protein phosphatases, and (iii) at the level of phosphorylation-independent, allosteric kinase activation (Figure 1A). Activation loop phosphorylation increases the AMPK activity by about 100-fold, while allosteric regulation changes AMPK activity up to ten-fold in mammalian cells and about two-fold in recombinant, bacterially produced AMPK [24,31–33]. AMP activates, and ATP inhibits, AMPK through all three mechanisms. ADP more weakly protects against activation loop dephosphorylation, does not allosterically activate AMPK [33–36], and it may not stimulate activation loop phosphorylation [33,37], although the latter is controversial [36].

The two main mammalian AMPK activation loop-phosphorylating kinases are the tumor suppressor LKB1 in complex with STRAD and MO25, and Ca^{2+}/calmodulin-dependent protein kinase kinase β (CaMKK2) [38–42]. While CaMKK2 mediates Ca^{2+}-dependent AMPK phosphorylation, AMP binding to the γ subunit increases activation loop phosphorylation through LKB1 by inducing a conformation that stabilizes formation of a complex between myristoylated AMPK, Axin, and LKB1/STRAD/MO25 [37,43]. However, the structural details of this interaction remain unknown. In addition, activation loop phosphorylation is also modulated by phosphorylation of the ST loop [15–17] and by ubiquitination of AMPK [44] and LKB1 [45].

In addition to adenine nucleotides, glucose, glycogen, and nicotinamide adenine dinucleotides are also important energy metabolites. Glucose has recently been identified as an important AMPK activity regulator, but it does so without direct AMPK binding [43,46]. In contrast, both glycogen and NADPH and NADH can directly bind AMPK: glycogen at the CBM [22,47], and in a reconstituted system, NADPH and NADH at the adenine nucleotide sensor site CBS3 [34,48]. However, the physiological relevance of the glycogen [47,49,50] and NADPH/NADH [34,48] interactions for AMPK activity regulation remains unclear.

Finally, a number of pharmacological activators bind AMPK at a unique site at the interface between CBM and KD (so called allosteric drug and metabolite [ADaM] site), as first shown for Merck compound 991 [51], and derivatives of the Abbot compound A769662 [52]. Binding greatly stabilizes the association of the highly dynamic CBM with the KD [53], an interaction that is also modulated by CBM phosphorylation and carbohydrate binding [49,53]. ADaM site agonists activate AMPK both directly and through increased protection against activation loop dephosphorylation, whose structural details will be covered in detail in a separate article in this issue.

Besides activity regulation, the level of AMPK is regulated by ubiquitination and proteasomal degradation in brown adipose tissue [54], testis [55], certain cancers [55,56], and in the presence of high levels of glucose [57].

3.1. The γ Subunit Contains Three Functional Adenine Nucleotide Binding Sites

The structure of the yeast and mammalian AMPK core scaffolds revealed a disk-shaped γ subunit composed of four CBS sites. Each CBS consists of a strand-helix-strand-strand-helix fold (β1-α1-β2-β3-α2) with long intervening loops (Figure 4A). β1 is often incomplete, but where present, it forms a three-stranded sheet with the two central β-strands (β2 and β3). The β-sheet of one CBS packs parallel with the sheet of a neighboring CBS. The interface between the two sheets forms two clefts, one on the top flat side and one on the bottom flat side of the disk, which are the binding sites for adenine nucleotides (Figure 4C,D). Therefore, each binding site requires a tandem CBS pair to form a functional unit termed the Bateman domain (CBS1 + CBS2 = Bateman domain 1, CBS3 + CBS4 = Bateman domain 2). The structures of the core complexes in the presence of AMP [24,58], ADP [34], or ATP [24,58] revealed adenine nucleotide binding at three sites in mammalian AMPK: CBS1, CBS3, and CBS4.

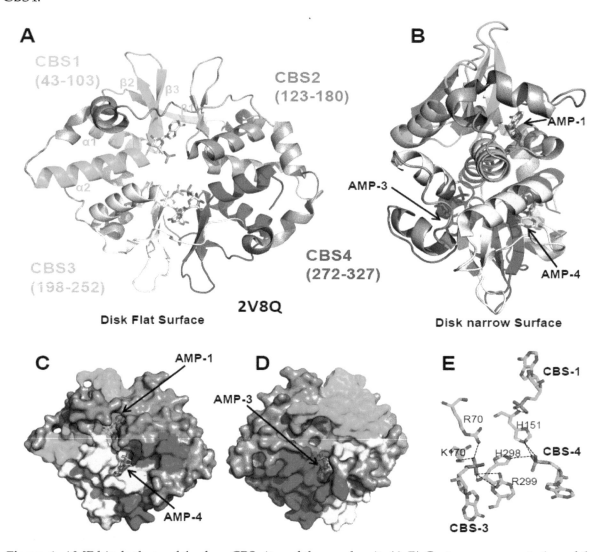

Figure 4. AMP binds three of the four CBS sites of the γ subunit. (**A,B**) Cartoon representation of the γ subunit in two different orientations. AMP molecules are shown in stick representation. The four CBS sites are shown in different colors with the secondary structure elements of CBS1 labeled. (**C,D**) Surface representation of the front and back sides of the disk flat surfaces illustrating the AMP-occupied binding pockets 1, 3, and 4, and the empty CBS2 pocket. (**E**) The phosphate groups (orange) of the three AMP molecules (cyan C atoms) coordinately interact with a set of polar γ subunit residues (green C atoms); O: red, N: blue.

3.2. CBS3 Is the Adenine Nucleotide Sensor Site

While the structure of the core complexes revealed how adenine nucleotides bind the γ subunit, they did not provide information on how the binding signal is transduced to the KD in the α subunit. In 2011, the Gamblin and Carling groups crystallized an AMPK complex containing rat α1, human β2 CTD, and rat γ1 [34]. While this complex is not regulated by protection against activation loop dephosphorylation [51], it retained direct AMPK activation by AMP and ADP. The structure revealed that the α-linker that connects AID and α-CTD directly bound the γ subunit [34], which has been validated in all subsequent AMP-bound AMPK complex structures with a resolved α-linker. A segment of the linker, termed regulatory subunit-interacting motif 2 (αRIM2) [27,59], interacts with AMP at CBS3, suggesting that αRIM2 functions as an adenine nucleotide sensor, and that it mediates the transduction of the adenine-binding signal to the KD [27,34,59]. This function has been validated by several experimental approaches. First, the mutation of either of the two key αRIM2 residues (E362 and R363 in rat α1 and human α2; E364 and R365 in human α1) abolished or largely reduced both AMP-dependent direct AMPK activation [27,49,51] and AMP-dependent protection against activation loop dephosphorylation [49]. Second, AMP increases, and ATP decreases the interaction between isolated α-linker and core AMPK in a reconstituted system, and the AMP increase requires intact E364 and R365 [49]. Third, the AMP-mimetic synthetic AMPK activator C2 activates AMPK α-isotype-selectively (it fully activates α1-containing complexes, but only partially α2 complexes), and this selectivity can be fully reversed by a swap of the αRIM2 regions [60,61].

4. If CBS3 Is the Sensor Site, What Are the Roles of CBS1 and CBS4?

Of the three functional CBS sites, only CBS3 interacts with the α subunit, an interaction that is directly modulated by AMP and ATP. In contrast, CBS1 and CBS4 do not interact with any part of the α- or β-subunit. Moreover, CBS4 binds AMP very tightly [24,58] and it is unlikely to exchange AMP under physiological conditions, yet mutations in CBS4 abolish regulation by AMP [36,58], while mutations in CBS1 have either no [58] or only a small [36] effect on AMPK regulation. Important insight came from a mutational study. When CBS1, CBS4, and the ATP-binding site in the KD are mutated, so that CBS3 remains the only functional adenine nucleotide binding site, it binds AMP only very weakly and with 10–100 times lower affinity than ATP [48]. Since the cellular ATP concentrations are much higher than AMP and ADP concentrations, CBS3 by itself would remain almost completely ATP-bound under both normal and energy stress conditions. However, the phosphates of adenine nucleotides bound to CBS1, 3, and 4 coordinately bind a set of charged and polar amino acids (Figure 4E), so that binding to one site affects binding to the other two sites. Through these coordinated interactions, AMP bound at CBS4, together with additional interactions from αRIM2, stabilizes AMP at CBS3. This increases CBS3's affinity for AMP by two orders of magnitude, and its AMP/ATP binding preference by two to three orders of magnitude [48], allowing CBS3 to sensitively detect physiological energy stress versus non-stress adenine nucleotide levels. Conversely, both CBS3 and CBS1 strongly stabilize AMP-binding at CBS4, so that under physiological conditions CBS4 remains essentially non-exchangeably AMP-bound and CBS1 largely ATP-bound [48].

5. AMP-Binding at CBS3 Destabilizes an Inhibitory AID–KD Interaction

The KD is followed by the AID, a small 48 amino acid domain that inhibits kinase activity about tenfold in the context of a KD–AID fragment [62,63]. Crystal structures of the fission yeast [21] and human [49] AMPK KD–AID fragments revealed a three-helical AID, whose C-terminal helix (α3) directly binds the hinge between the KD N- and C-lobes at the backside of the KD (Figures 2B and 5A). In contrast, in structures of active, AMP-bound AMPK [34,49], the AID is rotated away from the KD and bound to the γ subunit (see structure overlay in Figure 5B). The AID–KD interaction arrests the KD in a unique inactive conformation, in which the ATP binding K47, the Mg^{2+}-binding DFG loop, and the substrate-binding catalytic loop are misaligned, and H139 of the regulatory spine is out of register [21,49] (so called "HRD-out" conformation [30]; Figure 5C). The inhibitory function of the KD–AID interaction was further validated by the mutation of interface residues in either the KD or the AID, all of which made AMPK constitutively active [21,49]. Conversely, binding of the AID to the γ subunit, as seen in structures of AMP-bound AMPK [34,49], allows the KD to adopt the active conformation [27,34,51,59] (Figure 5D). Consistently, mutations in AID-interacting γ subunit residues make AMPK constitutively inactive [59].

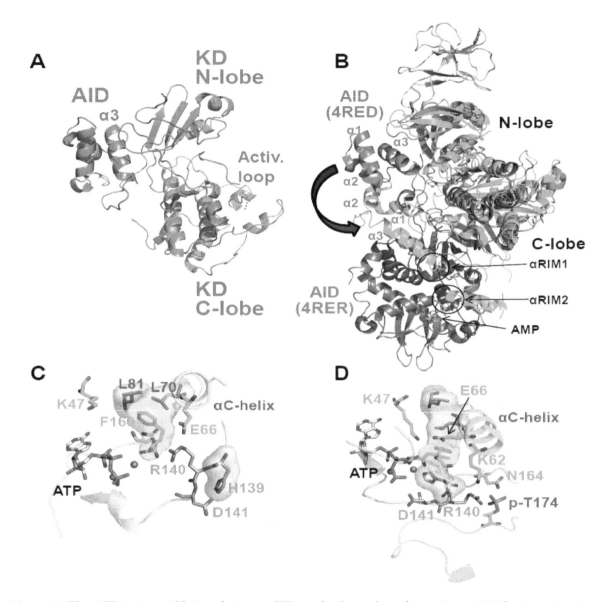

Figure 5. The AID is in equilibrium between KD- and γ-bound conformations. (**A**) Cartoon structure of the human α_1 KD-AID complex. (**B**) Overlay of the inactive KD-AID structure with the structure of active holo-AMPK (α subunit: green; β- and γ-subunits: grey). The arrow indicates the repositioning of the AID in the active structure. (**C,D**) Catalytic center of the inactive (**C**) and active (**D**) AMPK conformation. Stick plus translucent surface presentations indicate the regulatory spine residues L70, L81, H139, and F160. Mg^{2+}-ATP was modeled into both structures for orientation, even though it cannot bind to the inactive structure shown in panel C. Spheres: Mg^{2+} ions.

A Highly Conserved Interaction Network Links αRIM2/CBS3 and AID-αRIM1/CBS2 Binding

The structure of AMP-bound AMPK α_1-β_2CTD-γ_1 [34] first revealed the AID conformation in active AMPK, in which the border of the AID and the N-terminus of the α-linker, termed αRIM1, binds the γ-subunit at the unoccupied CBS2 [27,34,51,59]. Mutational analysis by Ja-Wei Wu's group provided a molecular pathway to link αRIM2 binding of AMP-occupied CBS3 to direct AMPK kinase activation. They first showed that αRIM1/CBS2 interface amino acids corresponding to human α1 I335/M3364 and F342/Y343, and human γ1 R171 and F179 are required for AMP-mediated relief of AMPK autoinhibition [59]. In active, AMP-bound holo-AMPK, the direct interaction of γ1 K170 with both AMP/CBS3 and αRIM2 α1 E364 positions three key residues at the αRIM1 interface. First, the residue following K170, R171, forms Van der Waals interactions and a backbone hydrogen bond with αRIM1 α1 F342. Second, the K170-interacting residues K174 and F175 form Van der Waals bonds with F342 and both Van der Waals and π-stacking interactions with γ1 F179. The latter is the linchpin of the interface and directly interacts with all four αRIM1 residues that are required for the relief of AMPK autoinhibition (I335, M336, Y343, and F342; Figure 6). Similarly, E364, R171, and F179 are also all required for the relief of AMPK autoinhibition [59]. The mutational analysis thus provides strong support that this AMP-stabilized interaction network that is seen in all active structures of holo-AMPK is responsible for shifting the AID equilibrium from the inactive, KD-bound conformation to the active, γ/CBS2-bound conformation.

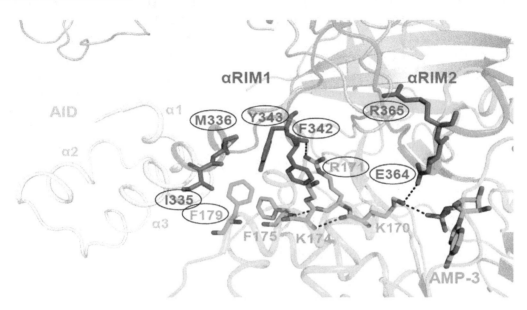

Figure 6. αRIM2/CBS3 and AID-αRIM1/CBS2 interactions are linked. Structure of human AMP-bound AMPK $\alpha_1\beta_2\gamma_1$ (4RER) with key residues shown in a stick presentation; the α-linker is shown in magenta, the γ subunit in cyan, and the AID in light green. AMP bound at CBS3 and αRIM2 E364 directly interact with γ1 K170, which positions the αRIM1-binding residues R171, and indirectly through K174 and F175, F179, thus stabilizing the AID-γ subunit interaction. Consistently, mutations of the αRIM1/γ subunit (and αRIM2/CBS3) interface residues highlighted by oval outlines (human α_1: F342D/Y343D, I335D/M336D, E364, R365; γ_1: R171A, F179D) are constitutively AMP-non-responsive. Dashed lines indicate hydrogen bonds.

ATP binding is thought to disrupt this network. In the structure of the core AMPK complex co-crystallized with ATP [58], ATP was bound to CBS4 and CBS1, which sterically interfered with nucleotide binding at CBS3, and caused rearrangement and disruption of the interaction network [58,59]. However, the physiological relevance of this structure remains unclear, since under physiological conditions, CBS4 does not seem to exchange AMP (see above; [24,48,58]). Therefore, a final understanding of how ATP disrupts the CBS3–α-linker–AID network will require the structure of the holo-AMPK complex, including the α-linker, in ATP-bound conformation.

ADaM site ligands, while not focus of this review, directly activate AMPK by a completely different mechanism. Through binding of both the CBM and the KD [51,52] and stabilization of the CBM–KD interaction [53], the N-terminus of the β-linker at the CBM border adopts a helix that packs parallel to the αC-helix, and it has therefore been named C-interacting helix [51]. This suggested that ADaM site ligands may activate AMPK by stabilizing αC through induced formation of the C-interacting helix, reminiscent of the regulatable αC stabilization of several other protein kinases [64]. Support for this model came from the mutation of H233 in the C-interacting helix, which reduced activation by the ADaM site ligand 991 [51], and by direct demonstration through hydrogen/deuterium exchange mass spectrometry (HDX-MS) that 991 binding strongly and selectively stabilizes αC [48].

6. Regulation of Activation Loop Accessibility

A major regulatory mechanism for AMPK activation by AMP and ADP is the protection of activation loop p-T172 (human $\alpha1$ T174) against dephosphorylation. p-T172 protection can be demonstrated in a cell-free, reconstituted system independent of the phosphatase used (e.g., PP2C, PP2A, λ-phosphatase), and AMP does not, or only slightly inhibit the dephosphorylation of a different substrate, casein, by PP2Cα [32]. Therefore, reduced dephosphorylation is not due to phosphatase inhibition, but to an AMP/ADP-induced change in the activation loop accessibility. The crystal structure of AMP-bound, phosphorylated AMPK α_1–β_2CTD–γ_1 (PDB: 4CFH) first demonstrated that the activation loop directly interacts with the core of AMPK [34]. Specifically, the stable β-CTD directly bound and stabilized the activation loop (Figure 7). The authors therefore proposed that the core shields the activation loop from phosphatase access. In agreement, mutation of the activation loop-interacting $\beta2$ H235 increased p-T172 dephosphorylation in the context of holo-AMPK [34]. However, the construct used in the structure was not regulated by protection against activation loop dephosphorylation [51], indicating that additional parts of AMPK, likely either the β-linker and/or the CBM, were also required for AMP-mediated, and probably ADaM site ligand-mediated protection against activation loop dephosphorylation. Consistently, in structures in which the β-linker is largely resolved (e.g., $\beta2$-linker in 4RER [49], $\beta1$-linker in 5ISO [65]), p-T172 is clearly protected by the β-linker, especially in the case of the $\beta2$-linker. Finally, how can the activation loop in AMP-bound conformation be largely inaccessible to protein phosphatases without affecting accessibility to the T172-phosphorylating upstream protein kinases? Answers to these fundamental questions will likely require the structure of holo-AMPK in the alternative, ATP-bound state and analysis of AMPK's conformational landscape and dynamics in solution.

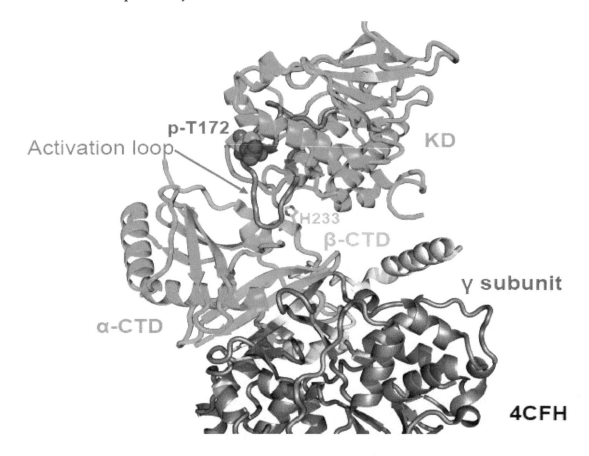

Figure 7. The β-CTD binds and stabilizes the activation loop. Structure of AMP-bound, phosphorylated AMPK α_1–β_2CTD–γ_1 (PDB: 4CFH). The activation loop is highlighted in orange, and p-T172 is shown in sphere presentation.

7. Conclusions and Future Directions

AMPK is a molecular machine consisting of the adenine nucleotide-binding core (γ subunit plus α- and β-CTDs), the catalytic KD, and at least four dynamic domains (AID, CBM, and the α- and β-linkers). We propose that adenine nucleotides, ADaM site ligands, and CBM phosphorylation affect the conformation of the KD through induced movements of the dynamic domains, while phosphorylation of activation loop and S/T loop modulate the KD conformation directly. Through concerted efforts, the mechanism of direct, allosteric AMPK activation through AID movement and αC stabilization is relatively well understood. However, the structural basis of direct inhibition by ATP, of activation loop accessibility regulation through ligands and possibly phosphorylation, and of the AMP-induced interaction with Axin and the LKB1 complex all remain poorly understood. The most important future challenges in AMPK structural biology will therefore be the determination of the structures of holo-AMPK in its inhibited, ATP-bound conformation, and in complex with Axin and LKB/STRAD/MO25.

Author Contributions: The manuscript was written by K.M. with input from all authors.

Abbreviations

ADaM site	Allosteric drug and metabolite-binding site
AID	Autoinhibitory domain
AMPK	AMP-activated protein kinase
αRIM	α-regulatory subunit interaction motif
CaMKK2	Ca^{2+}/calmodulin-dependent protein kinase kinase β
CBM	Carbohydrate-binding module
CBS	Cystathionine β-synthetase
CTD	C-terminal domain
HDX-MS	Hydrogen deuterium exchange mass spectrometry
KD	Kinase domain
LKB1	Liver kinase B1
MO25	Mouse protein-25
PP2A	Protein phosphatase 2A
PP2C	Protein phosphatase 2C
STRAD	STE20-related kinase adaptor

References

1. Yuan, H.X.; Xiong, Y.; Guan, K.L. Nutrient sensing, metabolism, and cell growth control. *Mol. Cell* **2013**, *49*, 379–387. [CrossRef] [PubMed]
2. Garcia, D.; Shaw, R.J. AMPK: Mechanisms of Cellular Energy Sensing and Restoration of Metabolic Balance. *Mol. Cell* **2017**, *66*, 789–800. [CrossRef] [PubMed]
3. Hardie, D.G. AMP-activated protein kinase: An energy sensor that regulates all aspects of cell function. *Genes Dev.* **2011**, *25*, 1895–1908. [CrossRef] [PubMed]
4. Hardie, D.G. Keeping the home fires burning: AMP-activated protein kinase. *J. R. Soc. Interface* **2018**, *15*, 20170774. [CrossRef] [PubMed]
5. Steinberg, G.R.; Kemp, B.E. AMPK in Health and Disease. *Physiol. Rev.* **2009**, *89*, 1025–1078. [CrossRef] [PubMed]
6. Hardie, D.G.; Schaffer, B.E.; Brunet, A. AMPK: An Energy-Sensing Pathway with Multiple Inputs and Outputs. *Trends Cell Biol.* **2016**, *26*, 190–201. [CrossRef] [PubMed]

7.   Hardie, D.G. AMPK: A target for drugs and natural products with effects on both diabetes and cancer. *Diabetes* **2013**, *62*, 2164–2172. [CrossRef] [PubMed]

8.   Hardie, D.G. Targeting an energy sensor to treat diabetes. *Science* **2017**, *357*, 455–456. [CrossRef] [PubMed]

9.   Hardie, D.G.; Ross, F.A.; Hawley, S.A. AMP-activated protein kinase: A target for drugs both ancient and modern. *Chem. Biol.* **2012**, *19*, 1222–1236. [CrossRef] [PubMed]

10. Guigas, B.; Viollet, B. Targeting AMPK: From Ancient Drugs to New Small-Molecule Activators. *EXS* **2016**, *107*, 327–350. [PubMed]

11. Cokorinos, E.C.; Delmore, J.; Reyes, A.R.; Albuquerque, B.; Kjobsted, R.; Jorgensen, N.O.; Tran, J.L.; Jatkar, A.; Cialdea, K.; Esquejo, R.M.; et al. Activation of Skeletal Muscle AMPK Promotes Glucose Disposal and Glucose Lowering in Non-human Primates and Mice. *Cell Metab.* **2017**, *25*, 1147–1159.e10. [CrossRef] [PubMed]

12. Myers, R.W.; Guan, H.P.; Ehrhart, J.; Petrov, A.; Prahalada, S.; Tozzo, E.; Yang, X.; Kurtz, M.M.; Trujillo, M.; Gonzalez Trotter, D.; et al. Systemic pan-AMPK activator MK-8722 improves glucose homeostasis but induces cardiac hypertrophy. *Science* **2017**, *357*, 507–511. [CrossRef] [PubMed]

13. Steneberg, P.; Lindahl, E.; Dahl, U.; Lidh, E.; Straseviciene, J.; Backlund, F.; Kjellkvist, E.; Berggren, E.; Lundberg, I.; Bergqvist, I.; et al. PAN-AMPK activator O304 improves glucose homeostasis and microvascular perfusion in mice and type 2 diabetes patients. *JCI Insight* **2018**, *3*. [CrossRef] [PubMed]

14. Ross, F.A.; MacKintosh, C.; Hardie, D.G. AMP-activated protein kinase: A cellular energy sensor that comes in 12 flavours. *FEBS J.* **2016**, *283*, 2987–3001. [CrossRef] [PubMed]

15. Hawley, S.A.; Ross, F.A.; Gowans, G.J.; Tibarewal, P.; Leslie, N.R.; Hardie, D.G. Phosphorylation by Akt within the ST loop of AMPK-alpha1 down-regulates its activation in tumour cells. *Biochem. J.* **2014**, *459*, 275–287. [CrossRef] [PubMed]

16. Hurley, R.L.; Barre, L.K.; Wood, S.D.; Anderson, K.A.; Kemp, B.E.; Means, A.R.; Witters, L.A. Regulation of AMP-activated protein kinase by multisite phosphorylation in response to agents that elevate cellular cAMP. *J. Biol. Chem.* **2006**, *281*, 36662–36672. [CrossRef] [PubMed]

17. Suzuki, T.; Bridges, D.; Nakada, D.; Skiniotis, G.; Morrison, S.J.; Lin, J.D.; Saltiel, A.R.; Inoki, K. Inhibition of AMPK catabolic action by GSK3. *Mol. Cell* **2013**, *50*, 407–419. [CrossRef] [PubMed]

18. Littler, D.R.; Walker, J.R.; Davis, T.; Wybenga-Groot, L.E.; Finerty, P.J., Jr.; Newman, E.; Mackenzie, F.; Dhe-Paganon, S. A conserved mechanism of autoinhibition for the AMPK kinase domain: ATP-binding site and catalytic loop refolding as a means of regulation. *Acta Crystallogr. Sect. F Struct. Biol. Cryst. Commun.* **2010**, *66*, 143–151. [CrossRef] [PubMed]

19. Nayak, V.; Zhao, K.; Wyce, A.; Schwartz, M.F.; Lo, W.S.; Berger, S.L.; Marmorstein, R. Structure and dimerization of the kinase domain from yeast Snf1, a member of the Snf1/AMPK protein family. *Structure* **2006**, *14*, 477–485. [CrossRef] [PubMed]

20. Handa, N.; Takagi, T.; Saijo, S.; Kishishita, S.; Takaya, D.; Toyama, M.; Terada, T.; Shirouzu, M.; Suzuki, A.; Lee, S.; et al. Structural basis for compound C inhibition of the human AMP-activated protein kinase alpha2 subunit kinase domain. *Acta Crystallogr. D Biol. Crystallogr.* **2011**, *67*, 480–487. [CrossRef] [PubMed]

21. Chen, L.; Jiao, Z.H.; Zheng, L.S.; Zhang, Y.Y.; Xie, S.T.; Wang, Z.X.; Wu, J.W. Structural insight into the autoinhibition mechanism of AMP-activated protein kinase. *Nature* **2009**, *459*, 1146–1149. [CrossRef] [PubMed]

22. Polekhina, G.; Gupta, A.; van Denderen, B.J.; Feil, S.C.; Kemp, B.E.; Stapleton, D.; Parker, M.W. Structural basis for glycogen recognition by AMP-activated protein kinase. *Structure* **2005**, *13*, 1453–1462. [CrossRef] [PubMed]

23. Amodeo, G.A.; Rudolph, M.J.; Tong, L. Crystal structure of the heterotrimer core of Saccharomyces cerevisiae AMPK homologue SNF1. *Nature* **2007**, *449*, 492–495. [CrossRef] [PubMed]

24. Xiao, B.; Heath, R.; Saiu, P.; Leiper, F.C.; Leone, P.; Jing, C.; Walker, P.A.; Haire, L.; Eccleston, J.F.; Davis, C.T.; et al. Structural basis for AMP binding to mammalian AMP-activated protein kinase. *Nature* **2007**, *449*, 496–500. [CrossRef] [PubMed]

25. Townley, R.; Shapiro, L. Crystal structures of the adenylate sensor from fission yeast AMP-activated protein kinase. *Science* **2007**, *315*, 1726–1729. [CrossRef] [PubMed]

26. Jin, X.; Townley, R.; Shapiro, L. Structural insight into AMPK regulation: ADP comes into play. *Structure* **2007**, *15*, 1285–1295. [CrossRef] [PubMed]

27. Chen, L.; Xin, F.J.; Wang, J.; Hu, J.; Zhang, Y.Y.; Wan, S.; Cao, L.S.; Lu, C.; Li, P.; Yan, S.F.; et al. Conserved regulatory elements in AMPK. *Nature* **2013**, *498*, E8–E10. [CrossRef] [PubMed]

28. Kornev, A.P.; Haste, N.M.; Taylor, S.S.; Eyck, L.F. Surface comparison of active and inactive protein kinases identifies a conserved activation mechanism. *Proc. Natl. Acad. Sci. USA* **2006**, *103*, 17783–17788. [CrossRef] [PubMed]

29. Kornev, A.P.; Taylor, S.S. Dynamics-Driven Allostery in Protein Kinases. *Trends Biochem. Sci.* **2015**, *40*, 628–647. [CrossRef] [PubMed]

30. Meharena, H.S.; Chang, P.; Keshwani, M.M.; Oruganty, K.; Nene, A.K.; Kannan, N.; Taylor, S.S.; Kornev, A.P. Deciphering the structural basis of eukaryotic protein kinase regulation. *PLoS Biol.* **2013**, *11*, e1001680. [CrossRef] [PubMed]

31. Sanders, M.J.; Ali, Z.S.; Hegarty, B.D.; Heath, R.; Snowden, M.A.; Carling, D. Defining the mechanism of activation of AMP-activated protein kinase by the small molecule A-769662, a member of the thienopyridone family. *J. Biol. Chem.* **2007**, *282*, 32539–32548. [CrossRef] [PubMed]

32. Davies, S.P.; Helps, N.R.; Cohen, P.T.; Hardie, D.G. 5′-AMP inhibits dephosphorylation, as well as promoting phosphorylation, of the AMP-activated protein kinase. Studies using bacterially expressed human protein phosphatase-2C alpha and native bovine protein phosphatase-2AC. *FEBS Lett.* **1995**, *377*, 421–425. [PubMed]

33. Gowans, G.J.; Hawley, S.A.; Ross, F.A.; Hardie, D.G. AMP is a true physiological regulator of AMP-activated protein kinase by both allosteric activation and enhancing net phosphorylation. *Cell Metab.* **2013**, *18*, 556–566. [CrossRef] [PubMed]

34. Xiao, B.; Sanders, M.J.; Underwood, E.; Heath, R.; Mayer, F.V.; Carmena, D.; Jing, C.; Walker, P.A.; Eccleston, J.F.; Haire, L.F.; et al. Structure of mammalian AMPK and its regulation by ADP. *Nature* **2011**, *472*, 230–233. [CrossRef] [PubMed]

35. Carling, D.; Clarke, P.R.; Zammit, V.A.; Hardie, D.G. Purification and characterization of the AMP-activated protein kinase. Copurification of acetyl-CoA carboxylase kinase and 3-hydroxy-3-methylglutaryl-CoA reductase kinase activities. *Eur. J. Biochem.* **1989**, *186*, 129–136. [PubMed]

36. Oakhill, J.S.; Steel, R.; Chen, Z.P.; Scott, J.W.; Ling, N.; Tam, S.; Kemp, B.E. AMPK is a direct adenylate charge-regulated protein kinase. *Science* **2011**, *332*, 1433–1435. [CrossRef] [PubMed]

37. Zhang, Y.L.; Guo, H.; Zhang, C.S.; Lin, S.Y.; Yin, Z.; Peng, Y.; Luo, H.; Shi, Y.; Lian, G.; Zhang, C.; et al. AMP as a low-energy charge signal autonomously initiates assembly of AXIN-AMPK-LKB1 complex for AMPK activation. *Cell Metab.* **2013**, *18*, 546–555. [CrossRef] [PubMed]

38. Hawley, S.A.; Boudeau, J.; Reid, J.L.; Mustard, K.J.; Udd, L.; Makela, T.P.; Alessi, D.R.; Hardie, D.G. Complexes between the LKB1 tumor suppressor, STRAD alpha/beta and MO25 alpha/beta are upstream kinases in the AMP-activated protein kinase cascade. *J. Biol.* **2003**, *2*, 28. [CrossRef] [PubMed]

39. Hurley, R.L.; Anderson, K.A.; Franzone, J.M.; Kemp, B.E.; Means, A.R.; Witters, L.A. The Ca2+/calmodulin-dependent protein kinase kinases are AMP-activated protein kinase kinases. *J. Biol. Chem.* **2005**, *280*, 29060–29066. [CrossRef] [PubMed]

40. Shaw, R.J.; Kosmatka, M.; Bardeesy, N.; Hurley, R.L.; Witters, L.A.; DePinho, R.A.; Cantley, L.C. The tumor suppressor LKB1 kinase directly activates AMP-activated kinase and regulates apoptosis in response to energy stress. *Proc. Natl. Acad. Sci. USA* **2004**, *101*, 3329–3335. [CrossRef] [PubMed]

41. Woods, A.; Dickerson, K.; Heath, R.; Hong, S.P.; Momcilovic, M.; Johnstone, S.R.; Carlson, M.; Carling, D. Ca2+/calmodulin-dependent protein kinase kinase-beta acts upstream of AMP-activated protein kinase in mammalian cells. *Cell Metab.* **2005**, *2*, 21–33. [CrossRef] [PubMed]

42. Woods, A.; Johnstone, S.R.; Dickerson, K.; Leiper, F.C.; Fryer, L.G.; Neumann, D.; Schlattner, U.; Wallimann, T.; Carlson, M.; Carling, D. LKB1 is the upstream kinase in the AMP-activated protein kinase cascade. *Curr. Biol.* **2003**, *13*, 2004–2008. [CrossRef] [PubMed]

43. Zhang, C.S.; Jiang, B.; Li, M.; Zhu, M.; Peng, Y.; Zhang, Y.L.; Wu, Y.Q.; Li, T.Y.; Liang, Y.; Lu, Z.; et al. The lysosomal v-ATPase-Ragulator complex is a common activator for AMPK and mTORC1, acting as a switch between catabolism and anabolism. *Cell Metab.* **2014**, *20*, 526–540. [CrossRef] [PubMed]

44. Deng, M.; Yang, X.; Qin, B.; Liu, T.; Zhang, H.; Guo, W.; Lee, S.B.; Kim, J.J.; Yuan, J.; Pei, H.; et al. Deubiquitination and Activation of AMPK by USP10. *Mol. Cell* **2016**, *61*, 614–624. [PubMed]

45. Lee, S.W.; Li, C.F.; Jin, G.; Cai, Z.; Han, F.; Chan, C.H.; Yang, W.L.; Li, B.K.; Rezaeian, A.H.; Li, H.Y.; et al. Skp2-dependent ubiquitination and activation of LKB1 is essential for cancer cell survival under energy stress. *Mol. Cell* **2015**, *57*, 1022–1033. [CrossRef] [PubMed]

46. Zhang, C.S.; Hawley, S.A.; Zong, Y.; Li, M.; Wang, Z.; Gray, A.; Ma, T.; Cui, J.; Feng, J.W.; Zhu, M.; et al. Fructose-1,6-bisphosphate and aldolase mediate glucose sensing by AMPK. *Nature* **2017**, *548*, 112–116. [CrossRef] [PubMed]

47. Polekhina, G.; Gupta, A.; Michell, B.J.; van Denderen, B.; Murthy, S.; Feil, S.C.; Jennings, I.G.; Campbell, D.J.; Witters, L.A.; Parker, M.W.; et al. AMPK beta subunit targets metabolic stress sensing to glycogen. *Curr. Biol.* **2003**, *13*, 867–871. [CrossRef]

48. Gu, X.; Yan, Y.; Novick, S.J.; Kovich, A.; Goswami, D.; Ke, J.; Tan, M.H.E.; Wang, L.; Li, X.; de Waal, P.; et al. Deconvoluting AMP-dependent kinase (AMPK) adenine nucleotide binding and sensing. *J. Biol. Chem.* **2017**, *292*, 12653–12666. [CrossRef] [PubMed]

49. Li, X.; Wang, L.; Zhou, X.E.; Ke, J.; de Waal, P.W.; Gu, X.; Tan, M.H.; Wang, D.; Wu, D.; Xu, H.E.; et al. Structural basis of AMPK regulation by adenine nucleotides and glycogen. *Cell Res.* **2015**, *25*, 50–66. [CrossRef] [PubMed]

50. McBride, A.; Ghilagaber, S.; Nikolaev, A.; Hardie, D.G. The glycogen-binding domain on the AMPK beta subunit allows the kinase to act as a glycogen sensor. *Cell Metab.* **2009**, *9*, 23–34. [CrossRef] [PubMed]

51. Xiao, B.; Sanders, M.J.; Carmena, D.; Bright, N.J.; Haire, L.F.; Underwood, E.; Patel, B.R.; Heath, R.B.; Walker, P.A.; Hallen, S.; et al. Structural basis of AMPK regulation by small molecule activators. *Nat. Commun.* **2013**, *4*, 3017. [CrossRef] [PubMed]

52. Calabrese, M.F.; Rajamohan, F.; Harris, M.S.; Caspers, N.L.; Magyar, R.; Withka, J.M.; Wang, H.; Borzilleri, K.A.; Sahasrabudhe, P.V.; Hoth, L.R.; et al. Structural Basis for AMPK Activation: Natural and Synthetic Ligands Regulate Kinase Activity from Opposite Poles by Different Molecular Mechanisms. *Structure* **2014**, *22*, 1161–1172. [CrossRef] [PubMed]

53. Gu, X.; Bridges, M.D.; Yan, Y.; de Waal, P.; Zhou, X.E.; Suino-Powell, K.M.; Xu, H.E.; Hubbell, W.L.; Melcher, K. Conformational heterogeneity of the allosteric drug and metabolite (ADaM) site in AMP-activated protein kinase (AMPK). *J. Biol. Chem.* **2018**, *239*, 16994–17007. [CrossRef] [PubMed]

54. Qi, J.; Gong, J.; Zhao, T.; Zhao, J.; Lam, P.; Ye, J.; Li, J.Z.; Wu, J.; Zhou, H.M.; Li, P. Downregulation of AMP-activated protein kinase by Cidea-mediated ubiquitination and degradation in brown adipose tissue. *EMBO J.* **2008**, *27*, 1537–1548. [CrossRef] [PubMed]

55. Pineda, C.T.; Ramanathan, S.; Fon Tacer, K.; Weon, J.L.; Potts, M.B.; Ou, Y.H.; White, M.A.; Potts, P.R. Degradation of AMPK by a cancer-specific ubiquitin ligase. *Cell* **2015**, *160*, 715–728. [CrossRef] [PubMed]

56. Vila, I.K.; Yao, Y.; Kim, G.; Xia, W.; Kim, H.; Kim, S.J.; Park, M.K.; Hwang, J.P.; Gonzalez-Billalabeitia, E.; Hung, M.C.; et al. A UBE2O-AMPKalpha2 Axis that Promotes Tumor Initiation and Progression Offers Opportunities for Therapy. *Cancer Cell* **2017**, *31*, 208–224. [CrossRef] [PubMed]

57. Lee, J.O.; Lee, S.K.; Kim, N.; Kim, J.H.; You, G.Y.; Moon, J.W.; Jie, S.; Kim, S.J.; Lee, Y.W.; Kang, H.J.; et al. E3 ubiquitin ligase, WWP1, interacts with AMPKalpha2 and down-regulates its expression in skeletal muscle C2C12 cells. *J. Biol. Chem.* **2013**, *288*, 4673–4680. [CrossRef] [PubMed]

58. Chen, L.; Wang, J.; Zhang, Y.Y.; Yan, S.F.; Neumann, D.; Schlattner, U.; Wang, Z.X.; Wu, J.W. AMP-activated protein kinase undergoes nucleotide-dependent conformational changes. *Nat. Struct. Mol. Biol.* **2012**, *19*, 716–718. [CrossRef] [PubMed]

59. Xin, F.J.; Wang, J.; Zhao, R.Q.; Wang, Z.X.; Wu, J.W. Coordinated regulation of AMPK activity by multiple elements in the alpha-subunit. *Cell Res.* **2013**, *23*, 1237–1240. [CrossRef] [PubMed]

60. Hunter, R.W.; Foretz, M.; Bultot, L.; Fullerton, M.D.; Deak, M.; Ross, F.A.; Hawley, S.A.; Shpiro, N.; Viollet, B.; Barron, D.; et al. Mechanism of action of compound-13: An alpha1-selective small molecule activator of AMPK. *Chem. Biol.* **2014**, *21*, 866–879. [CrossRef] [PubMed]

61. Langendorf, C.G.; Ngoei, K.R.; Scott, J.W.; Ling, N.X.; Issa, S.M.; Gorman, M.A.; Parker, M.W.; Sakamoto, K.; Oakhill, J.S.; Kemp, B.E. Structural basis of allosteric and synergistic activation of AMPK by furan-2-phosphonic derivative C2 binding. *Nat. Commun.* **2016**, *7*, 10912. [CrossRef] [PubMed]

62. Crute, B.E.; Seefeld, K.; Gamble, J.; Kemp, B.E.; Witters, L.A. Functional domains of the alpha1 catalytic subunit of the AMP-activated protein kinase. *J. Biol. Chem.* **1998**, *273*, 35347–35354. [CrossRef] [PubMed]

63. Pang, T.; Xiong, B.; Li, J.Y.; Qiu, B.Y.; Jin, G.Z.; Shen, J.K.; Li, J. Conserved alpha-helix acts as autoinhibitory sequence in AMP-activated protein kinase alpha subunits. *J. Biol. Chem.* **2007**, *282*, 495–506. [CrossRef] [PubMed]

64. Palmieri, L.; Rastelli, G. alphaC helix displacement as a general approach for allosteric modulation of protein kinases. *Drug Discov. Today* **2013**, *18*, 407–414. [CrossRef] [PubMed]
65. Willows, R.; Sanders, M.J.; Xiao, B.; Patel, B.R.; Martin, S.R.; Read, J.; Wilson, J.R.; Hubbard, J.; Gamblin, S.J.; Carling, D. Phosphorylation of AMPK by upstream kinases is required for activity in mammalian cells. *Biochem. J.* **2017**, *474*, 3059–3073. [CrossRef] [PubMed]

9

AMPK Activation Reduces Hepatic Lipid Content by Increasing Fat Oxidation In Vivo

Marc Foretz [1,2,3,*], Patrick C. Even [4] and Benoit Viollet [1,2,3,*]

[1]   INSERM, U1016, Institut Cochin, Département d'Endocrinologie Métabolisme et Diabète, 24, rue du Faubourg Saint Jacques, 75014 Paris, France

[2]   CNRS, UMR8104, 75014 Paris, France

[3]   Université Paris Descartes, Sorbonne Paris Cité, 75014 Paris, France

[4]   UMR PNCA, AgroParisTech, INRA, Université Paris-Saclay, 75005 Paris, France; patrick.even@agroparistech.fr

*   Correspondence: marc.foretz@inserm.fr (M.F.); benoit.viollet@inserm.fr (B.V.)

Abstract: The energy sensor AMP-activated protein kinase (AMPK) is a key player in the control of energy metabolism. AMPK regulates hepatic lipid metabolism through the phosphorylation of its well-recognized downstream target acetyl CoA carboxylase (ACC). Although AMPK activation is proposed to lower hepatic triglyceride (TG) content via the inhibition of ACC to cause inhibition of de novo lipogenesis and stimulation of fatty acid oxidation (FAO), its contribution to the inhibition of FAO in vivo has been recently questioned. We generated a mouse model of AMPK activation specifically in the liver, achieved by expression of a constitutively active AMPK using adenoviral delivery. Indirect calorimetry studies revealed that liver-specific AMPK activation is sufficient to induce a reduction in the respiratory exchange ratio and an increase in FAO rates in vivo. This led to a more rapid metabolic switch from carbohydrate to lipid oxidation during the transition from fed to fasting. Finally, mice with chronic AMPK activation in the liver display high fat oxidation capacity evidenced by increased [C^{14}]-palmitate oxidation and ketone body production leading to reduced hepatic TG content and body adiposity. Our findings suggest a role for hepatic AMPK in the remodeling of lipid metabolism between the liver and adipose tissue.

Keywords: AMPK; liver; lipid metabolism; fatty acid oxidation; indirect calorimetry

1. Introduction

AMP-activated protein kinase (AMPK) is a phylogenetically conserved serine/threonine protein kinase viewed as a fuel gauge monitoring systemic and cellular energy status which plays a crucial role in protecting cellular function under energy-restricted conditions [1]. AMPK is a heterotrimeric protein consisting of a catalytic α-subunit and two regulatory subunits, β and γ, with each subunit existing as at least two isoforms. AMPK is activated in response to a variety of metabolic stresses that typically change the cellular AMP:ATP ratio caused by increasing ATP consumption or reducing ATP production, as seen following glucose deprivation and inhibition of mitochondrial oxidative phosphorylation as well as exercise and muscle contraction. Activation of AMPK initiates metabolic changes to reprogram metabolism by switching cells from an anabolic to a catabolic state, shutting down the ATP-consuming synthetic pathways and restoring energy balance. This regulation involves AMPK-dependent phosphorylation of key regulators of many important pathways [2,3].

One of the first identified AMPK targets is acetyl CoA carboxylase (ACC), playing a role in the control of fatty acid metabolism via the regulation of malonyl-CoA synthesis [4]. Malonyl-CoA is both a critical precursor of biosynthesis of fatty acids and an inhibitor of fatty acid uptake into mitochondria via the transport system involving carnitine palmitoyltransferase-1. By inhibiting ACC and lowering the concentration of its reaction product malonyl-CoA, AMPK activation is expected to coordinate the partitioning of fatty acids between oxidative and biosynthetic pathways by increasing fatty acid oxidation (FAO) capacity and inhibiting de novo lipogenesis (DNL), respectively. For these reasons, AMPK has emerged as a promising therapeutic target to treat metabolic disorders that occur in conditions such as nonalcoholic fatty liver disease (NAFLD). There is now literature precedence demonstrating the impact of hepatic AMPK activation in the setting of NAFLD [5]. In addition, recent advances in the development of allosteric and isoform-biased small-molecule AMPK activators have reinforced the potential for the pharmacological activation of AMPK as a treatment modality for hepatic steatosis [6–9]. Recent evidences showed that regulation of hepatic lipogenesis by AMPK activation mainly resides in the phosphorylation and inactivation of ACC, but not in the control of lipogenic gene expression [7,8,10]. Accordingly, genetic mouse models of hepatic AMPK deficiency and ACC with knock-in phosphorylation mutations confirmed the importance of the activation of AMPK and phosphorylation of ACC for the improvement of fatty liver disease induced by AMPK-activating drugs [7,8,11]. These studies also provided in vitro and in vivo evidence for the contribution of both hepatic FAO and DNL in the reduction of hepatic triglyceride (TG) accumulation mediated through pharmacological AMPK activation. However, one study recently questioned the effect of liver-specific activation of AMPK on FAO rates in vivo [10]. In that study, by using a genetic mouse model expressing in the liver a gain-of-function AMPKγ1 mutant, Woods et al. demonstrated that the effect of hepatic AMPK activation in the protection against hepatic steatosis is largely dependent on the suppression of de novo lipogenesis, but not on the stimulation of hepatic fatty acid oxidation [10]. Therefore, in the present study, we examined the impact of AMPK activation in the liver on hepatic lipid metabolism and determined its effect on FAO rates in vivo, measured by indirect calorimetry.

2. Results

As a first step to elucidating the impact of AMPK activation in the liver on hepatic lipid metabolism in vivo, we generated a mouse model in which AMPK activation specifically in the liver is achieved by expression of a constitutively active AMPK. Mice were injected intravenously with an adenovirus expressing a constitutively active form of AMPKα2 (Ad AMPK-CA) or GFP as a control (Ad GFP). This resulted in AMPK-CA expression restricted to the liver and undetectable in all other tissues (Figure 1A and data not shown). High levels of hepatic AMPK-CA expression were maintained until day 8, with no change in endogenous AMPKα expression (Figure 1A). ACC protein levels were low in Ad AMPK-CA livers, but the phospho-ACC/total ACC ratio was twice that in control livers, demonstrating an increase in ACC phosphorylation and therefore AMPK activation following AMPK-CA expression (Figure 1B). Decreased hepatic malonyl CoA levels in Ad AMPK-CA compared to Ad GFP livers also confirmed inhibition of ACC activity (Figure 1C). The levels of carnitine palmitoyltransferase (CPT)-1a and -2 mRNA expression were similar in the liver of Ad GFP and AMPK-CA mice (Figure 1D). There were no significant changes in body weight and food intake during the week following the injection of AMPK-CA or GFP adenoviruses (Figure 1E,F).

We studied the metabolic consequences of AMPK activation in the liver by monitoring energy expenditure and respiratory exchange ratio (RER) during a 22-h fasting period, determined by indirect calorimetry. The values of RER provide an approximation of carbohydrate and lipid oxidation to generate energy, ranging from 1.0 to approximately 0.7, respectively. In fed mice, the RER associated with the total and resting metabolism rates was lower in Ad AMPK-CA than in Ad GFP mice. During fasting, Ad AMPK-CA mice reached maximal rates of lipid oxidation after only 3 h of fasting, whereas such rates were not achieved until 12 h in Ad GFP mice (Figure 2, upper panel). Thus, AMPK activation in the liver enhances lipid oxidation, leading to a more rapid metabolic switch from carbohydrate to lipid oxidation during the transition from fed to fasting. Thereafter, RER stabilized at the same values in Ad GFP and Ad AMPK-CA mice, suggesting that the rate of lipid oxidation reached the same maximum intensity in both groups. Total and resting metabolic rates and spontaneous activity were similar in Ad GFP and Ad AMPK-CA mice (Figure 2, middle and lower panels). All mice exhibited a period of intense activity during the night period between 00:00 and 06:00 h. According to previous observations of fed mice, this hyperactivity was probably related to the fact that the mice were fasted and seeking for food. Analysis of the changes in total metabolism and RER induced by bursts of spontaneous activity that occurred during the light period (i.e., when RER was lower in Ad AMPK-CA than in control Ad GFP mice) showed that the utilization of glucose and lipids by the working muscles was very similar in both groups (same changes in RER). These observations agree with the conclusion that the rapid mobilization and utilization of lipids in Ad AMPK-CA mice in response to fasting is probably specific to the constitutive activation of AMPK in the liver.

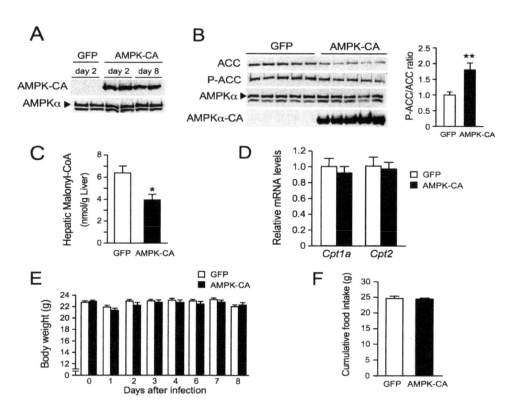

Figure 1. Effects of the expression of an active form of AMP-activated protein kinase (AMPK) in the liver on body weight and food intake. Ten-week-old male C57BL/6J mice received injections of adenovirus (Ad) expressing the green fluorescent protein (GFP) or a constitutively active form of AMPKα2 (AMPK-CA) and were studied for the indicated times after adenovirus injection and in the indicated nutritional state. (A) Western blot analysis of liver lysates with antibodies raised against pan-AMPKα and myc-tagged AMPK-CA was performed on days 2 and day 8 after adenovirus administration; (B) Western blot analysis of liver lysates from fed mice 48 h after the injection of Ad GFP or Ad AMPK-CA, with the antibodies indicated. Each lane represents a liver sample from an individual mouse. The panel on the right shows Ser79 phosphorylated acetyl CoA carboxylase/ total acetyl CoA carboxylase (P-ACC/ACC) ratios from the quantification of immunoblot images ($n = 5$); (C) Hepatic malonyl-CoA levels in 8 h-fasted mice 48 h after the injection of Ad GFP or Ad AMPK-CA ($n = 5$); (D) Effect of AMPK activation in the liver on the expression of *Cpt1a* and *Cpt2* genes. Total RNA was isolated from the liver of 24 h-fasted mice 48 h after the injection of Ad GFP or Ad AMPK-CA ($n = 6$). The expression of *Cpt1a* and *Cpt2* genes was assessed by real-time quantitative RT-PCR. Relative mRNA levels are expressed as fold-activation relative to levels in Ad GFP livers; (E) Body weight changes and (F) cumulative food intake measured for 8 days after adenovirus administration ($n = 11$–12 per group). Data are means ± standard error of mean (SEM). * $p < 0.05$, ** $p < 0.01$ versus Ad GFP mice by unpaired two-tailed Student's *t*-test (**B–D,F**) or by one-way ANOVA with Bonferroni post-hoc test (**E**).

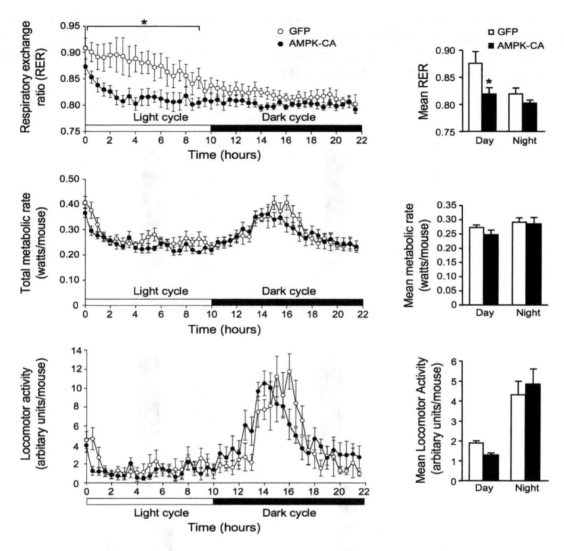

Figure 2. Effects of the expression of an active form of AMPK in the liver on respiratory exchange ratio. Whole-animal indirect calorimetry was used to assess oxygen consumption (VO_2) and carbon dioxide production (VCO_2) in mice infected with Ad GFP or Ad AMPK-CA for 48 h. Fed adenovirus-infected mice were placed in a metabolic chamber at 10:00 h. They were kept in the cage for 22 h, with free access to water but no food. Upper panel: The respiratory exchange ratio (RER = VO_2/VCO_2) was calculated from VO_2 and VCO_2 data and plotted at 15-min intervals. An RER of 1.0 is expected for glucose oxidation and an RER of 0.7 corresponds to lipid oxidation. The right panel shows mean RER results for Ad GFP and Ad AMPK-CA mice (n = 6 per group) during light or dark periods. Middle panel: Total metabolic rate. The right panel shows mean metabolic rates for Ad GFP and Ad AMPK-CA mice (n = 6 per group) during light and dark periods. Lower panel: Locomotor activity. The right panel shows the mean locomotor activity results for Ad GFP- and Ad AMPK-CA mice (n = 6 per group) during light and dark periods. Data are means ± SEM. * p < 0.05 versus Ad GFP mice by one-way ANOVA with Bonferroni post-hoc test.

We then investigated whether AMPK activation mediated increased hepatic fatty acid oxidation, by measuring the rate of β-oxidation through assays of [^{14}C]-palmitoyl-CoA oxidation in the liver (Figure 3A). Rates of palmitoyl-CoA oxidation were ~25% higher in Ad AMPK-CA mice than in control Ad GFP mice. Indirect support for the increase in FAO is provided by the increase in plasma ketone bodies in Ad AMPK-CA mice (Figure 3B) and a corresponding decrease in plasma triglyceride (TG) and free fatty acid (FFA) concentrations (Figure 3C,D). To determine whether AMPK activation increased fatty acid utilization, [C^{14}]-palmitate was injected into Ad AMPK-CA and Ad GFP mice and its incorporation into lipids measured. AMPK activation was associated with an increase in hepatic fatty acid uptake of ~25% (Figure 3E). These findings were correlated with increased expression of the fatty acid transporters *Slc27a4* (fatty acid transport protein 4, *Fatp4*), *Cd36* (fatty acid translocase, *Fat*), and *Fabp4* (fatty acid binding protein 4) (Figure 3F).

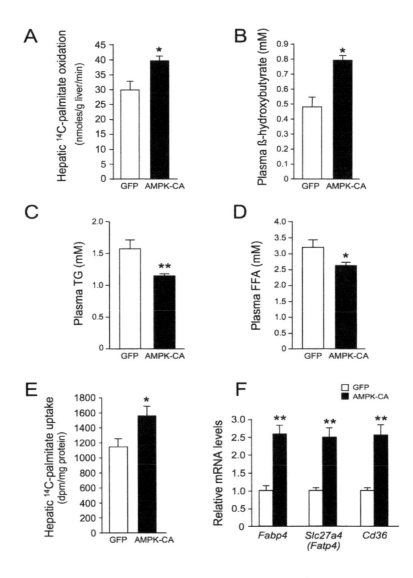

Figure 3. Long-term adenovirus-mediated expression of an active form of AMPK in the liver increases hepatic lipid oxidation and fatty acid uptake. Ten-week-old male C57BL/6J mice received injections of Ad GFP or Ad AMPK CA and were studied at the indicated times after adenovirus injection and in the indicated nutritional state. (**A**) Hepatic [1-^{14}C]-palmitate oxidation in fed mice 48 h after the injection of Ad GFP or Ad AMPK-CA (n = 4); (**B**) Plasma β-hydroxybutyrate levels in 24 h-fasted mice 48 h after the injection of Ad GFP or Ad AMPK-CA (n = 6); (**C**) Plasma triglyceride (TG) and (**D**) plasma free fatty acid (FFA) levels in overnight-fasted mice 8 days after the injection of Ad GFP or Ad AMPK-CA (n = 12); (**E**) Hepatic [1-^{14}C]-palmitate uptake in 24 h-fasted mice 48 h after the injection of Ad GFP or Ad AMPK-CA (n = 5); (**F**) Effect of AMPK activation in the liver on the expression of the fatty acid transporters. Total RNA was isolated from the liver of 24 h-fasted mice 48 h after the injection of Ad GFP or Ad AMPK-CA (n = 5). The expression of *Slc27a4* (*Fatp4*), *Cd36*, and *Fabp4* genes was assessed by real-time quantitative RT-PCR. Relative mRNA levels are expressed as fold-activation relative to levels in Ad GFP livers. Data are means ± SEM. * $p < 0.05$, ** $p < 0.01$ versus Ad GFP-infected mice by unpaired two-tailed Student's t-test.

Long-term (8 days) expression of AMPK-CA in liver was sufficient to modify hepatic lipid content and lowered TG levels by ~45% and cholesterol levels by ~10% (Figure 4A,B). This is in line with low abundance of lipid droplets in hepatocytes from Ad AMPK-CA compared to Ad GFP mice revealed by liver ultrastructure changes using transmission electron microscopy (Figure 4C). The increase in hepatic β-oxidation was related to systemic changes in adiposity and resulted in a significant decrease in body fat mass (Figure 5A). This decrease was confirmed by the careful weighing of adipose tissue from epidydimal and inguinal fat pads (Figure 5B). The epidydimal fat pads were much smaller, as was adipocyte diameter (Figure 5C–E). As a result, plasma leptin concentration, a marker of adiposity, was halved in Ad AMPK-CA mice (Figure 5F).

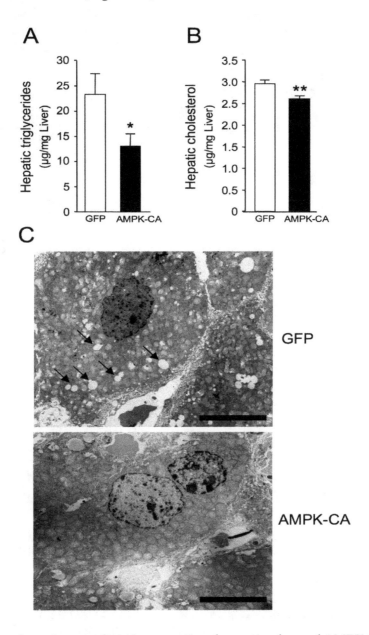

Figure 4. Long-term adenovirus-mediated expression of an active form of AMPK in the liver reduces hepatic lipid accumulation. Ten-week-old male C57BL/6J mice received injections of Ad GFP or Ad AMPK-CA. Fed mice were studied on day 8 after adenovirus administration. (**A**) Liver triglyceride content and (**B**) liver cholesterol content (n = 9–10). Data are means ± SEM. * $p < 0.05$, ** $p < 0.01$ versus Ad GFP-infected mice by unpaired two-tailed Student's t-test; (**C**) Representative images of transmission electron microscopy showing the ultrastructure change in Ad GFP and Ad AMPK-CA livers. Scale bar: 10 μm. Black arrowheads in insets depict lipid droplets.

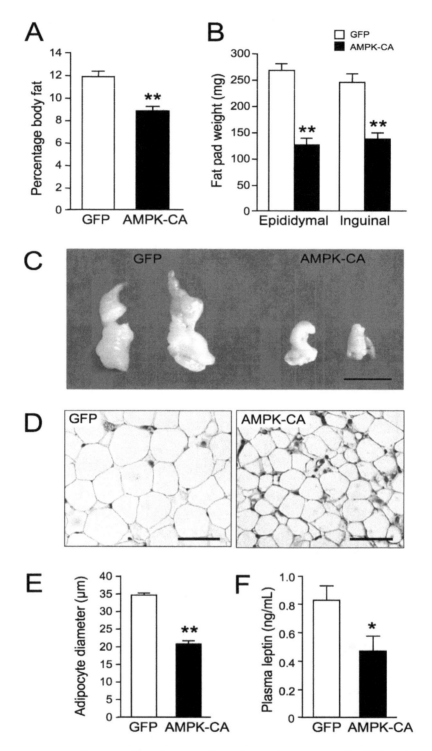

Figure 5. Long-term adenovirus-mediated expression of an active form of AMPK in the liver diminishes peripheral adiposity. Ten-week-old male C57BL/6J mice received injections of Ad GFP or Ad AMPK-CA. Fed mice were studied on day 8 after adenovirus administration. (**A**) Body fat content was measured by dual X-ray absorptiometry ($n = 10$ per group); (**B**) Epididymal and inguinal subcutaneous fat-pad weight ($n = 10$ per group); (**C**) Representative epididymal white fat pads fixed in formalin. Scale bar: 1 cm; (**D**) Representative hematoxylin-and-eosin-stained sections of epididymal adipose tissues. Scale bars: 50 μm; (**E**) Mean adipocyte size in epididymal white adipose tissues. The diameter of at least 200 cells per sample was determined ($n = 4$ mice per group); (**F**) Plasma leptin levels in fed mice ($n = 10$ per group). Data are means $\pm$ SEM. * $p < 0.05$, ** $p < 0.001$ versus Ad GFP mice by unpaired two-tailed Student's t-test.

3. Discussion

Activation of AMPK has been reported to reduce hepatic lipid content in many preclinical studies, yet the importance of hepatic FAO and DNL for its TG-lowering effect has been unclear [6–10]. Here, we report that AMPK activation in the liver is capable of significant reduction in liver TG through the stimulation of fatty acid utilization, as evidenced by a reduction of RER and increased palmitate oxidation and ketone body production. These results are reminiscent of the acute effect of the direct AMPK activator A-769662, showing a concurrent drop in RER in fed rats and leading to the reduction in liver TGs after chronic treatment of obese mice [9]. Importantly, it has been demonstrated that A-769662 acts in an AMPK-dependent manner to induce fat utilization [12]. We also recently confirmed that A-769662 was capable to restore hepatic fatty acid oxidation after chronic treatment of a fatty liver mouse model [8]. Further support for a significant role of lipid oxidation following hepatic AMPK activation recently came from a study investigating the therapeutic beneficial of the β1-biased activator PF-06409577 in a high-fat-fed mouse model, where the contribution of de novo lipogenesis is essentially negligible for hepatic TG accumulation [7]. Acute or 42 days' dosing with PF-06409577 resulted in a large increase in circulating β-hydroxybutyrate and lower hepatic TG, an effect that was lost in mice lacking AMPK specifically in the liver [7]. In addition to the impact on FAO, activation of AMPK in the liver has also been largely documented as a source of inhibition of DNL [6–10]. Thus, our results substantially contribute to the current view that following hepatic AMPK activation, lowering of hepatic TGs may arise through the capacity of AMPK to combine between both inhibition of TG synthesis and stimulation of lipid utilization [5,7,8]. Definitive evidence for a dual effect of hepatic AMPK activation on lipid synthesis and utilization is provided by in vitro and in vivo studies with AMPK-deficient mouse models and primary culture of hepatocytes treated with various pharmacological activators of AMPK [7,8]. The balance and contribution between inhibition of DNL and stimulation of FAO may depend on the source of hepatic TGs at the origin of the development of hepatic steatosis. Consistent with this notion, pharmacological AMPK-induced inhibition of DNL has been suggested to play a significant role in the improvement of hepatic steatosis of animal models where DNL mainly contributes to hepatic TG accumulation [7,8]. Similarly, transgenic mice expressing specifically in the liver a naturally occurring gain-of-function AMPKγ1 mutant were completely protected against hepatic steatosis when fed a high-fructose diet, known to increase hepatic lipogenesis [10]. In that study, the effect of AMPK activation was relying exclusively on the inhibition of DNL, because no difference in FAO and RER was detected, despite hepatic AMPK activation. However, it is possible that mice fed a high-sucrose diet preferentially oxidize carbohydrates as their primary source of energy and this obscures the effect of AMPK activation on fat oxidation due to the competition between glucose and fat for substrate oxidation [13]. Intriguingly, in the same study, these mice expressing a gain-of-function AMPKγ1 mutant in the liver failed to stimulate FAO and reduce hepatic lipids when fed a high-fat diet [10]. These results contrast with the effectiveness of the AMPKα2-CA mutant used in the present study and various direct AMPK activators in stimulating hepatic FAO and reducing hepatic TG accumulation in vivo [7,8]. What causes this discrepancy is unclear, but we speculate that basal AMPK activity increased by mutation of the AMPKγ1 subunit is insufficient to fully phosphorylate and inactivate ACC and therefore presumably to stimulate FAO in vivo. Given the observation of the lowering in AMPK activity in the liver of high-fat-fed mice and fatty liver mouse models [8,14–16], AMPK activation probably needs to reach a higher threshold before the stimulation of FAO can be effective [8], providing an alternative explanation for the absence of a significant effect on hepatic lipid content in mice expressing the gain-of-function AMPKγ1 mutant on a high-fat diet.

AMPK has been proposed as a potential pharmacological target for the treatment of NAFLD due to its capacity to increase FAO and inhibit DNL in the liver [5]. One mechanism by which AMPK regulates the partitioning of fatty acids between oxidative and biosynthetic pathways is accomplished by the phosphorylation and inactivation of ACC, the rate-controlling enzyme for the synthesis of malonyl-CoA, which is both a critical precursor for biosynthesis of fatty acids and a potent inhibitor of long-chain fatty acyl-CoA transport into mitochondria for β-oxidation. This is supported by the observation of increased fatty acid synthesis and reduced FAO in the liver of mice lacking AMPK phosphorylation sites on ACC1/ACC2 [11]. In addition, these mice are resistant to the inhibition of lipogenesis in vivo induced by the AMPK-activating drugs metformin and A-769662 [11]. The effects of metformin and the direct AMPK activator PF-06409577 on lipid synthesis are abolished in hepatocytes isolated from these mice as well as liver AMPK-deficient mice [7,8,17]. The direct AMPK activator A-769662 also failed to increase fatty acid oxidation in these hepatocytes with mutation at the AMPK phosphorylation sites on ACC isoforms [11]. Thus, the action of AMPK in the improvement of hepatic steatosis is likely mediated through the phosphorylation of ACC to increase FAO and suppress DNL [11]. Recent studies performed in mice and humans treated with pharmacological inhibitors of ACC support the concept that direct inhibition of ACC is a promising therapeutic option for the management of fatty liver disease [18,19].

We have previously shown that short-term (48 h) expression of AMPK-CA in the liver paradoxically induced a concomitant hepatic lipid accumulation and increase in fatty acid oxidation [20]. Interestingly, a similar phenotype is observed during the physiological response to fasting, where hepatic TG contents rise significantly [21]. We hypothesized that in response to short-term AMPK activation, the hepatic lipid oxidation capacity is overloaded by the uptake of mobilized fatty acids from adipose tissue, which are stored temporarily as TGs in the liver until they are oxidized [20]. As anticipated, we report here that long-term (8 days) expression of AMPK-CA finally leads to a decrease in hepatic lipid content, but also in a reduction of body adiposity. However, body weight of Ad AMPK-CA mice was not significantly altered due to the small amount of fat mass loss compared to total body weight. Our data are corroborated with the effect of chronic treatment with the AMPK activators metformin, AICAR, or A-769662 in mice, which are associated with reduced fatty liver and fat pad weight [9,22–24], although no change in body composition was reported in diet-induced obese (DIO) mice treated with the AMPK β1-biased activator PF-06409577 [7]. Overall, these observations suggest a role for hepatic AMPK in the remodeling of lipid metabolism through crosstalk between liver and adipose tissue. However, the nature of the hepatic signal triggering the mobilization of fatty acids from adipose tissue to the liver remains to be elucidated. One possibility is the secretion of liver-derived proteins known as hepatokines, which could act on adipose tissue to stimulate lipolysis. FGF21 and Angptl3 are reasonable candidates playing important roles in the regulation of lipid metabolism [25,26]. Interestingly, FGF21 expression is induced by metformin and AICAR in hepatocytes [27].

In conclusion, chronic AMPK activation in the liver increases lipid oxidation, thereby decreasing hepatic lipid content and body adiposity, suggesting a role for hepatic AMPK in the remodeling of lipid metabolism between the liver and adipose tissue. Overall, our data emphasizes the potential therapeutic implications for hepatic AMPK activation in vivo.

4. Material and Methods

4.1. Reagents and Antibodies

Adenovirus expressing GFP and a myc epitope-tagged constitutively active form of AMPKα2 (AMPK-CA) were generated as previously described [20]. Primary antibodies directed against total AMPKα (#2532), total acetyl-CoA carboxylase (ACC) (#3676), and ACC phosphorylated at Ser79 (#3661) were purchased from Cell Signaling Technology (Danvers, MA, USA) and myc epitope tag (clone 9E10) from Sigma (Saint-Quentin-Fallavier, France). HRP-conjugated secondary antibodies were purchased from Calbiochem (Burlington, MA, USA).

4.2. Animals

Animal studies were approved by the Paris Descartes University ethics committees (no. CEEA34.BV.157.12) and performed under a French authorization to experiment on vertebrates (no. 75-886) in accordance with the European guidelines. C57BL/6J mice were obtained from Harlan France (Gannat, France). All mice were maintained in a barrier facility under a 12-h light/12-h dark cycle with free access to water and standard mouse diet (in terms of energy: 65% carbohydrate, 11% fat, 24% protein).

4.3. Metabolic Parameters

Blood was collected into heparin-containing tubes, and centrifuged to obtain plasma. Plasma leptin levels were assessed using mouse ELISA kit (Crystal Chem, Elk Grove Village, IL, USA). Plasma triglyceride, free fatty acid, and β-hydroxybutyrate levels were determined enzymatically (Dyasis, Grabels, France).

4.4. Liver Triglyceride, Cholesterol, and Malonyl-CoA Contents

For the extraction of total lipids from the liver, a portion of frozen tissue was homogenized in acetone (500 µL/50 mg tissue) and incubated on a rotating wheel overnight at 4 °C. Samples were centrifuged at 4 °C for 10 min at $5000 \times g$, and the triglyceride and cholesterol concentrations of the supernatants were determined with enzymatic colorimetric assays (Diasys, Grabels, France). Hepatic malonyl CoA ester content was measured using a modified high-performance liquid chromatography method [28].

4.5. Indirect Calorimetry

Mice were placed in a metabolic cage from 10:00 h until 08:00 h the next day (22 h). The metabolic cage was continuously connected to an open-circuit, indirect calorimetry system controlled by a computer running a data acquisition and analysis program, as previously described [29]. Mice were housed with free access to water but no food. Air flow through the chamber was regulated at 0.5 L/min by a mass flow-meter, and temperature was maintained close to thermoneutrality (30 °C ± 1 °C). Oxygen consumption (VO_2) and carbon dioxide production (VCO_2) were recorded at one-second intervals. Spontaneous activity was measured by means of 3 piezo-electric force transducers positioned in triangle under the metabolic cage, with sampling of the electrical signal at 100 Hz. Data were averaged every 10 s and stored on a hard disk for further processing. Computer-assisted processing of respiratory exchanges and spontaneous activity signals was performed to extract the respiratory exchanges specifically associated with spontaneous activity (Kalman filtering method) [29]. This separation provided information about total, resting, and activity-related O_2 consumption and CO_2 production. The respiratory exchange ratio (RER) was calculated as the ratio of VCO_2 produced/VO_2 consumed.

4.6. Assessment of Fatty Acid Oxidation in Liver Homogenates

The rate of mitochondrial palmitate oxidation was measured in fresh liver homogenate from fed mice anesthetized with a xylamine/ketamine mixture via intraperitoneal injection according to a modified version of the method described by Yu et al. [30]. The rate of palmitate oxidation was assessed by collecting and counting the radiolabeled acid-soluble metabolites (ASMs) produced from the oxidation of [1-^{14}C]-palmitate. Briefly, a portion of liver (200 mg) was homogenized in 19 volumes of ice-cold buffer containing 250 mM sucrose, 1 mM EDTA, and 10 mM Tris–HCl pH 7.4. For the assessment of palmitate oxidation, 75 µL of liver homogenate was incubated with 425 µL of reaction mixture (pH 7.4) in a 25 mL flask. The reaction mixture contained 100 mM sucrose, 10 mM Tris–HCl, 80 mM KCl, 5 mM K_2HPO_4, 1 mM $MgCl_2$, 0.2 mM EDTA, 1 mM dithiothreitol, 5.5 mM ATP, 1 mM NAD, 0.03 mM cytochrome C, 2 mM L-carnitine, 0.5 mM malate, and 0.1 mM coenzyme A. The reaction was started by adding 120 µM palmitate plus 1.7 µCi [1-^{14}C]-palmitate (56 mCi/mmol) complexed with fatty acid-free bovine serum albumin in a 5:1 molar ratio. Each homogenate was incubated in triplicate in the presence or absence of 75 µM antimycin A plus 10 µM rotenone to inhibit mitochondrial β-oxidation. After 30 min of incubation at 37 °C in a shaking water bath, the reaction was stopped by adding 200 μL of ice-cold 3 M perchloric acid. The radiolabeled ASMs produced from the oxidation of [1-^{14}C]-palmitate were assayed in the supernatants of the acid precipitate. ASM radioactivity was determined by liquid scintillation counting. Mitochondrial β-oxidation was calculated as the difference between the total β-oxidation rate and the peroxisomal β-oxidation rate, which was determined following incubation of the homogenate with antimycin A and rotenone. Data are expressed in nanomoles of radiolabeled ASM produced per gram of liver per hour.

4.7. Palmitate Uptake by the Liver

The in vivo uptake of palmitate by the liver was assessed by injection of 10 μCi [1-^{14}C]-palmitate (56 mCi/mmol) complexed with 1% fatty acid-free bovine serum albumin in a final volume of 200 μL PBS via the inferior vena cava in anesthetized 24 h-fasted mice by the intraperitoneal injection of a xylamine/ketamine mixture. Four minutes after injection, the superior vena cava was clamped and the hepatic portal vein was sectioned. A needle was inserted into the inferior vena cava toward the liver, and 10 mL of ice-cold PBS was injected under pressure with a syringe. At the end of this procedure, the liver was pale and the fluid emerging from the portal vein was clear. The liver was removed and used for lipid extraction and for the measurement of radioactivity by scintillation counting. Rates of palmitate uptake are expressed as disintegrations per minute (dpm) per gram of protein per hour.

4.8. Fat Mass and Histomorphometry

The total body fat content of mice was determined by dual energy X-ray absorptiometry (Lunar PIXImus2 mouse densitometer; GE Healthcare, Chicago, IL, USA), in accordance with the manufacturer's instructions. Body weight was determined and the left and right epididymal and inguinal white fat pads were harvested and weighed. Epididymal fat pads were then fixed in 10% neutral buffered formalin and embedded in paraffin. Tissues were cut into 4-μm sections and stained with hematoxylin and eosin. For the determination of adipocyte size, photomicrographs of the stained sections were obtained at ×100 magnification. Mean adipocyte diameter was calculated from measurements of at least 200 cells per sample.

4.9. Injection of Recombinant Adenovirus

Male C57BL/6J mice were anesthetized with isoflurane before the injection (between 9:00 and 10:00 h) into the penis vein of 1×10^9 pfu of either Ad GFP or Ad AMPK-CA in a final volume of 200 μL of sterile 0.9% NaCl. Mice were sacrificed 48 h or 8 days after adenovirus injection, as indicated in the figure legends. For the eight-day studies, mouse weight and food intake were measured daily.

4.10. Isolation of Total mRNA and Quantitative RT-PCR Analysis

Total RNA from mouse liver tissue was extracted using Trizol (Invitrogen, Carlsbad, CA, USA), and single-strand cDNA was synthesized from 5 μg of total RNA with random hexamer primers (Applied Biosystems, Foster City, CA, USA) and Superscript II (Life Technologies, Carlsbad, CA, USA). Real-time RT-PCR reactions were carried out in a final volume of 20 μL containing 125 ng of reverse-transcribed total RNA, 500 nM of primers, and 10 μL of 2× PCR mix containing Sybr Green (Roche, Meylan, France). The reactions were performed in 96-well plates in a LightCycler 480 instrument (Roche) with 40 cycles. The relative amounts of the mRNAs studied were determined by means of the second-derivative maximum method, with LightCycler 480 analysis software and 18S RNA as the invariant control for all studies. The sense and antisense PCR primers used, respectively, were as follows: for *Cpt1a*, 5'-AGATCAATCGGACCCTAGACAC-3', 5'-CAGCGAGTAGCGCATAGTCA-3'; for *Cpt2*, 5'-CAGCACAGCATCGTACCCA-3', 5'-TCCCAATG CCGTTCTCAAAAT-3'; for *Cd36*, 5'-TGGCTAAATGAGACTGGGACC-3', 5'-ACATCACCACTCCAATCCCAAG-3'; for *Slc27a4* (*Fatp4*), 5'-GCACACTCAGCCGCCTGCTTCA-3', 5'-TCACAGCTTCTCCTCGCCTGCCTG-3'; for *Fabp4*, 5'-GT

GATGCCTTTGTGGGAACCT-3′, 5′-ACTCTTGTGGAAGTCGCCT-3′; for 18S, 5′-GTAACCCGTTGAA CCCCATT-3′, 5′-CCATCCAATCGGTAGTAGCG-3′.

4.11. Western Blot Analysis

After the indicated incubation time in the figure legends, cultured hepatocytes were lysed in ice-cold lysis buffer containing 50 mM Tris, pH 7.4, 1% Triton X-100, 150 mM NaCl, 1 mM EDTA, 1 mM EGTA, 10% glycerol, 50 mM NaF, 5 mM sodium pyrophosphate, 1 mM Na_3VO_4, 25 mM sodium-β-glycerophosphate, 1 mM DTT, 0.5 mM PMSF and protease inhibitors (Complete Protease Inhibitor Cocktail; Roche). Lysates were sonicated on ice for 15 s to shear DNA and reduce viscosity. The tissues were homogenized in ice-cold lysis buffer using a ball-bearing homogenizer (Retsch, Eragny, France). The homogenate was centrifuged for 10 min at $10,000 \times g$ at 4 °C, and the supernatants were removed for determination of total protein content with a BCA protein assay kit (Thermo Fisher Scientific, Waltham, MA, USA). Fifty micrograms of protein from the supernatant were separated on 7.5% or 10% SDS-PAGE gels and transferred to nitrocellulose membranes. The membranes were blocked for 30 min at 37 °C with Tris-buffered saline supplemented with 0.05% NP40 and 5% nonfat dry milk. Immunoblotting was performed following standard procedures, and the signals were detected by chemiluminescence reagents (Thermo). X-ray films were scanned, and band intensities were quantified by Image J (NIH) densitometry analysis.

4.12. Transmission Electron Microscopy

Livers were fixed in 3% glutaraldehyde, 0.1 M sodium phosphate buffer (pH 7.4) for 24 h at 4 °C, postfixed with 1% osmium tetroxide, dehydrated with 100% ethanol, and embedded in epoxy resin. For ultrastructure analysis, ultrathin slices (70–100 nm thick) were cut from the resin blocks with a Reichert Ultracut S ultramicrotome (Reichert Technologies, Depew, NY, USA), stained with lead citrate and uranyl acetate, and examined in a transmission electron microscope (model 1011; JEOL, Tokyo, Japan) at the Cochin Institute electron microscopy facility.

4.13. Statistical Analysis

Results are expressed as means ± standard error of mean (SEM). Comparisons between groups were made by unpaired two-tailed Student's t-test or ANOVA for multiple comparisons where appropriate. Differences between groups were considered statistically significant when $p < 0.05$.

Author Contributions: M.F. designed and performed experiments, interpreted data, and wrote the manuscript. P.C.E. performed and analyzed indirect calorimetry experiments. B.V. interpreted the data and wrote the manuscript.

Acknowledgments: The authors thank Alain Schmitt (Cellular imaging: Electron microscopy facility at Institut Cochin, Paris, France) for transmission electron microscopy pictures and Jason Dyck (University of Alberta, Edmonton, Canada) for malonyl-CoA assays.

References

1. Hardie, D.G. AMP-activated protein kinase: Maintaining energy homeostasis at the cellular and whole-body levels. *Annu. Rev. Nutr.* **2014**, *34*, 31–55. [CrossRef] [PubMed]
2. Hardie, D.G.; Lin, S.C. AMP-activated protein kinase—Not just an energy sensor. *F1000Res* **2017**, *6*, 1724. [CrossRef] [PubMed]
3. Day, E.A.; Ford, R.J.; Steinberg, G.R. AMPK as a Therapeutic Target for Treating Metabolic Diseases. *Trends Endocrinol. Metab.* **2017**, *28*, 545–560. [CrossRef] [PubMed]

4. Viollet, B.; Foretz, M.; Guigas, B.; Horman, S.; Dentin, R.; Bertrand, L.; Hue, L.; Andreelli, F. Activation of AMP-activated protein kinase in the liver: A new strategy for the management of metabolic hepatic disorders. *J. Physiol.* **2006**, *574*, 41–53. [CrossRef] [PubMed]

5. Smith, B.K.; Marcinko, K.; Desjardins, E.M.; Lally, J.S.; Ford, R.J.; Steinberg, G.R. Treatment of nonalcoholic fatty liver disease: Role of AMPK. *Am. J. Physiol. Endocrinol. Metab.* **2016**, *311*, E730–E740. [CrossRef] [PubMed]

6. Gomez-Galeno, J.E.; Dang, Q.; Nguyen, T.H.; Boyer, S.H.; Grote, M.P.; Sun, Z.; Chen, M.; Craigo, W.A.; van Poelje, P.D.; MacKenna, D.A.; et al. A Potent and Selective AMPK Activator That Inhibits de Novo Lipogenesis. *ACS Med. Chem. Lett.* **2010**, *1*, 478–482. [CrossRef] [PubMed]

7. Esquejo, R.M.; Salatto, C.T.; Delmore, J.; Albuquerque, B.; Reyes, A.; Shi, Y.; Moccia, R.; Cokorinos, E.; Peloquin, M.; Monetti, M.; et al. Activation of Liver AMPK with PF-06409577 Corrects NAFLD and Lowers Cholesterol in Rodent and Primate Preclinical Models. *EBioMedicine* **2018**, *31*, 122–132. [CrossRef] [PubMed]

8. Boudaba, N.; Marion, A.; Huet, C.; Pierre, R.; Viollet, B.; Foretz, M. AMPK Re-Activation Suppresses Hepatic Steatosis but its Downregulation Does Not Promote Fatty Liver Development. *EBioMedicine* **2018**, *28*, 194–209. [CrossRef] [PubMed]

9. Cool, B.; Zinker, B.; Chiou, W.; Kifle, L.; Cao, N.; Perham, M.; Dickinson, R.; Adler, A.; Gagne, G.; Iyengar, R.; et al. Identification and characterization of a small molecule AMPK activator that treats key components of type 2 diabetes and the metabolic syndrome. *Cell Metab.* **2006**, *3*, 403–416. [CrossRef] [PubMed]

10. Woods, A.; Williams, J.R.; Muckett, P.J.; Mayer, F.V.; Liljevald, M.; Bohlooly, Y.M.; Carling, D. Liver-Specific Activation of AMPK Prevents Steatosis on a High-Fructose Diet. *Cell Rep.* **2017**, *18*, 3043–3051. [CrossRef] [PubMed]

11. Fullerton, M.D.; Galic, S.; Marcinko, K.; Sikkema, S.; Pulinilkunnil, T.; Chen, Z.P.; O'Neill, H.M.; Ford, R.J.; Palanivel, R.; O'Brien, M.; et al. Single phosphorylation sites in Acc1 and Acc2 regulate lipid homeostasis and the insulin-sensitizing effects of metformin. *Nat. Med.* **2013**, *19*, 1649–1654. [CrossRef] [PubMed]

12. Hawley, S.A.; Fullerton, M.D.; Ross, F.A.; Schertzer, J.D.; Chevtzoff, C.; Walker, K.J.; Peggie, M.W.; Zibrova, D.; Green, K.A.; Mustard, K.J.; et al. The ancient drug salicylate directly activates AMP-activated protein kinase. *Science* **2012**, *336*, 918–922. [CrossRef] [PubMed]

13. Randle, P.J.; Garland, P.B.; Hales, C.N.; Newsholme, E.A. The glucose fatty-acid cycle. Its role in insulin sensitivity and the metabolic disturbances of diabetes mellitus. *Lancet* **1963**, *1*, 785–789. [CrossRef]

14. Lindholm, C.R.; Ertel, R.L.; Bauwens, J.D.; Schmuck, E.G.; Mulligan, J.D.; Saupe, K.W. A high-fat diet decreases AMPK activity in multiple tissues in the absence of hyperglycemia or systemic inflammation in rats. *J. Physiol. Biochem.* **2013**, *69*, 165–175. [CrossRef] [PubMed]

15. Muse, E.D.; Obici, S.; Bhanot, S.; Monia, B.P.; McKay, R.A.; Rajala, M.W.; Scherer, P.E.; Rossetti, L. Role of resistin in diet-induced hepatic insulin resistance. *J. Clin. Investig.* **2004**, *114*, 232–239. [CrossRef] [PubMed]

16. Yu, X.; McCorkle, S.; Wang, M.; Lee, Y.; Li, J.; Saha, A.K.; Unger, R.H.; Ruderman, N.B. Leptinomimetic effects of the AMP kinase activator AICAR in leptin-resistant rats: Prevention of diabetes and ectopic lipid deposition. *Diabetologia* **2004**, *47*, 2012–2021. [CrossRef] [PubMed]

17. Hawley, S.A.; Ford, R.J.; Smith, B.K.; Gowans, G.J.; Mancini, S.J.; Pitt, R.D.; Day, E.A.; Salt, I.P.; Steinberg, G.R.; Hardie, D.G. The Na$^+$/Glucose Cotransporter Inhibitor Canagliflozin Activates AMPK by Inhibiting Mitochondrial Function and Increasing Cellular AMP Levels. *Diabetes* **2016**, *65*, 2784–2794. [CrossRef] [PubMed]

18. Harriman, G.; Greenwood, J.; Bhat, S.; Huang, X.; Wang, R.; Paul, D.; Tong, L.; Saha, A.K.; Westlin, W.F.; Kapeller, R.; et al. Acetyl-CoA carboxylase inhibition by ND-630 reduces hepatic steatosis, improves insulin sensitivity, and modulates dyslipidemia in rats. *Proc. Natl. Acad. Sci. USA* **2016**, *113*, E1796–E1805. [CrossRef] [PubMed]

19. Kim, C.W.; Addy, C.; Kusunoki, J.; Anderson, N.N.; Deja, S.; Fu, X.; Burgess, S.C.; Li, C.; Ruddy, M.; Chakravarthy, M.; et al. Acetyl CoA Carboxylase Inhibition Reduces Hepatic Steatosis but Elevates Plasma Triglycerides in Mice and Humans: A Bedside to Bench Investigation. *Cell Metab.* **2017**, *26*, 576. [CrossRef] [PubMed]

20. Foretz, M.; Ancellin, N.; Andreelli, F.; Saintillan, Y.; Grondin, P.; Kahn, A.; Thorens, B.; Vaulont, S.; Viollet, B. Short-term overexpression of a constitutively active form of AMP-activated protein kinase in the liver leads to mild hypoglycemia and fatty liver. *Diabetes* **2005**, *54*, 1331–1339. [CrossRef] [PubMed]

21. Lin, X.; Yue, P.; Chen, Z.; Schonfeld, G. Hepatic triglyceride contents are genetically determined in mice: Results of a strain survey. *Am. J. Physiol. Gastrointest. Liver Physiol.* **2005**, *288*, G1179–G1189. [CrossRef] [PubMed]

22. Borgeson, E.; Wallenius, V.; Syed, G.H.; Darshi, M.; Lantero Rodriguez, J.; Biorserud, C.; Ragnmark Ek, M.; Bjorklund, P.; Quiding-Jarbrink, M.; Fandriks, L.; et al. AICAR ameliorates high-fat diet-associated pathophysiology in mouse and ex vivo models, independent of adiponectin. *Diabetologia* **2017**, *60*, 729–739. [CrossRef] [PubMed]

23. Lin, H.Z.; Yang, S.Q.; Chuckaree, C.; Kuhajda, F.; Ronnet, G.; Diehl, A.M. Metformin reverses fatty liver disease in obese, leptin-deficient mice. *Nat. Med.* **2000**, *6*, 998–1003. [CrossRef] [PubMed]

24. Henriksen, B.S.; Curtis, M.E.; Fillmore, N.; Cardon, B.R.; Thomson, D.M.; Hancock, C.R. The effects of chronic AMPK activation on hepatic triglyceride accumulation and glycerol 3-phosphate acyltransferase activity with high fat feeding. *Diabetol. Metab. Syndr.* **2013**, *5*, 29. [CrossRef] [PubMed]

25. Hotta, Y.; Nakamura, H.; Konishi, M.; Murata, Y.; Takagi, H.; Matsumura, S.; Inoue, K.; Fushiki, T.; Itoh, N. Fibroblast growth factor 21 regulates lipolysis in white adipose tissue but is not required for ketogenesis and triglyceride clearance in liver. *Endocrinology* **2009**, *150*, 4625–4633. [CrossRef] [PubMed]

26. Shimamura, M.; Matsuda, M.; Kobayashi, S.; Ando, Y.; Ono, M.; Koishi, R.; Furukawa, H.; Makishima, M.; Shimomura, I. Angiopoietin-like protein 3, a hepatic secretory factor, activates lipolysis in adipocytes. *Biochem. Biophys. Res. Commun.* **2003**, *301*, 604–609. [CrossRef]

27. Nygaard, E.B.; Vienberg, S.G.; Orskov, C.; Hansen, H.S.; Andersen, B. Metformin stimulates FGF21 expression in primary hepatocytes. *Exp. Diabetes Res.* **2012**, *2012*, 465282. [CrossRef] [PubMed]

28. Dyck, J.R.; Barr, A.J.; Barr, R.L.; Kolattukudy, P.E.; Lopaschuk, G.D. Characterization of cardiac malonyl-CoA decarboxylase and its putative role in regulating fatty acid oxidation. *Am. J. Physiol.* **1998**, *275*, H2122–H2129. [CrossRef] [PubMed]

29. Even, P.C.; Mokhtarian, A.; Pele, A. Practical aspects of indirect calorimetry in laboratory animals. *Neurosci. Biobehav. Rev.* **1994**, *18*, 435–447. [CrossRef]

30. Yu, X.X.; Drackley, J.K.; Odle, J. Rates of mitochondrial and peroxisomal beta-oxidation of palmitate change during postnatal development and food deprivation in liver, kidney and heart of pigs. *J. Nutr.* **1997**, *127*, 1814–1821. [CrossRef] [PubMed]

10

AMP-activated Protein Kinase Controls Immediate Early Genes Expression Following Synaptic Activation through the PKA/CREB Pathway

Sébastien Didier [†], Florent Sauvé [†], Manon Domise, Luc Buée, Claudia Marinangeli and Valérie Vingtdeux *

Université de Lille, Inserm, Centre Hospitalo-Universitaire de Lille, UMR-S1172—JPArc—Centre de Recherche Jean-Pierre AUBERT, F-59000 Lille, France; sebastien.didier1@gmail.com (S.D.); florent.sauve@inserm.fr (F.S.); manon.domise@inserm.fr (M.D.); luc.buee@inserm.fr (L.B.); c.marinangeli83@gmail.com (C.M.)
* Correspondence: valerie.vingtdeux@inserm.fr
† These authors contributed equally to this work.

Abstract: Long-term memory formation depends on the expression of immediate early genes (IEGs). Their expression, which is induced by synaptic activation, is mainly regulated by the $3',5'$-cyclic AMP (cAMP)-dependent protein kinase/cAMP response element binding protein (cAMP-dependent protein kinase (PKA)/ cAMP response element binding (CREB)) signaling pathway. Synaptic activation being highly energy demanding, neurons must maintain their energetic homeostasis in order to successfully induce long-term memory formation. In this context, we previously demonstrated that the expression of IEGs required the activation of AMP-activated protein kinase (AMPK) to sustain the energetic requirements linked to synaptic transmission. Here, we sought to determine the molecular mechanisms by which AMPK regulates the expression of IEGs. To this end, we assessed the involvement of AMPK in the regulation of pathways involved in the expression of IEGs upon synaptic activation in differentiated primary neurons. Our data demonstrated that AMPK regulated IEGs transcription via the PKA/CREB pathway, which relied on the activity of the soluble adenylyl cyclase. Our data highlight the interplay between AMPK and PKA/CREB signaling pathways that allows synaptic activation to be transduced into the expression of IEGs, thus exemplifying how learning and memory mechanisms are under metabolic control.

Keywords: AMPK; synaptic activation; PKA; CREB; soluble Adenylyl cyclase; Immediate early genes; transcription

1. Introduction

Long-term memory formation as well as long lasting forms of synaptic plasticity depend on the expression of new genes and proteins. These activity-regulated genes, referred to as immediate early genes (IEGs), encode for transcription factors and proteins that have the potential to transduce synaptic activity directly into immediate changes of neural function. They include, for example, *Arc/Arg3.1*, *Egr1/Zif268*, and *c-Fos*. These genes are indirect markers of neuronal activity and are used to map neuronal networks and circuits engaged in information processing and plasticity [1]. For instance, Arc (activity-regulated cytoskeleton-associated protein) is a cytosolic protein found in post-synaptic densities that regulates the endocytosis of AMPA receptors [2], Notch signaling, spine density, and morphology [3] through actin remodeling [4]. *Arc* knock-out (KO) mice display impairments in the formation of long-term memories while short-term memory is not affected [5]. Egr1/Zif268 and c-Fos interact with an array of other transcription factors to regulate gene expression. *Egr1* and *c-Fos* KO animals display deficits in complex behavioral tasks and memories [6,7].

Signaling pathways involved in activity-driven regulation of transcription and translation have been the object of many studies, however, not all the components have been elucidated. One of the most studied mediators of these transcriptional changes is the transcription factor $3',5'$-cyclic AMP (cAMP) response element-binding (CREB) protein [8]. Indeed, many of the IEGs contain cAMP response elements (CRE) and thus are regulated by the transcription factor CREB. CREB signaling is regulated by phosphorylation on its Ser[133], a key regulatory site where phosphorylation ensures the transcriptional function of CREB [9,10]. While several signaling pathways and kinases are known to induce CREB phosphorylation, the most important CREB kinase is the $3',5'$-cyclic AMP (cAMP)-dependent protein kinase (PKA). PKA activity, in turn, is known to be regulated upstream by signaling pathways leading to the increase of intracellular cAMP levels, and thus by the activity of adenylyl cyclases (ACs), the best characterized of which being the G protein-coupled receptors (GPCRs) [11].

Altogether, these processes are induced by synaptic activation and in particular by glutamatergic neurotransmission. Importantly, glutamatergic transmission is a highly energy-consuming process [12,13]. Within neurons, energy levels are regulated by the AMP-activated protein kinase (AMPK). AMPK is a Ser/Thr protein kinase, which is an important intracellular energy sensor and regulator. AMPK is composed of a catalytic subunit α and two regulatory subunits β and γ [14]. AMPK activity is regulated by the intracellular levels of adenine nucleotides AMP and ATP [15,16] and by the phosphorylation of its α subunit on Thr[172] [17–19]. Interestingly, we recently reported that AMPK was necessary to maintain energy levels in neurons during synaptic activation [20]. Indeed, following glutamatergic synaptic stimulation we showed that AMPK activity was necessary to up-regulate glycolysis and mitochondrial respiration in order to maintain ATP levels within neurons. Failure to maintain energy homeostasis, through AMPK inhibition, prevented IEGs protein expression, synaptic plasticity, and hence long-term memory formation. This evidence strongly suggested that AMPK might act as a gatekeeper inside the neurons to allow signal transduction only in conditions where energy supplies are sufficient.

The goal of the present study was to determine the signaling pathway regulated by AMPK that allows the expression of IEGs. To this end, synaptic activation was induced in primary neurons, and AMPK and PKA signaling pathways were studied in these conditions. Our results showed that both signaling pathways were required for the expression of IEGs to occur. Interestingly, we also showed that the soluble adenylyl cyclase (AC) was responsible for PKA activation. Finally, inhibition of AMPK led to a downregulation of PKA pathway activation. Altogether, these data show how AMPK and PKA pathways interplay to regulate the expression of IEGs following synaptic activation.

2. Results

2.1. AMPK Activity is Required for Synaptic Activity-Induced IEGs Transcriptional Regulation

In order to determine the signaling pathway regulated by AMPK that allows for the expression of IEGs, we used primary neuronal cultures at 15 days in vitro (DIV) in which glutamatergic synaptic activation was induced using bicuculline and 4-aminopyridine (Bic/4-AP) as previously described [20–22]. As we recently showed, synaptic activation (SA) in this model led to the rapid activation of AMPK, as indicated by the increased phosphorylation of AMPK at Thr[172], and of its direct target, the Acetyl-CoA carboxylase (ACC), at Ser[79]. Additionally, after 2 h of SA, a significant increase of the IEGs Arc, EgrI, and c-Fos expression was observed (Figure 1a, b). Further, AMPK inhibition using Compound C (Cc) prevented the expression of IEGs following SA (Figure 1c). These data demonstrated, as we previously reported [20], that proper AMPK activation is necessary for the expression of IEGs.

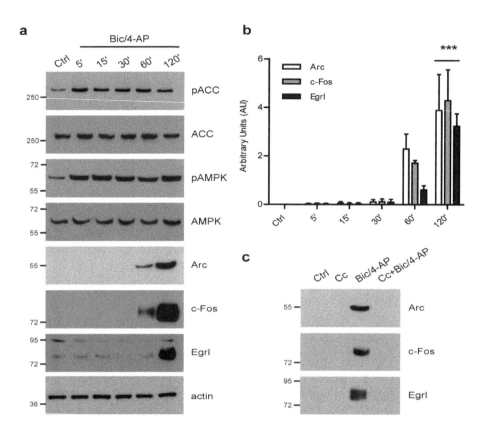

Figure 1. AMP-activated protein kinase (AMPK) is required for the expression of immediate early genes (IEGs) following glutamatergic activation. (**a**) Primary neurons at 15 days in vitro (DIV) treated with bicuculline and 4-aminopyridine (Bic/4-AP) (50µM/2.5mM) for the indicated time were subjected to immunoblotting with anti- phospho-AMPK (pAMPK), phospho-acetyl-CoA carboxylase (pACC), acetyl-CoA carboxylase (ACC), total AMPK, Arc, c-Fos, EgrI, and actin antibodies. Results are representative of at least four experiments. (**b**) Quantification of Western blot (WB) as in (**a**) showing Arc, c-Fos, and EgrI expression. Results show mean ± SD ($n = 4$). One-way ANOVA followed by Bonferroni's post-hoc test were used for evaluation of statistical significance, * $p < 0.05$, *** $p < 0.001$ compared to control condition. (**c**) Primary neurons at 15 DIV were pre-treated for 20 min in presence or absence of the AMPK inhibitor Compound C (Cc, 10 µM) prior to being treated with Bic/4-AP (50µM/2.5mM, 2 h). Cell lysates were subjected to immunoblotting with anti-Arc, c-Fos, EgrI, and actin antibodies. Results are representative of at least four experiments.

IEGs protein expression relies on the transcription of new genes, however, it was also proposed that it could result from the translation of a pre-existing pool of messenger RNA (mRNA) that is dendritically localized [23]. Therefore, we next thought to determine whether the expression of IEGs in our system was dependent on new mRNA expression or whether a pre-existing pool of mRNA could be sufficient to allow for the expression of IEGs following SA. To this end, translation was inhibited using anisomycin A and transcription inhibited using the RNA polymerase inhibitor actinomycin D. Both anisomycin A and actinomycin D repressed the expression of IEGs' proteins induced by SA (Figure 2a), showing that both *de novo* translation and transcription were necessary for the expression of IEGs. Altogether, these data implied that the expression of IEGs required new mRNA synthesis following SA. Indeed, Bic/4-AP stimulation led to a significant up-regulation of *Arc, c-Fos,* and *EgrI* mRNA (Figure 2b–d). We next assessed whether AMPK was required for this increased transcription to occur. Pre-treatment with the AMPK inhibitor Cc prevented the expression of IEGs, demonstrating that AMPK repression led to an inhibition of the activity-mediated IEG's mRNA levels of induction (Figure 2b–d). Altogether, these results showed that AMPK activity is involved in the transcriptional regulation of IEGs.

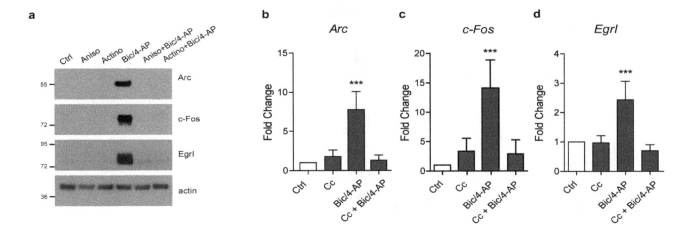

Figure 2. Expression of IEGs required *de novo* messenger RNA (mRNA) transcription and translation. (**a**) Primary neurons at 15 DIV co-treated with Bic/4-AP (50μM/2.5 mM) and the translation inhibitor anisomycin A (Aniso, 25 μM) or the transcription inhibitor actinomycin D (Actino, 10 μM) for 2 h were subjected to immunoblotting with anti-Arc, c-Fos, EgrI, and actin antibodies. Results are representative of at least four experiments. Results demonstrate that both translation and transcription are required for the expression of IEGs following synaptic activation (SA). (**b–d**) mRNA levels of *Arc*, *c-Fos*, and *EgrI* were determined by quantitative PCR in primary neurons at 15 DIV treated with Bic/4-AP (50μM/2.5mM, 30 min) after 20 min with or without pre-treatment with the AMPK inhibitor Compound C (Cc, 10 μM). Results show mean ± SD ($n = 4$–6). One-way ANOVA followed by Bonferroni's post-hoc test were used for evaluation of statistical significance, *** $p < 0.001$.

2.2. PKA Pathway Is Activated Following SA and Is Required for The Expression of IEGs

As the main pathway involved in the regulation of IEGs transcription is the PKA/CREB pathway, we questioned whether AMPK could cross-talk with this signaling pathway. We first assessed the activation of the PKA/CREB pathway following SA. To this end, we used an anti-phospho-PKA substrate antibody that detects proteins containing a phosphorylated Ser/Thr residue within the consensus sequence for PKA, thus giving an indirect readout of PKA activation status. Bic/4-AP stimulation led to a rapid and sustained activation of PKA, as observed using the anti-phospho-PKA substrate antibody as well as to the phosphorylation of CREB at Ser[133], a direct target of PKA (Figure 3a–c). Altogether, these data demonstrated that the PKA pathway was rapidly activated following SA and led to the activation of CREB.

We next determined whether the PKA pathway was required for the expression of IEGs following SA. To this end, primary neurons were pre-treated with the pharmacological PKA inhibitor H89, prior to being treated with Bic/4-AP (Figure 3d–f). Additionally, as H89 was reported to display off-target effects, to further validate the implication of PKA, we also used PKI 14–22 amide, a specific PKA peptide inhibitor (PKI) (Figure 3g–i). Results showed that both H89 and PKI prevented the PKA-substrate and CREB phosphorylation induced by SA (Figure 3d–i). Furthermore, our results showed that PKA inhibition by H89 or PKI led to a significant reduction of the expression of IEGs (Figure 3j,k). Consistent with previous reports, the present results show that PKA activation during SA is required for the expression of IEGs.

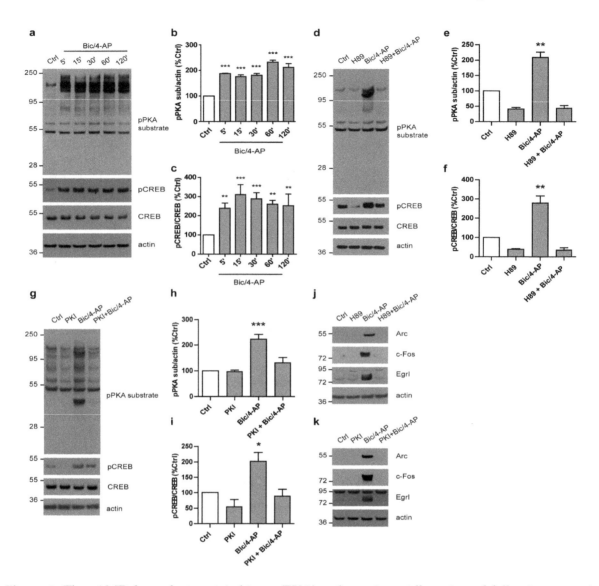

Figure 3. The cAMP-dependent protein kinase (PKA) pathway is rapidly activated following synaptic activation and is required for the expression of IEGs. (3′,5′-cyclic AMP = cAMP) (**a**) Primary neurons at 15 DIV treated with Bic/4-AP (50µM/2.5mM) for the indicated times were subjected to immunoblotting with anti- phospho-PKA substrate (pPKA sub), phospho-CREB (pCREB), total CREB, and actin antibodies. (cAMP response element binding = CREB) (**b**, **c**) Quantification of WB as in (a) showing the ratios pPKA sub/actin (**b**) and pCREB/CREB (**c**) expressed as a percentage of control ($n = 3$). (**d**–**g**) Primary neurons at 15 DIV treated with Bic/4-AP (50µM/2.5mM, 10 min) with or without 20 min pre-treatment with the PKA inhibitors H89 (20 µM, **d**–**f**) or PKA peptide inhibitor (PKI) (50 µM, **g**–**i**) were subjected to immunoblotting with anti- pPKA sub, pCREB, total CREB, and actin antibodies (**d**,**i**). Quantification of WB as in (d) and (g) showing the ratios pPKA sub/actin (**e**,**h**) and pCREB/CREB (**f**,**i**) expressed as a percentage of control ($n = 4$). (**j**,**k**) Primary neurons at 15 DIV treated with Bic/4-AP (50µM/2.5mM, 2 h) with or without 20 min pre-treatment with the PKA inhibitors H89 (20 µM, **j**) or PKI (50 µM, **k**) were subjected to immunoblotting with anti-Arc, cFos, EgrI, and actin antibodies. Results are representative of at least four experiments. Results show mean ± SD. One-way ANOVA followed by Bonferroni's post-hoc test were used for evaluation of statistical significance. * $p < 0.05$, ** $p < 0.01$, *** $p < 0.001$.

2.3. PKA Activation Following Synaptic Activation is Mediated by the Soluble AC

PKA is activated by the second messenger cAMP that is produced from ATP by AC. We next determined which of the ACs were responsible for PKA activation. To this end, neurons were pre-treated with various inhibitors of ACs, including inhibitors directed against the membrane bound

ACs, (SQ22536 or NKY80), or the specific inhibitor of the soluble AC (sAC) KH7 before Bic/4-AP stimulation [24]. Results showed that only KH7 inhibited PKA-substrate and CREB phosphorylation following SA (Figure 4a,c,d) and hence inhibited the expression of IEGs (Figure 4b). Thus, PKA was activated following SA-regulated expression of IEGs via the sAC activity, since only KH7 pre-treatment repressed PKA activation and the expression of IEGs.

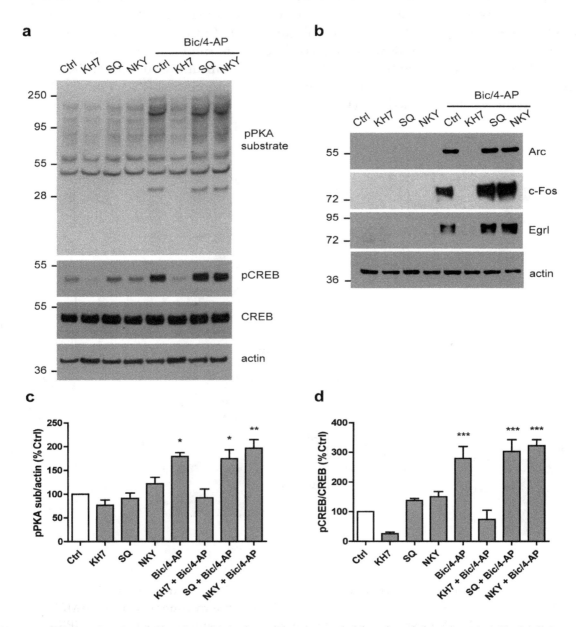

Figure 4. PKA activation following SA is dependent on soluble adenylyl cyclase (sAC). (**a**) Primary neurons at 15 DIV treated with Bic/4-AP (50μM/2.5mM, 10 min) with or without 20 min pre-treatment with the adenylyl cyclase (AC) inhibitors KH7 (20 μM), SQ22536 (SQ, 20 μM), and NKY80 (NKY, 20 μM) were subjected to immunoblotting with anti- phospho-PKA substrate (pPKA sub), phospho-CREB (pCREB), total CREB, and actin antibodies. (Ctrl was without pre-treatment) (**b**) Quantification of WB as in (a) showing the ratios pPKA sub/actin (c) and pCREB/CREB (d) expressed as a percentage of control ($n = 4$). (**c**) Primary neurons at 15 DIV treated with Bic/4-AP (2 h) with or without 20 min pre-treatment with the AC inhibitors KH7 (20 μM), SQ (20 μM), and NKY (20 μM) were subjected to immunoblotting with anti- Arc, c-Fos, EgrI, and actin antibodies. Results are representative of at least four experiments. Results show mean ± SD. One-way ANOVA followed by Bonferroni's post-hoc test were used for evaluation of statistical significance. * $p < 0.05$, ** $p < 0.01$, *** $p < 0.001$.

2.4. AMPK Regulates PKA Activation Following SA

To determine whether AMPK could be involved in the regulation of the PKA pathway, neurons were pre-treated with Cc as described above. Interestingly, Cc-pre-treatment prohibited PKA activation mediated by Bic/4-AP, as both PKA substrate and CREB were no longer phosphorylated (Figure 5a–c). Further experiments using short hairpin RNA (shRNA) directed against AMPK were performed. AMPK expression was down-regulated in primary neurons using shRNA directed against the α1 and α2 AMPK catalytic subunits (shAMPK) (Figure 5d,e). In these conditions, shAMPK reduced the phosphorylation of ACC following SA, confirming its inhibitory effect on AMPK signaling (Figure 5d). Moreover, following SA, shAMPK led to a significant reduction of PKA-substrate and CREB phosphorylation as compared to the control non-targeting shRNA (shNT) (Figure 5d,f,g), thus validating the results obtained with Cc.

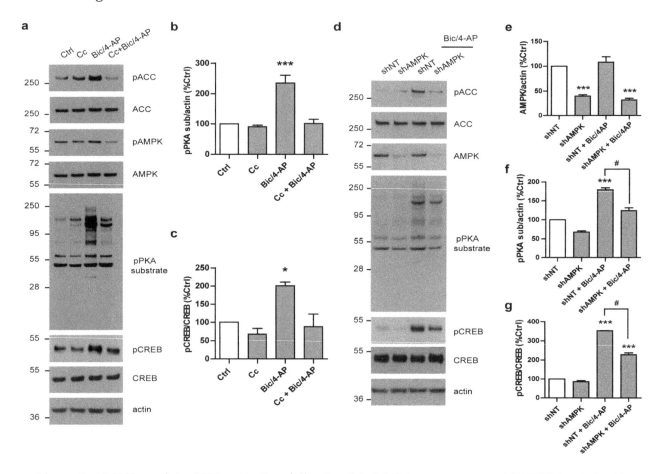

Figure 5. AMPK regulates PKA activation following SA. (**a**) Primary neurons at 15 DIV treated with Bic/4-AP (50µM/2.5mM, 10 min) with or without 20 min pre-treatment with the AMPK inhibitor Compound C (Cc, 10 µM) were subjected to immunoblotting with anti- phospho-AMPK (pAMPK), phospho-ACC (pACC), phospho-PKA substrate (pPKA sub), phospho-CREB (pCREB), total AMPK, ACC, CREB, and actin antibodies. (**b,c**) Quantification of WB as in (a) showing the ratios pPKA sub/actin (**b**) and pCREB/CREB (**c**) expressed as a percentage of control ($n = 6$). (**d**) 15 DIV primary neurons transduced for seven days with control non-targeting short hairpin RNA (shRNA) non-targeting shRNA (shNT) or with AMPK shRNA (shAMPK) were stimulated with Bic/4-AP (10 min) and subjected to immunoblotting with anti- pACC, pPKA sub, pCREB,total AMPK, ACC, CREB, and actin antibodies. (**e,f,g**) Quantification of WB as in (d) showing the ratios AMPK/actin (**e**), pPKA sub/actin (**f**), and pCREB/CREB (**g**) expressed as percentage of control ($n = 3$). Results show mean ± SD. One-way ANOVA followed by Bonferroni's post-hoc test were used for evaluation of statistical significance. * $p < 0.05$, ** $p < 0.01$, *** $p < 0.001$ as compared to Ctrl (**b,c**) or shNT (**e,f,g**), # $p < 0.05$ as compared to shNT + Bic/4-AP condition (**e,f**).

Altogether, these results show for the first time that AMPK activation cross-talks with the PKA pathway to regulate the expression of IEGs following SA.

3. Discussion

Changes in the expression of IEGs is an important process mediated by synaptic activity that is necessary for the conversion of short-term memory to long-term memory. With the present study, we extended on our previous data to determine the mechanism by which AMPK activity following synaptic activation led to the expression of IEGs. Here, we showed that SA led to the activation of the PKA/CREB pathway in an AMPK-dependent manner.

Whether AMPK directly or indirectly regulated the PKA/CREB pathway remains to be explored. However, in a previous report, we demonstrated that AMPK was required during SA to maintain intracellular ATP levels [20], therefore, it is possible that AMPK indirectly regulated the PKA pathway via controlling ATP levels. ATP, indeed, is converted by AC into cAMP, the second messenger that regulates PKA. Therefore, it is possible that the drop of ATP levels, due to AMPK inhibition, could lead to a parallel decrease of cAMP production, and eventually to a decrease of the signaling systems dependent on PKA.

Interestingly, our data showed that membrane-bound AC are not responsible for PKA activation following SA. Rather, it is the unconventional sAC (ADCY10) that is involved. sAC is distributed throughout the cytoplasm and in cellular organelles including the nucleus and mitochondria. Its functions are distinct from those of the transmembrane AC. For instance, it is insensible to G-proteins and forskolin regulation. However, in neuronal cells sAC activity can also be activated by intracellular Ca^{2+} elevations that increase its affinity for ATP [23] but also by bicarbonate anions ($HCO_3{}^-$) that increase the enzyme's V_{max}. Importantly, $HCO_3{}^-$ can be metabolically generated within the cells under the action of carbonic anhydrases (CA), hence sAC activity can be modulated by metabolically generated $HCO_3{}^-$ within the mitochondria [25–27]. Thus, mitochondrial metabolism regulated by AMPK could be another level of regulation of sAC, and hence cAMP production. Finally, we cannot exclude the possibility that AMPK could also regulate in a more direct fashion the sAC, through phosphorylation for instance. Finally, recent results have reported that activation of the mitochondrial cannabinoid receptor (mtCB1) caused inhibition of mitochondrial sAC, which resulted in reduction of PKA-dependent regulation of mitochondrial respiration, and eventually amnesic effects [28].

It is also interesting to note that AMPK was reported to be regulated by phosphorylation on $Ser^{485/491}$ on its catalytic subunits, respectively to $\alpha1$ and $\alpha2$. This phosphorylation occurs in response to agents that elevate intracellular cAMP, such as forskolin and isobutylmethylxanthine, and is likely to be mediated by PKA. These agents, however, act via membrane bound AC. Therefore, further investigations would be required to determine whether the sAC could also regulate PKA-mediated phosphorylation of AMPK. Interestingly, this phosphorylation of AMPK could be implicated in attenuating its activity given that it is associated with a down-regulation of its phosphorylation on Thr^{172} [29]. Further, in adipocytes, PKA was found to phosphorylate Ser^{173} on the AMPK α subunit to regulate lipolysis in response to PKA-activating signals [30]. Altogether these studies show that PKA can negatively regulate AMPK activity. It is therefore possible that PKA activation could in return repress AMPK activity, which could be an interesting mechanism to recover a basal AMPK activity state following SA.

Importantly, our data ([20], this study) suggest that neuronal energetic status may influence the formation of long-term memory. The hypothesis that AMPK influences these processes by maintaining ATP levels raises the question of long-term memory formation in an energetic stress environment. Metabolic disorders such as obesity and diabetes are characterized by peripheral metabolic dysfunction, but also cognitive deficits [31], elevated neurodegenerative disease risk, especially for Alzheimer's disease [32], and have recently been associated with central metabolic perturbations [33]. Interestingly, other studies have shown that several neurodegenerative diseases, including Alzheimer's disease, are not only associated with hypometabolism, but also to an activation of AMPK [34].

In conclusion, our study adds a player in the induction of signaling pathways involved in the regulation of the expression of IEGs, and hence memory formation. Altogether, our data suggest that through energy levels regulation, AMPK might indirectly control the activity of other signaling pathways, including those regulated by the second messenger cAMP.

4. Materials and Methods

4.1. Chemicals and Reagents/Antibodies

Antibodies directed against AMPKα (1/1000, Rabbit), ACC (1/1000, Rabbit), phospho-Ser^{79}ACC (1/1000, Rabbit), phospho-PKA substrate (RRXS*/T*) (1/2000, Rabbit), phospho-AMPK substrate (1/1000, Rabbit), and phospho-Ser^{133}CREB (1/1000, Rabbit) were obtained from Cell Signaling technology (Danvers, MA, USA). Anti phospho-Thr^{172}AMPKα (1/1000, Rabbit), CREB (1/500, Rabbit), Arc (1/500, Mouse), c-Fos (1/500, Mouse), and EgrI (1/500, Rabbit) antibodies were from Santa-Cruz (Dallas, TX, USA). Anti-actin (1/15 000, Mouse) antibody was from BD Bioscience (Franklin Lakes, NJ, USA). HRP-coupled secondary antibodies directed against the primary antibodies' hosts were obtained from Cell Signaling technology. Bicuculline (Bic), H 89, PKI 14-22 amide, NKY 80, SQ 22536, and KH 7 were purchased from Tocris (Bristol, UK), 4-aminopyridine (4-AP) was purchased from Sigma (St Louis, MO, USA), and Compound C (Cc) was from Santa Cruz (Dallas, TX, USA).

4.2. Primary Neuronal Cell Culture and Treatments

All animal experiments were performed according to procedures approved by the local Animal Ethical Committee following European standards for the care and use of laboratory animals (agreement APAFIS#4689-2016032315498524 v5 from CEEA75, Lille, France; approved on Oct 11, 2016). Primary neurons were prepared as previously described [35]. Briefly, fetuses at stage E18.5 were obtained from pregnant C57BL/6J wild-type female mice (The Jackson Laboratory, Bar Harbor, ME, USA). Forebrains were dissected in ice-cold dissection medium composed of Hanks' balanced salt solution (HBSS) (Invitrogen, Carlsbad, CA, USA) supplemented with 0.5 % w/v D-glucose (Sigma, St Louis, MO, USA) and 25 mM Hepes (Invitrogen, Carlsbad, CA, USA). Neurons were dissociated and isolated in ice-cold dissection medium containing 0.01 % w/v papain (Sigma, St Louis, MO, USA), 0.1 % w/v dispase (Sigma, St Louis, MO, USA), and 0.01 % w/v DNaseI (Roche, Rotkreuz, Switzerland), and by incubation at 37 °C for 15 min. Cells were spun down at 220 x g for 5 min at 4 °C, resuspended in Neurobasal medium supplemented with 2% B27, 1 mM NaPyr, 100 units/mL penicillin, 100 µg/ml streptomycin, and 2 mM Glutamax (Invitrogen, Carlsbad, CA, USA). For Western blots experiments, 12-well plates were seeded with 500,000 neurons per well and for RT-qPCR experiments, 6-well plates were seeded with 1,000,000 neurons per well. Fresh medium was added every 3 days (1:3 of starting volume). Cells were then treated and collected between DIV 14 to 17. For shRNA transduction, shRNA vectors from the TRC-Mm1.0 (Mouse) library, shAMPK α1 (CloneID:TRCN0000024000) shAMPK α2 (CloneID:TRCN0000024046), and non-targeting control shRNA (RHS6848) were obtained from Dharmacon, Lafayette, CO, USA. For the lentiviral production, HEK 293T cells were transfected for 72 h before collecting the supernatant as previously described [20]. Supernatant was concentrated using Amicon® Ultra 15-mL Centrifugal Filters (EMD Millipore, Burlington, MA, USA). Primary neuronal cultures were transduced with both AMPK α1 and AMPK α2 shRNA or the non-targeting shRNA at DIV 7, 7 days before performing experimentation.

4.3. Immunoblotting

For Western blot (WB) analysis, 15 µg of proteins from total cell lysates were separated in 8–16% Tris-Glycine gradient gels and transferred to nitrocellulose membranes. Membranes were then blocked in 5% fat-free milk in Tris Buffer Saline-0.01% Tween-20, and incubated with specific primary antibodies overnight at 4 °C. Proteins were thereafter detected via the use of Horseradish Peroxidase-conjugated secondary antibodies and electrochemiluminescence detection system (ThermoFisher Scientific,

Waltham, MA, USA). The Western blot bands corresponding to proteins of interest, or smears for phosphorylated-PKA substrate, were analyzed using the FIJI software v1.51n [36].

4.4. Quantitative Real-Time RT-PCR for the Expression of IEGs

Total RNA was isolated using the NucleoSpin® RNA kit (Macherey-Nagel, Düren, Germany) according to the manufacturer's instructions. One microgram of total RNA was reverse-transcribed using the Applied Biosystems High-Capacity cDNA reverse transcription kit (ThermoFisher Scientific, Watham, MA, USA). Real-time quantitative reverse transcription polymerase chain reaction (qRT-PCR) analyses were performed using Power SYBR Green PCR Master Mix (ThermoFisher Scientific, Watham, MA, USA) on a StepOneTM Real-Time PCR System (ThermoFisher Scientific, Watham, MA, USA) using the following primers: *β-actin* forward: 5'-CTAAGGCCAACCGTGAAAAG-3', reverse: 5'-ACCAGAGGCATACAGGGACA-3'; *Arc* forward: 5'-GGTGAGCTGAAGCCACAAAT-3', reverse: 5'-TTCACTGGTATGAATCACTGCTG-3'; *Egr1* forward: 5'-AAGACACCCCCCCATGAA-C-3', reverse: 5'-CTCATCCGAGCGAGAAAAGC-3'; and *c-Fos* forward: 5'-CGAAGGGAACGGAATAAG-3', reverse: 5'-CTCTGGGAAGCCAAGGTC-3'. The thermal cycler conditions were as follows: hold for 10 min at 95 °C, followed by 45 cycles of a two-step PCR consisting of a 95 °C step for 15 s followed by a 60 °C step for 25 s. Amplifications were carried out in triplicate, and the relative expression of target genes was determined by the $\Delta\Delta CT$ method using β-actin for normalization.

4.5. Statistical Analyses

All statistical analyses were performed using GraphPad Prism (Prism 5.0d, GraphPad Software Inc, La Jolla, CA, USA).

Author Contributions: Conceptualization, V.V. and C.M.; Investigation, S.D., F.S., M.D., C.M., and V.V.; Validation, S.D. and C.M.; Formal Analysis, S.D., F.S., and C.M.; Resources, L.B; Visualization, F.S.; Supervision, V.V.; Writing–Original Draft, C.M. and V.V.; Writing – Review and Editing, All authors; Funding Acquisition, L.B. and V.V.

Acknowledgments: We thank the animal core facility (animal facilities of Université de Lille-Inserm) of "Plateformes en Biologie Santé de Lille" as well as C. Degraeve, M. Besegher-Dumoulin, J. Devassine, R. Dehaynin, and D. Taillieu for animal care.

Abbreviations

AC	Adenylyl cyclase
ACC	Acetyl-CoA carboxylase
AMPK	AMP-activated protein kinase
Arc	Activity-regulated cytoskeleton-associated protein
Bic/4-AP	Bicuculline/4-aminopyridine
cAMP	3',5'-cyclic AMP
Cc	Compound C
CREB	cAMP response element binding
DIV	Days in vitro
GPCRs	G protein-coupled receptors
IEGs	Immediate Early Genes
KO	Knock-out
PKA	cAMP-dependent protein kinase
SA	Synaptic activation
sAC	Soluble adenylyl cyclase
shRNA	Short hairpin RNA

References

1. Guzowski, J.F.; Timlin, J.A.; Roysam, B.; McNaughton, B.L.; Worley, P.F.; Barnes, C.A. Mapping behaviorally relevant neural circuits with immediate-early gene expression. *Curr. Opin. Neurobiol.* **2005**, *15*, 599–606. [CrossRef] [PubMed]

2. Shepherd, J.D.; Rumbaugh, G.; Wu, J.; Chowdhury, S.; Plath, N.; Kuhl, D.; Huganir, R.L.; Worley, P.F. *Arc/Arg3.1* mediates homeostatic synaptic scaling of AMPA receptors. *Neuron* **2006**, *52*, 475–484. [CrossRef] [PubMed]

3. Peebles, C.L.; Yoo, J.; Thwin, M.T.; Palop, J.J.; Noebels, J.L.; Finkbeiner, S. Arc regulates spine morphology and maintains network stability in vivo. *Proc. Natl. Acad. Sci. USA* **2010**, *107*, 18173–18178. [CrossRef] [PubMed]

4. Fukazawa, Y.; Saitoh, Y.; Ozawa, F.; Ohta, Y.; Mizuno, K.; Inokuchi, K. Hippocampal LTP is accompanied by enhanced F-actin content within the dendritic spine that is essential for late LTP maintenance in vivo. *Neuron* **2003**, *38*, 447–460. [CrossRef]

5. Plath, N.; Ohana, O.; Dammermann, B.; Errington, M.L.; Schmitz, D.; Gross, C.; Mao, X.; Engelsberg, A.; Mahlke, C.; Welzl, H.; et al. *Arc/Arg3.1* is essential for the consolidation of synaptic plasticity and memories. *Neuron* **2006**, *52*, 437–444. [CrossRef] [PubMed]

6. Jones, M.W.; Errington, M.L.; French, P.J.; Fine, A.; Bliss, T.V.; Garel, S.; Charnay, P.; Bozon, B.; Laroche, S.; Davis, S. A requirement for the immediate early gene Zif268 in the expression of late LTP and long-term memories. *Nat. Neurosci.* **2001**, *4*, 289–296. [CrossRef] [PubMed]

7. Paylor, R.; Johnson, R.S.; Papaioannou, V.; Spiegelman, B.M.; Wehner, J.M. Behavioral assessment of c-fos mutant mice. *Brain Res.* **1994**, *651*, 275–282. [CrossRef]

8. Alberini, C.M. Transcription factors in long-term memory and synaptic plasticity. *Physiol. Rev.* **2009**, *89*, 121–145. [CrossRef] [PubMed]

9. Gonzalez, G.A.; Montminy, M.R. Cyclic AMP stimulates somatostatin gene transcription by phosphorylation of CREB at serine 133. *Cell* **1989**, *59*, 675–680. [CrossRef]

10. Naqvi, S.; Martin, K.J.; Arthur, J.S. CREB phosphorylation at Ser133 regulates transcription via distinct mechanisms downstream of cAMP and MAPK signalling. *Biochem. J.* **2014**, *458*, 469–479. [CrossRef] [PubMed]

11. Pierce, K.L.; Premont, R.T.; Lefkowitz, R.J. Seven-transmembrane receptors. *Nat. Rev. Mol. Cell. Biol.* **2002**, *3*, 639–650. [CrossRef] [PubMed]

12. Attwell, D.; Laughlin, S.B. An energy budget for signaling in the grey matter of the brain. *J. Cereb. Blood Flow MeTable* **2001**, *21*, 1133–1145. [CrossRef] [PubMed]

13. Harris, J.J.; Jolivet, R.; Attwell, D. Synaptic energy use and supply. *Neuron* **2012**, *75*, 762–777. [CrossRef] [PubMed]

14. Kahn, B.B.; Alquier, T.; Carling, D.; Hardie, D.G. AMP-activated protein kinase: Ancient energy gauge provides clues to modern understanding of metabolism. *Cell. MeTable* **2005**, *1*, 15–25. [CrossRef] [PubMed]

15. Hardie, D.G.; Salt, I.P.; Hawley, S.A.; Davies, S.P. AMP-activated protein kinase: An ultrasensitive system for monitoring cellular energy charge. *Biochem. J.* **1999**, *338*, 717–722. [CrossRef] [PubMed]

16. Scott, J.W.; Hawley, S.A.; Green, K.A.; Anis, M.; Stewart, G.; Scullion, G.A.; Norman, D.G.; Hardie, D.G. CBS domains form energy-sensing modules whose binding of adenosine ligands is disrupted by disease mutations. *J. Clin. Invest.* **2004**, *113*, 274–284. [CrossRef] [PubMed]

17. Stapleton, D.; Mitchelhill, K.I.; Gao, G.; Widmer, J.; Michell, B.J.; Teh, T.; House, C.M.; Fernandez, C.S.; Cox, T.; Witters, L.A.; Kemp, B.E. Mammalian AMP-activated protein kinase subfamily. *J. Biol. Chem.* **1996**, *271*, 611–614. [CrossRef] [PubMed]

18. Hawley, S.A.; Davison, M.; Woods, A.; Davies, S.P.; Beri, R.K.; Carling, D.; Hardie, D.G. Characterization of the AMP-activated protein kinase kinase from rat liver and identification of threonine 172 as the major site at which it phosphorylates AMP-activated protein kinase. *J. Biol. Chem.* **1996**, *271*, 27879–27887. [CrossRef] [PubMed]

19. Stein, S.C.; Woods, A.; Jones, N.A.; Davison, M.D.; Carling, D. The regulation of AMP-activated protein kinase by phosphorylation. *Biochem. J.* **2000**, *345*, 437–443. [CrossRef] [PubMed]

20. Marinangeli, C.; Didier, S.; Ahmed, T.; Caillerez, R.; Domise, M.; Laloux, C.; Bégard, S.; Carrier, S.; Colin, M.; Marchetti, P.; et al. AMP-Activated Protein Kinase Is Essential for the Maintenance of Energy Levels during Synaptic Activation. *iScience* **2018**, *9*, 1–13. [CrossRef] [PubMed]

21. Hardingham, G.E.; Fukunaga, Y.; Bading, H. Extrasynaptic NMDARs oppose synaptic NMDARs by triggering CREB shut-off and cell death pathways. *Nat. Neurosci.* **2002**, *5*, 405–414. [CrossRef] [PubMed]

22. Hoey, S.E.; Williams, R.J.; Perkinton, M.S. Synaptic NMDA receptor activation stimulates alpha-secretase amyloid precursor protein processing and inhibits amyloid-beta production. *J. Neurosci.* **2009**, *29*, 4442–4460. [CrossRef] [PubMed]

23. Steward, O.; Farris, S.; Pirbhoy, P.S.; Darnell, J.; Driesche, S.J. Localization and local translation of Arc/Arg3.1 mRNA at synapses: Some observations and paradoxes. *Front. Mol. Neurosci.* **2014**, *7*. [CrossRef] [PubMed]

24. Bitterman, J.L.; Ramos-Espiritu, L.; Diaz, A.; Levin, L.R.; Buck, J. Pharmacological distinction between soluble and transmembrane adenylyl cyclases. *J. Pharmacol. Exp. Ther.* **2013**, *347*, 589–598. [CrossRef] [PubMed]

25. Tresguerres, M.; Buck, J.; Levin, L.R. Physiological carbon dioxide, bicarbonate, and pH sensing. *Pflugers. Arch.* **2010**, *460*, 953–964. [CrossRef] [PubMed]

26. Acin-Perez, R.; Salazar, E.; Brosel, S.; Yang, H.; Schon, E.A.; Manfredi, G. Modulation of mitochondrial protein phosphorylation by soluble adenylyl cyclase ameliorates cytochrome oxidase defects. *EMBO Mol. Med.* **2009**, *1*, 392–406. [CrossRef] [PubMed]

27. Acin-Perez, R.; Salazar, E.; Kamenetsky, M.; Buck, J.; Levin, L.R.; Manfredi, G. Cyclic AMP produced inside mitochondria regulates oxidative phosphorylation. *Cell. MeTable* **2009**, *9*, 265–276. [CrossRef] [PubMed]

28. Hebert-Chatelain, E.; Desprez, T.; Serrat, R.; Bellocchio, L.; Soria-Gomez, E.; Busquets-Garcia, A.; Pagano Zottola, A.C.; Delamarre, A.; Cannich, A.; Vincent, P.; et al. A cannabinoid link between mitochondria and memory. *Nature* **2016**, *539*, 555–559. [CrossRef] [PubMed]

29. Hurley, R.L.; Barré, L.K.; Wood, S.D.; Anderson, K.A.; Kemp, B.E.; Means, A.R.; Witters, L.A. Regulation of AMP-activated protein kinase by multisite phosphorylation in response to agents that elevate cellular cAMP. *J. Biol. Chem.* **2006**, *281*, 36662–36672. [CrossRef] [PubMed]

30. Djouder, N.; Tuerk, R.D.; Suter, M.; Salvioni, P.; Thali, R.F.; Scholz, R.; Vaahtomeri, K.; Auchli, Y.; Rechsteiner, H.; Brunisholz, R.A.; et al. PKA phosphorylates and inactivates AMPKalpha to promote efficient lipolysis. *EMBO J.* **2010**, *29*, 469–481. [CrossRef] [PubMed]

31. Bischof, G.N.; Park, D.C. Obesity and Aging: Consequences for Cognition, Brain Structure, and Brain Function. *Psychosom. Med.* **2015**, *77*, 697–709. [CrossRef] [PubMed]

32. Kivipelto, M.; Ngandu, T.; Fratiglioni, L.; Viitanen, M.; Kåreholt, I.; Winblad, B.; Helkala, E.L.; Tuomilehto, J.; Soininen, H.; Nissinen, A. Obesity and vascular risk factors at midlife and the risk of dementia and Alzheimer disease. *Arch. Neurol.* **2005**, *62*, 1556–1560. [CrossRef] [PubMed]

33. Hwang, J.J.; Jiang, L.; Hamza, M.; Sanchez Rangel, E.; Dai, F.; Belfort-DeAguiar, R.; Parikh, L.; Koo, B.B.; Rothman, D.L.; Mason, G.; Sherwin, R.S. Blunted rise in brain glucose levels during hyperglycemia in adults with obesity and T2DM. *JCI Insight* **2017**, *2*. [CrossRef] [PubMed]

34. Vingtdeux, V.; Davies, P.; Dickson, D.W.; Marambaud, P. AMPK is abnormally activated in tangle- and pre-tangle-bearing neurons in Alzheimer's disease and other tauopathies. *Acta Neuropathol.* **2011**, *121*, 337–349. [CrossRef] [PubMed]

35. Domise, M.; Didier, S.; Marinangeli, C.; Zhao, H.; Chandakkar, P.; Buée, L.; Viollet, B.; Davies, P.; Marambaud, P.; Vingtdeux, V. AMP-activated protein kinase modulates tau phosphorylation and tau pathology in vivo. *Sci. Rep.* **2016**, *6*. [CrossRef] [PubMed]

36. Schindelin, J.; Arganda-Carreras, I.; Frise, E.; Kaynig, V.; Longair, M.; Pietzsch, T.; Preibisch, S.; Rueden, C.; Saalfeld, S.; Schmid, B.; et al. Fiji: an open-source platform for biological-image analysis. *Nat Methods.* **2012**, *9*, 676–682. [CrossRef] [PubMed]

11

Interactive Roles for AMPK and Glycogen from Cellular Energy Sensing to Exercise Metabolism

Natalie R. Janzen, Jamie Whitfield and Nolan J. Hoffman *

Exercise and Nutrition Research Program, Mary MacKillop Institute for Health Research, Australian Catholic University, Level 5, 215 Spring Street, Melbourne, Victoria 3000, Australia; natalie.janzen@acu.edu.au (N.R.J.); jamie.whitfield@acu.edu.au (J.W.)
* Correspondence: Nolan.Hoffman@acu.edu.au

Abstract: The AMP-activated protein kinase (AMPK) is a heterotrimeric complex with central roles in cellular energy sensing and the regulation of metabolism and exercise adaptations. AMPK regulatory β subunits contain a conserved carbohydrate-binding module (CBM) that binds glycogen, the major tissue storage form of glucose. Research over the past two decades has revealed that the regulation of AMPK is impacted by glycogen availability, and glycogen storage dynamics are concurrently regulated by AMPK activity. This growing body of research has uncovered new evidence of physical and functional interactive roles for AMPK and glycogen ranging from cellular energy sensing to the regulation of whole-body metabolism and exercise-induced adaptations. In this review, we discuss recent advancements in the understanding of molecular, cellular, and physiological processes impacted by AMPK-glycogen interactions. In addition, we appraise how novel research technologies and experimental models will continue to expand the repertoire of biological processes known to be regulated by AMPK and glycogen. These multidisciplinary research advances will aid the discovery of novel pathways and regulatory mechanisms that are central to the AMPK signaling network, beneficial effects of exercise and maintenance of metabolic homeostasis in health and disease.

Keywords: AMP-activated protein kinase; glycogen; exercise; metabolism; cellular energy sensing; energy utilization; liver; skeletal muscle; metabolic disease; glycogen storage disease

1. Introduction

The AMP-activated protein kinase (AMPK) is a heterotrimer composed of a catalytic α subunit and regulatory β and γ subunits, which becomes activated in response to a decrease in cellular energy status. Activation of AMPK results in metabolic adaptations such as increases in glucose uptake and glycolytic flux and fatty acid (FA) oxidation. AMPK activation simultaneously inhibits anabolic processes including protein and FA synthesis. AMPK can also translocate to the nucleus where it regulates transcription factors to increase energy production, meet cellular energy demands and inhibit cell growth and proliferation. Conversely, when energy levels are replete, AMPK activity returns to basal levels, allowing anabolic processes to resume. Given its central roles in cellular metabolic and growth signaling pathways, AMPK remains an appealing target for treating a range of pathologies associated with obesity and aging, including metabolic diseases such as obesity and type 2 diabetes (T2D).

In response to changes in energy supply and demand, glycogen, predominately stored in the liver and skeletal muscle, serves as an important source of energy to maintain metabolic homeostasis. Glycogen is synthesized by the linking of glucose monomers during periods of nutrient excess. In response to energy stress and decreased arterial glucose concentration, rising glucagon levels induce increased hepatic glucose output by promoting the breakdown of glycogen and the conversion of non-glucose substrates into glucose. The newly formed glucose is released into the bloodstream to help restore blood glucose levels. Skeletal muscle glycogen serves as an accessible source of glucose to form adenosine triphosphate (ATP) and to reduce equivalents via glycolytic and oxidative phosphorylation pathways during muscle contraction.

A significant body of evidence demonstrates that AMPK binds glycogen. This physical interaction is mediated by the carbohydrate-binding module (CBM) located within the AMPK β subunit and is thought to allow AMPK to function as a sensor of stored cellular energy. While glycogen is stored in multiple tissues throughout the body, this review will primarily focus on the physical interactions underlying the AMPK β subunit binding to glycogen and its potential functional links to glycogen storage dynamics in the liver and skeletal muscles, as these tissues are central to metabolic and exercise-regulated biological processes. In addition, as the majority of research on this topic has been undertaken in human and mouse model systems, studies in these species will be highlighted. Following a brief background on the regulation of AMPK and glycogen, this review will critically assess the recent advances and focus primarily on studies within the past two decades that have added to our understanding of the physical basis of AMPK-glycogen binding and its potential functional interactions in exercise and metabolism. Key remaining biological questions related to the interactive roles of AMPK and glycogen will be posed along with a discussion of research advancements that are feasible in the next decade with new technologies and experimental models to determine how AMPK-glycogen binding may be therapeutically targeted in health and disease.

2. Roles for AMPK and Glycogen in Metabolism

2.1. AMPK Activation and Signaling

Structural biology-based studies over the past decade have provided new insights into the molecular mechanisms by which AMPK activation is regulated by nucleotides, including changes in AMP:ATP and ADP:ATP ratios (i.e., adenylate energy charge) that occur in response to cellular energy stress [1]. The binding of AMP and ADP to the cystathionine-β-synthase (CBS) domains of the γ subunit promotes AMPK activation through several complementary mechanisms. The binding of AMP promotes AMPK association with liver kinase B1 (LKB1) and the scaffolding protein axin which enhances the effect of T172 phosphorylation [2,3], the primary phosphorylation and activation site in the AMPK α subunit, while simultaneously preventing its dephosphorylation by protein phosphatases [4–7]. This activation is mediated by myristoylation of the G2 site on the N-terminus of the β subunit (Figure 1), which promotes AMPK association with cellular membranes and LKB1 [2]. Furthermore, the binding of AMP, but not ADP, can cause the allosteric activation of AMPK without T172 phosphorylation [5,7–9]. The combination of allosteric activation by nucleotides and increased T172 phosphorylation by LKB1 can increase AMPK activity 1000-fold [10]. Additionally, the phosphorylation of T172 can be regulated by changes in intracellular Ca^{2+} concentrations via the upstream kinase calcium/calmodulin-dependent protein kinase kinase β (CAMKK2) in the absence of changes in adenylate energy charge [11–13].

Once activated, AMPK serves as a metabolic 'switch' to promote catabolic pathways and inhibit anabolic processes. For example, AMPK increases glucose uptake into skeletal muscle by phosphorylating and inhibiting Tre-2, BUB2, CDC16, 1 domain family, members 1 (TBC1D1) and 4 (TBC1D4), promoting glucose transporter 4 (GLUT4) vesicle translocation to the sarcolemmal membrane [14–18]. AMPK also functions in the regulation of lipids, acutely promoting lipid oxidation and inhibiting FA synthesis, primarily through phosphorylation and the inhibition of acetyl-CoA carboxylase (ACC) [19,20]. At the transcriptional level, AMPK phosphorylates and inhibits sterol regulatory element-binding protein 1, a transcription factor that regulates lipid synthesis [21]. Mitochondrial biogenesis is stimulated by AMPK activity through an increase in the peroxisome proliferator-activated receptor gamma coactivator 1-alpha (PGC-1α) transcription, thereby promoting oxidative metabolism [22]. AMPK can also inhibit anabolic pathways by phosphorylating the regulatory associated protein of the mechanistic target of rapamycin (mTOR) (Raptor) and the tuberous sclerosis complex 2 (TSC2), which in turn inactivates mTOR and prevents the phosphorylation of its substrates [23–25].

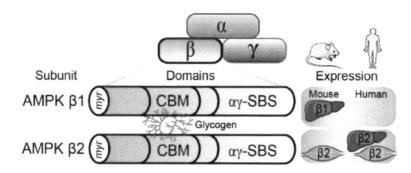

Figure 1. The AMPK is a heterotrimeric protein, consisting of a catalytic α subunit and regulatory β and γ subunits. The β subunit (β1 and β2 isoforms) possesses a glycogen-binding domain (CBM) that mediates AMPK's interaction with glycogen, an N-terminal myristoylation site (myr) and an αγ subunit binding sequence (αγ-SBS) involved in the heterotrimeric complex formation. Tissue expression of the β1 and β2 isoforms varies between humans and mice, as the β2 isoform is predominately expressed in both human liver and skeletal muscles, while mice predominately express the β1 isoform in the liver and the β2 isoform in the skeletal muscles.

2.2. The AMPK β Subunit and Carbohydrate-Binding Module

The AMPK β subunit exists in two isoforms (β1 and β2) and serves as a scaffolding subunit that binds to the AMPK catalytic α and regulatory γ subunits, playing an important role in the physical stability of the heterotrimer (Figure 1) [9,26,27]. In human and mouse skeletal muscles, the β2 isoform is predominantly expressed (Table 1) [28]. In contrast, the liver β subunit isoform expression differs across the mammalian species: the β1 isoform is predominantly expressed in mice, while β2 is predominantly expressed in humans [29,30]. However, despite isoform differences between species, both the β subunit isoforms contain the CBM which mediates physical AMPK-glycogen interaction and binding. Furthermore, the CBM is highly conserved between species, suggesting that the region possesses evolutionary significance and plays similar roles across species [31,32]. The CBM spans residues 68–163 of the β1 subunit and residues 67–163 of the β2 subunit [32,33] and is nearly identical in structure and sequence in both isoforms, with the major difference being the insertion of a threonine at residue 101 in the β2 CBM [34,35]. This insertion is believed to have occurred early in evolutionary history and provides the β2 CBM with a higher affinity for glycogen [33,34]. However, the reason for this divergence in β subunit isoforms is unknown.

Table 1. The AMPK β subunit isoform distribution in human and mouse tissues.

Tissue	B1	B2
Human vastus lateralis	ND	~100%
Human liver	ND	~100%
Mouse extensor digitorum longus	5%	95%
Mouse soleus	18%	78%
Mouse liver	100%	ND

Adapted from References [10,28–30,36]. ND, nondetectable.

2.3. Glycogen Dynamics

A number of proteins are associated with glycogen particles and function as regulators of glycogen synthesis, breakdown, particle size, and degree of branching. Glycogenin initiates glycogen formation and functions as the central protein of the glycogen particle [37]. Glycogen synthase (GS) is the rate-limiting enzyme in glycogen synthesis responsible for attaching UDP-glucose donors together in α-1,4 linkages, the linear links of the glycogen particle. As glucose-6-phosphate (G6P) is a precursor to UDP-glucose, its accumulation is a potent activator of GS, capable of overriding the inhibitory effects of phosphorylation mediated by proteins such as AMPK, glycogen synthase kinase 3, and protein kinase A [38,39]. As its name implies, glycogen branching enzyme (GBE) is responsible for introducing α-1,6 branch points to the growing glycogen particle. The rate-limiting enzyme of glycogen breakdown is glycogen phosphorylase (GP), which is known to be activated by elevated intracellular Ca^{2+}, epinephrine and cAMP concentrations [40,41]. When activated, GP degrades the α-1,4 links of glycogen particles and removes glycosyl units from the non-reducing ends of the glycogen particle [42]. The glycogen debranching enzyme (GDE) assists with the degradation of glycogen and is responsible for breaking the α-1,6 links to allow continued GP activity. Without GDE, GP can only degrade the outer tiers of glycogen particles and stops four glucose residues short of the α-1,6 branch point [39]. Further details regarding glycogen synthesis and breakdown are beyond the scope of this review, and readers are referred to other reviews covering this topic [39,43].

2.4. Glycogen Localization

Recently, there has been an increasing interest regarding the significance of glycogen's subcellular localization in skeletal muscles [44–48]. Glycogen can be concentrated beneath the sarcolemma (subsarcolemmal; SS), between the myofibers along the I band near the mitochondria and sarcoplasmic reticulum (intermyofibrillar; interMF), or within myofibers near the triad junction (intramyofibrillar; intraMF) [44,46]. Depletion of these different glycogen pools impacts muscle function and fatigue, such as impairing Ca^{2+} release and reuptake. Therefore, it has been hypothesized that these different pools of glycogen play a significant role in muscle contraction and fatigue beyond their role as an energy substrate [45,47,48].

Human skeletal muscles contain large stores of glycogen, which can exceed 100 mmol glucosyl units/kg wet weight (~500 mmol/kg dry weight) in the vastus lateralis muscle [49,50] and are primarily concentrated in the interMF space [49]. Conversely, rodents tend to have higher stores of glycogen in the liver compared to the skeletal muscle. For example, mice store about 120 μmol/g wet weight in liver and 15–20 μmol glucose/g wet weight in the type IIA flexor digitorum brevis muscle, with the highest concentrations in the intraMF pool [51]. Interestingly, while the relative contributions of intraMF glycogen to total glycogen are different between humans and mice, the intraMF content as a percentage of the total fiber volume is very similar between species [49,51]. Additionally, substrate utilization is different between species during exercise, as humans rely predominately on intramuscular stores and rodents rely on blood-borne substrates [52–55]. These differences in glycogen storage and utilization between humans and rodents are important considerations in the study design across species when assessing glycogen depletion and/or repletion.

3. Molecular Evidence of AMPK-Glycogen Binding

In 2003, it was first demonstrated that recombinant AMPK β1 CBM bound glycogen using a cell-free assay system [31]. Structural prediction and mutagenesis experiments targeting conserved residues within the CBM thought to mediate glycogen binding demonstrated that W100G and K126Q mutations abolished glycogen binding to the isolated β1 CBM, while W133L, S108E, and G147R mutations partially disrupted glycogen binding. Additionally, the AMPK heterotrimeric complex was found to bind glycogen more tightly than the β1 subunit in isolation; however, the reasons for this differential binding affinity remain unclear [31]. In support of these findings, cell-free assays have also revealed that glycogen has an inhibitory effect on AMPK activity [56]. Furthermore, mutation of critical residues in the β1 CBM (W100G, W133A, K126A, L148A, and T148A) ablated glycogen's inhibition. In these cell-free assays, glycogen with higher branch points had a greater inhibitory effect on AMPK, indicating that glycogen particle size has the capacity to influence AMPK-glycogen interactions [56]. It was also observed that glycogen particles co-localized with the β subunit of AMPK in the cytoplasm of CCL13 cells [57]. A follow-up structural-based study determined the CBM crystal structure in the presence of β-cyclodextrin and confirmed that AMPK indeed interacts with glycogen [32]. Additional experimental approaches such as immunogold cytochemistry have also shown that the AMPK α and β subunits of rat liver tissue are associated with the surface of glycogen particles in situ, providing further molecular evidence supporting this concept of physical AMPK-glycogen interaction [58].

In addition to its role in binding glycogen, the CBM of the β subunit also physically interacts with the kinase domain of the catalytic α subunit, forming a pocket, referred to as the allosteric drug and metabolite (ADaM) site. Small molecule AMPK activators such as A-769662, a β1 subunit specific activator, bind to this site and directly activate AMPK [27,59]. However, to date, any connection of A-769662's subunit specificity in relation to glycogen has been highly speculative and further research is required to establish potential direct links. The ADaM site is stabilized by autophosphorylation of S108 on the CBM and is dissociated when T172 on the α subunit is dephosphorylated [9]. Mutation of S108 to a phosphomimetic glutamic acid (S108E) resulted in reduced glycogen binding [31] and increased AMPK activity in response to AMP and A-769662, even in the presence of a non-phosphorylatable T172A mutation [8]. Conversely, mutation of S108 to a neutral alanine (S108A) had no effect on glycogen binding [31], but reduced AMPK activity in response to AMP and A-769662 [8]. Collectively, these findings infer that glycogen binding may inhibit AMPK activity by disrupting the interaction between the CBM and the kinase domain of the α subunit [1,9,56]. The inhibitory role of the β subunit T148 autophosphorylation on AMPK-glycogen binding has also been a focus of recent research. The mutation of T148 to a phosphomimetic aspartate (T148D) on the β1 subunit inhibits AMPK-glycogen binding in cellular systems [60]. The results from subsequent experiments in isolated rat skeletal muscle suggest that T148 is constitutively phosphorylated both at rest and following electrical stimulation, therefore, preventing glycogen from associating with the AMPK β2 subunit [61]. Further research is necessary to further elucidate the role of T148 in the context of AMPK-glycogen interactions.

Recent research has provided further structural insights into the affinity of the AMPK β subunits for carbohydrates. Isolated β2 CBM has a stronger affinity for carbohydrates than the β1 CBM, binding strongly to both branched and unbranched carbohydrates, with a preference for single α-1,6 branched carbohydrates [34]. One possible explanation for this difference is that a pocket is formed in the CBM by the T101 residue, which is unique to the β2 subunit, therefore, allowing binding to branched carbohydrates [33,34]. In addition, the β1 CBM possesses a threonine at residue 134 which may form a hydrogen bond with the neighboring W133, restricting the ability of the β1 CBM to accommodate carbohydrates, while the β2 CBM possesses a valine which does not bond with W133 [34]. This difference may explain the increased affinity of the β2 subunit for branched carbohydrates even though the 134 residue does not directly contact carbohydrates [34]. These findings indicate that the glycogen structure and branching affect AMPK binding, specifically to β1 subunits, which may dictate the inhibitory effect of glycogen observed in previous studies [34,56]. While the role of AMPK β isoform glycogen binding in the contexts of glycogen structure and branching has been investigated in vitro, it remains to be determined how these characteristics alter the dynamics of AMPK-glycogen binding in vivo.

4. Regulation of Cellular Energy Sensing by AMPK-Glycogen Binding

Several independent lines of evidence suggest that these physical AMPK and glycogen interactions also serve mechanistic functional roles in cellular energy sensing. A number of AMPK substrates are known to be directly involved in glycogen storage and breakdown, highlighting AMPK's role as an important regulator of glycogen metabolism. In vitro, AMPK regulates glycogen synthesis directly via the phosphorylation and inactivation of GS at site 2 [62]. In support of this finding, AMPK α2, but not α1, knockout (KO) mice display blunted phosphorylation of GS at site 2 and higher GS activity in response to stimulation by the AMPK activator 5-aminoimidazole-4-carboxamide ribonucleotide (AICAR) in skeletal muscle [63]. Paradoxically, chronic activation of AMPK also results in an accumulation of glycogen in skeletal and cardiac muscles [38]. While these divergent outcomes appear contradictory, it has been proposed that prolonged AMPK activation leads to glycogen accumulation by increasing glucose uptake and, subsequently, by increasing intracellular G6P, a known allosteric activator of GS. This hypothesis is further supported by recent independent findings using highly specific and potent pharmacological activators demonstrating that skeletal muscle AMPK activation results in increased skeletal muscle glucose uptake and glycogen synthesis in mice and non-human primates [64,65]. This accumulation of G6P overcomes the inhibition of GS by AMPK, thereby increasing GS activity [38]. Furthermore, AMPK activation also shifts fuel utilization towards FA oxidation post-exercise, allowing glucose to be utilized for glycogen resynthesis [66]. In addition to regulating GS activity, phosphorylation of GS at site 2 by AMPK causes GS to localize to the SS and interMF glycogen pools in humans [67]. These findings have been replicated in mouse models, as an R70Q mutation of the AMPK γ1 subunit results in the chronic activation of AMPK and glycogen accumulation in the skeletal muscle interMF region [68]. This has led to the suggestion that AMPK specifically senses and responds to interMF levels of glycogen [69]; however, further research is warranted to verify this hypothesis.

An increase in GP activity has also been observed to be associated with AMPK activation induced by AICAR treatment of isolated rat soleus muscles [70,71]. However, the ability to demonstrate a direct relationship between AMPK and GP activity has been limited by the identification of several AMPK-independent targets of AICAR, including phosphofructokinase, protein kinase C, and heat shock protein 90 [72]. Further research is therefore required to determine if this speculated relationship exists and elucidate the mechanism by which AMPK may regulate GP. In contrast, there is in vitro evidence that GDE binds to residues 68–123 of the AMPK β1 subunit [73]. Mutations in this region that disrupt glycogen binding (W100G and K128Q) do not affect the binding to GDE, indicating that GDE-AMPK binding is not likely mediated by glycogen [73]. AMPK's direct positive effect on glycogen accumulation, its known interaction with glycogen-associated proteins, and its ability to promote energy production through glucose uptake and fat oxidation when glycogen levels are low all support AMPK's role as a cellular energy sensor. Given the limited in vivo data currently available directly linking AMPK to glycogen-associated proteins, additional studies are necessary to further understand the potential direct binding partners and effects of AMPK on the glycogen-associated proteome.

It is important to consider additional factors that may impact physical and functional AMPK-glycogen interactions. In a proteomic screen utilizing purified glycogen from rat liver, AMPK was not included in the proteins detected to be associated with glycogen [74], and this has been replicated in a complementary study in adipocytes [75]. The authors suggested that this may be due to either AMPK protein below the level of detection being able to regulate glycogen or the predominance of the AMPK β1 subunit expression in the tissues studied, as this isoform has a lower affinity for glycogen compared to the β2 subunit [74,75]. In future studies interrogating AMPK and glycogen binding and functional interactions, considerations of the β subunit isoform expression and glycogen localization, as well as sample preparation and experimental variables that may limit the preservation and detection of AMPK-glycogen binding, are warranted in future studies to build upon this strong foundation of molecular and cellular evidence.

5. Linking AMPK and Glycogen to Exercise Metabolism in Physiological Settings

5.1. Regulation of Glycogen Storage by AMPK

In the fifteen years following the discovery of glycogen binding to the CBM on the β subunit, several studies utilizing AMPK isoform knockout (KO) mouse models have provided whole-body physiological evidence of AMPK's interactive functional roles with glycogen. Collectively, studies using AMPK α and β subunit KO mouse models have found that the ablation of AMPK alters liver and skeletal muscle glycogen content, supporting the role of AMPK in the regulation of tissue glycogen dynamics in vivo. Specifically, whole-body β2 KO mice have reduced basal glycogen levels in both liver and skeletal muscles associated with reduced muscle AMPK activity and attenuated maximal and submaximal running capacity compared to wild-type (WT) mice [76]. β2 KO mice also display reduced expression and activity of α1 and α2 subunits as well as compensatory upregulation of the β1 subunit in skeletal muscle [76]. Additional experiments utilizing this β2 KO model have demonstrated its negative impact on the whole-body and tissue metabolism and exercise capacity associated with attenuated AICAR-induced AMPK phosphorylation and glucose uptake in skeletal muscle [77]. As a result of these changes in the AMPK subunit expression and activity, it is difficult to elucidate the precise role of AMPK β2 in glycogen dynamics in this model. Similarly, muscle-specific AMPK β1/β2 KO mice display essentially no T172 phosphorylation in extensor digitorum longus (EDL) and the soleus muscle in response to electrical-stimulated contraction and have vastly reduced exercise capacity, carbohydrate utilization, and glucose uptake during treadmill running [55]. These defects were associated with reduced mitochondrial mRNA expression and reduced mitochondrial protein content [55]. Taken together, these findings suggest an important role of the β subunit in regulating AMPK activity and signaling, cellular glucose uptake and glycogen storage, mitochondrial function, and whole-body exercise capacity and metabolism.

In addition to mouse models targeting the AMPK β subunit(s), recent studies utilizing tissue-specific α1/α2 KO mice have provided support for the functional interactive roles of AMPK and glycogen. Liver-specific AMPK α1/α2 KO mice have an impaired ability to maintain euglycemia during exercise as a result of decreased hepatic glucose output due to decreased glycogenolysis [54]. Specifically, hepatic glycogen content was reduced in KO mice following both fasting and exercise. Phosphorylation of GS was unaffected in KO mice, but a decrease in UDP-glucose pyrophosphorylase 2 content was observed, suggesting reduced glycogen synthesis due to decreased glycogen precursors rather than an altered ability to synthesize glycogen [54]. In addition, when challenged with a long-term fast, these mice had reduced hepatic glycogenolysis and were unable to maintain liver ATP concentration without AMPK activity, providing further support of AMPK's role as an energy sensor [78]. Inducible muscle-specific α1/α2 KO mice have ablated skeletal muscle glycogen resynthesis and FA oxidation following exercise, even though glucose uptake was not affected, suggesting that AMPK functions as a switch to promote fat oxidation in order to preserve glucose for glycogen synthesis [79]. These findings indicate that AMPK can influence glycogen dynamics in physiological settings and that the ablation of AMPK activity reduces hepatic glucose output and is critical for skeletal muscle glycogen supercompensation following exercise. Collectively, studies using genetic models and pharmacological activators to date indicate that AMPK activation regulates glycogen synthesis in striated muscle (i.e., skeletal and cardiac muscles) secondary to increased glucose uptake and G6P accumulation, but not in the liver. Despite these important findings from AMPK transgenic mouse models, the precise role(s) of glycogen binding to the β subunits in the functional regulation of these physiological processes, as opposed to the ablation of the entire α subunits or β subunit(s) containing the CBM, remains to be elucidated.

5.2. Roles for Glycogen Availability in the Regulation of AMPK Activity

A series of physiological studies have demonstrated that low glycogen availability can amplify the AMPK signaling responses and adaptations to exercise. This was originally described in rat skeletal muscles in which AICAR treatment resulted in increased AMPK α2 activity and a markedly reduced glycogen synthase activity in a glycogen-depleted state compared to a glycogen-loaded state [80]. This observation was independent of adenine nucleotide concentrations and has subsequently been replicated in human skeletal muscle following exercise [81]. Additional studies of skeletal muscles have shown reductions in AMPK α1 and α2 association with glycogen, along with increased AMPK α2 activity and translocation to the nucleus following exercise in a glycogen-depleted state [82,83]. Furthermore, the consumption of a high-fat, low-carbohydrate diet followed by one day of a high-carbohydrate diet increases the resting skeletal muscle AMPK α activity in human skeletal muscle compared to a high-carbohydrate diet alone [84], supporting glycogen's inhibitory role on AMPK described in cell-free assays [56]. In a follow-up study, AMPK T172 phosphorylation was increased by exercise to a greater extent in the glycogen-depleted muscle than the normal glycogen repleted state [85]. Similarly, exercise in an overnight carbohydrate-fasted state resulted in increased AMPK T172 phosphorylation and the upregulation of signaling pathways involved in FA oxidation [86], while low glycogen stimulated peroxisome proliferator-activated receptor δ, a transcription factor that regulates fat utilization, in rat skeletal muscle following treadmill running [87]. Reduced glycogen availability is also associated with increases in the regulators of mitochondrial biogenesis, such as p53 and PGC-1α [88,89]. While none of these in vivo studies have directly assessed the functional role of AMPK-glycogen physical interaction, together they provide important physiological insights into how AMPK activity, subcellular localization, and signaling may be regulated by glycogen binding (Figure 2).

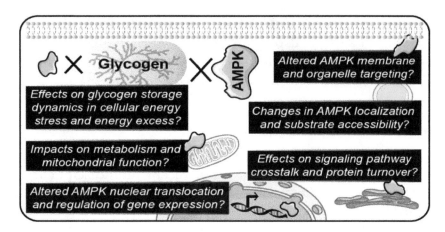

Figure 2. There are several potential alterations in cellular metabolism and signaling as a consequence of dysregulated AMPK-glycogen physical and functional interactions that represent key knowledge gaps in our current understanding and warrant further investigation in future studies. These potential alterations include changes in AMPK localization, translocation, substrates, and signaling pathway crosstalk, and subsequently, alterations in gene expression, cellular metabolism and glycogen storage.

5.3. Metabolic and Glycogen Storage Diseases as Models to Investigate AMPK-Glycogen Binding

Metabolic diseases such as insulin resistance and T2D are associated with impairments in AMPK activity, signaling, and glycogen storage dynamics. Obese patients with T2D have reduced skeletal muscle AMPK, ACC, and TBC1D4 phosphorylation following an acute bout of exercise [90]. In support of these findings, insulin resistance has been associated with suppressed AMPK activity in humans and mice [90,91], although results have been equivocal [92]. The liver-specific AMPK α1/α2 KO mice display an inability to maintain hepatic glucose output during exercise, highlighting the role of AMPK in maintaining euglycemia [54]. Skeletal muscle GS activity has also been demonstrated to be affected by insulin resistance and T2D, as there is increased phosphorylation of GS at site 2, the site phosphorylated by AMPK, which is not seen in healthy controls, resulting in nearly complete GS inactivation and dysregulation of glycogen synthesis [93]. Continued research in metabolic disease populations and rodent models can provide more insight into the significance of dysregulated AMPK and glycogen dynamics.

In addition, glycogen storage diseases provide pathophysiological models that can help provide additional insights into the influence of glycogen dynamics on AMPK. McArdle's disease is characterized by the accumulation of skeletal muscle glycogen due to a deficiency of GP. Individuals with McArdle's disease display higher muscle glycogen both at rest and following exercise compared to healthy controls, and an increased AMPK α2 activity and reduced GS activity in response to exercise [94]. Patients with McArdle's disease also demonstrate increased glucose clearance and ACC phosphorylation, indicating that AMPK activity is increased in order to maintain ATP concentration by promoting glucose uptake and FA oxidation [94]. The inability to break down glycogen, when coupled with retained, albeit reduced, glycogen synthesis, likely results in glycogen accumulation and the failure to utilize this energy source during exercise in this setting of the disease. A mouse model of McArdle's disease containing a p.R50X mutation, a nonsense mutation of nucleotide 148 in exon 1 of the GP gene, showed increased basal AMPK phosphorylation in the tibialis anterior and quadriceps muscles, associated with an increased GLUT4 content and increased AMPK-mediated glucose uptake compared to WT [95]. Following exhaustive exercise, McArdle mice display increased AMPK phosphorylation in the tibialis anterior and EDL muscles, while WT mice display no significant increase in AMPK activity [96]. While increased AMPK activity in McArdle patients and rodent models seems contrary to previous findings, the authors hypothesized that since McArdle disease results in an inability to break down glycogen, there is a subsequent increase in the AMPK activity in order to maintain an energy balance via increased glucose uptake [95,96]. Other rodent models have directly targeted muscle GS, which is affected in patients with Glycogen Storage Disease 0 [97]. Muscle-specific glycogen synthase knock-out models display increased AMPK phosphorylation [98] and markedly reduced glycogen content in skeletal muscle in the basal state, likely due to the retained capacity to break down but an inability to resynthesize glycogen [99,100].

6. Multidisciplinary Techniques and Models to Interrogate Roles for AMPK-Glycogen Interactions

While much remains to be discovered with regard to the molecular and cellular roles and physiological relevance of AMPK-glycogen binding, recent multidisciplinary technical research advances can be used to help address remaining knowledge gaps. For example, global mass spectrometry-based phosphoproteomics have recently revealed a repertoire of new AMPK substrates, providing additional evidence regarding the complexity and interconnection of the AMPK signaling network. A recent phosphoproteomic analysis mapping the human skeletal muscle exercise signaling network before and immediately following a single bout of intense aerobic exercise, in combination with phosphoproteomic analysis of AICAR-stimulated signaling in rat L6 myotubes, identified several novel AMPK substrates [101]. Other recent efforts have predicted and identified novel AMPK substrate phosphorylation sites via chemical genetic screening combined with peptide capture in whole cells [102], as well as affinity proteomics approach to analyzing hepatocyte proteins containing the substrate recognition motif targeted by AMPK phosphorylation [103]. Together, these complementary large-scale approaches have expanded the range of biological functions known to be regulated by AMPK. While additional substrates residing in different subcellular locations and organelles are continuing to be uncovered, the mechanisms underlying AMPK subcellular localization and targeting to substrates residing in these different organelles remains unknown. Future global, unbiased studies such as phosphoproteomics can help identify novel glycogen-associated AMPK substrates, post-translational regulation of glycogen regulatory machinery, AMPK subunit-specific regulation, and subcellular substrate targeting. Furthermore, omics-based approaches will reveal how AMPK-glycogen binding may impact other levels of biological regulation, such as the transcriptome, proteome, metabolome, and lipidome, in the contexts of exercise, metabolism, and beyond [104].

Novel AMPK fluorescence resonance energy transfer (FRET)-based sensors have recently revealed heterogeneous activity and tissue-specific roles for AMPK. These AMPK FRET sensors have permitted the spatiotemporal and dynamic assessment of AMPK activity in single cells [105], 3D cell cultures [106], and transgenic mice [107]. These biosensors build upon traditional methods to interrogate AMPK activity such as kinase assays and immunoblotting, which are limited to targeted measures of mean cellular protein phosphorylation and do not allow the spatiotemporal and dynamic assessment of AMPK activity. Electron microscopy-based approaches have also been used to visualize AMPK-glycogen association in fixed rat liver samples [58]. While improved microscopy technologies and sensors have been used to assess AMPK or glycogen localization, few studies have directly assessed AMPK-glycogen interactions. Utilizing these recent technical advancements will allow for the interrogation of AMPK-glycogen interactions and dynamics across species and physiologically relevant settings (Figure 2).

Despite the large body of research using in vitro models and physiological evidence indicating the potential functional roles for AMPK-glycogen binding, to date, there are no models that have been developed to disrupt and/or examine this physical binding directly in vivo. AMPK subunit KO models, while providing important insights into the functions of AMPK, are limited by the potential compensatory upregulation of other subunit isoforms or the disrupted stability of the AMPK heterotrimer complex (e.g., Reference [76]). In addition, directly assessing the function of β subunit glycogen binding is challenging when additional functions are altered in the presence of subunit deletion, as AMPK activity is impaired when the scaffolding β subunit is removed. The design of novel in vivo models in the future will be informed by previous molecular and cellular findings to allow direct interrogation of the functional relevance of the β subunits and CBM. Generation of novel animal models to specifically target physical AMPK-glycogen binding will provide important advances regarding its physiological significance and capability to be therapeutically targeted in vivo to modulate metabolism and the health benefits of exercise.

Finally, previous studies have primarily utilized centrifugation-based assays to detect and quantify physical AMPK-glycogen association. Novel biotechnological platforms and proximity assays will aid this investigation of AMPK-glycogen binding and AMPK's proximity to glycogen with improved sensitivity and specificity across molecular, cellular, and physiological models. Furthermore, newly developed kinase activity reporters [108] and other non-radioactive activity assays [109] will help provide new measures of intracellular AMPK activity dynamics and complement traditional surrogate measures such as immunoblot analyses of AMPK and ACC phosphorylation. Together these technological advances expand the repertoire of available tools to monitor the range of biological processes regulated by AMPK and further our understanding of the mechanisms and physiological significance underlying AMPK-glycogen interactions.

7. Potential Therapeutic Relevance of Targeting AMPK-Glycogen Binding

Consistent with the therapeutic relevance of the CBM, several lines of evidence demonstrate that the CBM may play a direct functional role in AMPK conformation and activation. The CBM contains the critical S108 autophosphorylation site required for drug-induced AMPK activation in the absence of AMP [8]. Although located on opposite sides of the AMPK heterotrimer, the CBM is conformationally connected to the regulatory AMPK γ subunit and its stabilization is affected by adenine nucleotide binding (e.g., AMP) to the CBS motifs [26]. Despite physical AMPK-glycogen interaction being mediated by the β subunit, mutations in the γ subunit also result in alterations in AMPK activation and glycogen metabolism. The γ2 subunit is known to contain mutations that cause constitutive AMPK activation, resulting in glycogen storage diseases in humans. These mutations result in glycogen accumulation with coexisting deleterious effects on cardiac electrical properties that are characteristic of familial hypertrophic cardiomyopathy and Wolff-Parkinson-White syndrome [110]. In addition, gain of function mutations in the AMPK γ3 subunit predominantly expressed in skeletal muscle result in excess glycogen storage [111] as well as improvements in metabolism via increased mitochondrial biogenesis [112]. Constitutive AMPK activation associated with these γ subunit mutations promotes glycogen synthesis by increasing glucose uptake. As mentioned above, the CBM interacts with the α subunit, forming the ADaM site and stabilizing the kinase domain of the α subunit in its active formation [9,27]. However, when glycogen binds to the CBM, this interaction is destabilized, altering the ADaM site and inhibiting the AMPK activity [9]. For example, isoform-specific allosteric inhibition of AMPK has been shown to be dependent on the β2 subunit CBM in glycogen-containing pancreatic beta cells [113]. The CBM, therefore, functions as both a critical element of AMPK activation as well as a site for the allosteric inhibition by glycogen, highlighting the therapeutic potential of new drugs targeting the ADaM site.

8. Conclusions

AMPK is a central regulator of cellular metabolism and, therefore, possesses significant therapeutic potential for the prevention and treatment of a range of metabolic diseases. A growing body of evidence demonstrates that AMPK physically binds glycogen and this interaction can alter the conformation of AMPK, and subsequently, its activity and downstream signaling. AMPK activity subsequently regulates glycogen metabolism. Recent research has described experimental and physiological settings that impact functional AMPK and glycogen interactions, including AMPK β isoform affinity, glycogen availability, and particle size. Despite our understanding of AMPK's relationship with glycogen, much remains to be elucidated. Further research using new technologies and experimental models can reveal additional mechanisms underlying AMPK and glycogen's interactive roles in cellular energy sensing, exercise, and metabolism. Together, these findings will help provide insights into the physiological and therapeutic relevance of targeting AMPK and glycogen binding in health and disease.

Acknowledgments: The authors acknowledge that all publications related to AMPK and glycogen could not be discussed in this review due to word limitations and the primary focus on studies published within the past two decades.

Abbreviations

ACC	Acetyl-CoA carboxylase
ADP	Adenosine diphosphate
AICAR	5-aminoimidazole-4-carboxamide ribonucleotide
AMP	Adenosine monophosphate
AMPK	AMP-activated protein kinase
ATP	Adenosine triphosphate
CaMKKβ	Calcium/calmodulin-dependent protein kinase β
cAMP	Cyclic AMP
CBM	Carbohydrate-binding module
CBS	Cystathionine-β-synthase domains
EDL	Extensor digitorum longus
FA	Fatty acid
FRET	Fluorescence resonance energy transfer
G6P	Glucose-6-phosphate
GBE	Glycogen branching enzyme
GDE	Glycogen debranching enzyme
GLUT4	Glucose transporter 4
GP	Glycogen phosphorylase
GS	Glycogen synthase
interMF	Intermyofibrillar
intraMF	Intramyofibrillar
KO	Knock-out
LKB1	Liver kinase B1
mTOR	Mechanistic target of rapamycin
PGC-1α	Peroxisome proliferator-activated receptor gamma coactivator 1-α
Raptor	Regulatory associated protein of mechanistic target of rapamycin
SS	Subsarcolemmal
TBCID1	Tre-2, BUB2, CDC16, 1 domain family, member 1

TBC1D4   Tre-2, BUB2, CDC16, 1 domain family, member 4

TSC2   Tuberous sclerosis complex 2

T2D   Type 2 diabetes

UGP2   UDP-glucose pyrophosphorylase 2

WT   Wild type

References

1. Garcia, D.; Shaw, R.J. AMPK: Mechanisms of cellular energy sensing and restoration of metabolic balance. *Mol. Cell* **2017**, *66*, 789–800. [CrossRef] [PubMed]
2. Oakhill, J.S.; Chen, Z.P.; Scott, J.W.; Steel, R.; Castelli, L.A.; Ling, N.; Macaulay, S.L.; Kemp, B.E. β-Subunit myristoylation is the gatekeeper for initiating metabolic stress sensing by AMP-activated protein kinase (AMPK). *Proc. Natl. Acad. Sci. USA* **2010**, *107*, 19237–19241. [CrossRef] [PubMed]
3. Zhang, Y.L.; Guo, H.; Zhang, C.S.; Lin, S.Y.; Yin, Z.; Peng, Y.; Luo, H.; Shi, Y.; Lian, G.; Zhang, C.; et al. AMP as a low-energy charge signal autonomously initiates assembly of AXIN-AMPK-LKB1 complex for AMPK activation. *Cell Metab.* **2013**, *18*, 546–555. [CrossRef] [PubMed]
4. Garcia-Haro, L.; Garcia-Gimeno, M.A.; Neumann, D.; Beullens, M.; Bollen, M.; Sanz, P. The PP1-R6 protein phosphatase holoenzyme is involved in the glucose-induced dephosphorylation and inactivation of AMP-activated protein kinase, a key regulator of insulin secretion, in MIN6 β cells. *FASEB J.* **2010**, *24*, 5080–5091. [CrossRef] [PubMed]
5. Gowans, G.J.; Hawley, S.A.; Ross, F.A.; Hardie, D.G. AMP is a true physiological regulator of AMP-activated protein kinase by both allosteric activation and enhancing net phosphorylation. *Cell Metab.* **2013**, *18*, 556–566. [CrossRef] [PubMed]
6. Joseph, B.K.; Liu, H.Y.; Francisco, J.; Pandya, D.; Donigan, M.; Gallo-Ebert, C.; Giordano, C.; Bata, A.; Nickels, J.T., Jr. Inhibition of AMP kinase by the protein phosphatase 2A heterotrimer, PP2APpp2r2d. *J. Biol. Chem.* **2015**, *290*, 10588–10598. [CrossRef] [PubMed]
7. Xiao, B.; Sanders, M.J.; Underwood, E.; Heath, R.; Mayer, F.V.; Carmena, D.; Jing, C.; Walker, P.A.; Eccleston, J.F.; Haire, L.F.; et al. Structure of mammalian AMPK and its regulation by ADP. *Nature* **2011**, *472*, 230–233. [CrossRef] [PubMed]
8. Scott, J.W.; Ling, N.; Issa, S.M.; Dite, T.A.; O'Brien, M.T.; Chen, Z.P.; Galic, S.; Langendorf, C.G.; Steinberg, G.R.; Kemp, B.E.; et al. Small molecule drug A-769662 and AMP synergistically activate naive AMPK independent of upstream kinase signaling. *Chem. Biol.* **2014**, *21*, 619–627. [CrossRef] [PubMed]
9. Li, X.; Wang, L.; Zhou, X.E.; Ke, J.; de Waal, P.W.; Gu, X.; Tan, M.H.; Wang, D.; Wu, D.; Xu, H.E.; et al. Structural basis of AMPK regulation by adenine nucleotides and glycogen. *Cell Res.* **2015**, *25*, 50–66. [CrossRef] [PubMed]
10. Kjobsted, R.; Hingst, J.R.; Fentz, J.; Foretz, M.; Sanz, M.N.; Pehmoller, C.; Shum, M.; Marette, A.; Mounier, R.; Treebak, J.T.; et al. AMPK in skeletal muscle function and metabolism. *FASEB J.* **2018**. [CrossRef] [PubMed]
11. Hawley, S.A.; Pan, D.A.; Mustard, K.J.; Ross, L.; Bain, J.; Edelman, A.M.; Frenguelli, B.G.; Hardie, D.G. Calmodulin-dependent protein kinase kinase-β is an alternative upstream kinase for AMP-activated protein kinase. *Cell Metab.* **2005**, *2*, 9–19. [CrossRef] [PubMed]
12. Jensen, T.E.; Rose, A.J.; Jorgensen, S.B.; Brandt, N.; Schjerling, P.; Wojtaszewski, J.F.; Richter, E.A. Possible CaMKK-dependent regulation of AMPK phosphorylation and glucose uptake at the onset of mild tetanic skeletal muscle contraction. *Am. J. Physiol. Endocrinol. Metab.* **2007**, *292*, E1308–E1317. [CrossRef] [PubMed]
13. Woods, A.; Dickerson, K.; Heath, R.; Hong, S.P.; Momcilovic, M.; Johnstone, S.R.; Carlson, M.; Carling, D. Ca^{2+}/calmodulin-dependent protein kinase kinase-β acts upstream of AMP-activated protein kinase in mammalian cells. *Cell Metab.* **2005**, *2*, 21–33. [CrossRef] [PubMed]
14. Kjobsted, R.; Munk-Hansen, N.; Birk, J.B.; Foretz, M.; Viollet, B.; Bjornholm, M.; Zierath, J.R.; Treebak, J.T.; Wojtaszewski, J.F. Enhanced muscle insulin sensitivity after contraction/exercise is mediated by AMPK. *Diabetes* **2017**, *66*, 598–612. [CrossRef] [PubMed]
15. Pehmoller, C.; Treebak, J.T.; Birk, J.B.; Chen, S.; Mackintosh, C.; Hardie, D.G.; Richter, E.A.; Wojtaszewski, J.F. Genetic disruption of AMPK signaling abolishes both contraction-and insulin-stimulated TBC1D1

phosphorylation and 14-3-3 binding in mouse skeletal muscle. *Am. J. Physiol. Endocrinol. Metab.* **2009**, *297*, E665–E675. [CrossRef] [PubMed]

16.   Vichaiwong, K.; Purohit, S.; An, D.; Toyoda, T.; Jessen, N.; Hirshman, M.F.; Goodyear, L.J. Contraction regulates site-specific phosphorylation of TBC1D1 in skeletal muscle. *Biochem. J.* **2010**, *431*, 311–320. [CrossRef] [PubMed]

17.   Whitfield, J.; Paglialunga, S.; Smith, B.K.; Miotto, P.M.; Simnett, G.; Robson, H.L.; Jain, S.S.; Herbst, E.A.F.; Desjardins, E.M.; Dyck, D.J.; et al. Ablating the protein TBC1D1 impairs contraction-induced sarcolemmal glucose transporter 4 redistribution but not insulin-mediated responses in rats. *J. Biol. Chem.* **2017**, *292*, 16653–16664. [CrossRef] [PubMed]

18.   Stockli, J.; Meoli, C.C.; Hoffman, N.J.; Fazakerley, D.J.; Pant, H.; Cleasby, M.E.; Ma, X.; Kleinert, M.; Brandon, A.E.; Lopez, J.A.; et al. The RabGAP TBC1D1 plays a central role in exercise-regulated glucose metabolism in skeletal muscle. *Diabetes* **2015**, *64*, 1914–1922. [CrossRef] [PubMed]

19.   Hardie, D.G.; Ross, F.A.; Hawley, S.A. AMPK: A nutrient and energy sensor that maintains energy homeostasis. *Nat. Rev. Mol. Cell Biol.* **2012**, *13*, 251–262. [CrossRef] [PubMed]

20.   Fullerton, M.D.; Galic, S.; Marcinko, K.; Sikkema, S.; Pulinilkunnil, T.; Chen, Z.P.; O'Neill, H.M.; Ford, R.J.; Palanivel, R.; O'Brien, M.; et al. Single phosphorylation sites in Acc1 and Acc2 regulate lipid homeostasis and the insulin-sensitizing effects of metformin. *Nat. Med.* **2013**, *19*, 1649–1654. [CrossRef] [PubMed]

21.   Li, Y.; Xu, S.; Mihaylova, M.M.; Zheng, B.; Hou, X.; Jiang, B.; Park, O.; Luo, Z.; Lefai, E.; Shyy, J.Y.; et al. AMPK phosphorylates and inhibits SREBP activity to attenuate hepatic steatosis and atherosclerosis in diet-induced insulin-resistant mice. *Cell Metab.* **2011**, *13*, 376–388. [CrossRef] [PubMed]

22.   Hawley, J.A.; Hargreaves, M.; Joyner, M.J.; Zierath, J.R. Integrative biology of exercise. *Cell* **2014**, *159*, 738–749. [CrossRef] [PubMed]

23.   Gwinn, D.M.; Shackelford, D.B.; Egan, D.F.; Mihaylova, M.M.; Mery, A.; Vasquez, D.S.; Turk, B.E.; Shaw, R.J. AMPK phosphorylation of raptor mediates a metabolic checkpoint. *Mol. Cell* **2008**, *30*, 214–226. [CrossRef] [PubMed]

24.   Inoki, K.; Zhu, T.; Guan, K.L. Tsc2 mediates cellular energy response to control cell growth and survival. *Cell* **2003**, *115*, 577–590. [CrossRef]

25.   Tee, A.R.; Fingar, D.C.; Manning, B.D.; Kwiatkowski, D.J.; Cantley, L.C.; Blenis, J. Tuberous sclerosis complex-1 and -2 gene products function together to inhibit mammalian target of rapamycin (mTOR)-mediated downstream signaling. *Proc. Natl. Acad. Sci. USA* **2002**, *99*, 13571–13576. [CrossRef] [PubMed]

26.   Gu, X.; Yan, Y.; Novick, S.J.; Kovach, A.; Goswami, D.; Ke, J.; Tan, M.H.E.; Wang, L.; Li, X.; de Waal, P.W.; et al. Deconvoluting AMP-activated protein kinase (AMPK) adenine nucleotide binding and sensing. *J. Biol. Chem.* **2017**, *292*, 12653–12666. [CrossRef] [PubMed]

27.   Xiao, B.; Sanders, M.J.; Carmena, D.; Bright, N.J.; Haire, L.F.; Underwood, E.; Patel, B.R.; Heath, R.B.; Walker, P.A.; Hallen, S.; et al. Structural basis of AMPK regulation by small molecule activators. *Nat. Commun.* **2013**, *4*, 3017. [CrossRef] [PubMed]

28.   Olivier, S.; Foretz, M.; Viollet, B. Promise and challenges for direct small molecule AMPK activators. *Biochem. Pharmacol.* **2018**. [CrossRef] [PubMed]

29.   Stephenne, X.; Foretz, M.; Taleux, N.; van der Zon, G.C.; Sokal, E.; Hue, L.; Viollet, B.; Guigas, B. Metformin activates AMP-activated protein kinase in primary human hepatocytes by decreasing cellular energy status. *Diabetologia* **2011**, *54*, 3101–3110. [CrossRef] [PubMed]

30.   Wu, J.; Puppala, D.; Feng, X.; Monetti, M.; Lapworth, A.L.; Geoghegan, K.F. Chemoproteomic analysis of intertissue and interspecies isoform diversity of AMP-activated protein kinase (AMPK). *J. Biol. Chem.* **2013**, *288*, 35904–35912. [CrossRef] [PubMed]

31.   Polekhina, G.; Gupta, A.; Michell, B.J.; van Denderen, B.; Murthy, S.; Feil, S.C.; Jennings, I.G.; Campbell, D.J.; Witters, L.A.; Parker, M.W.; et al. AMPK β subunit targets metabolic stress sensing to glycogen. *Curr. Biol.* **2003**, *13*, 867–871. [CrossRef]

32.   Polekhina, G.; Gupta, A.; van Denderen, B.J.; Feil, S.C.; Kemp, B.E.; Stapleton, D.; Parker, M.W. Structural basis for glycogen recognition by AMP-activated protein kinase. *Structure* **2005**, *13*, 1453–1462. [CrossRef] [PubMed]

33. Koay, A.; Woodcroft, B.; Petrie, E.J.; Yue, H.; Emanuelle, S.; Bieri, M.; Bailey, M.F.; Hargreaves, M.; Park, J.T.; Park, K.H.; et al. AMPK β subunits display isoform specific affinities for carbohydrates. *FEBS Lett.* **2010**, *584*, 3499–3503. [PubMed]

34. Mobbs, J.I.; Di Paolo, A.; Metcalfe, R.D.; Selig, E.; Stapleton, D.I.; Griffin, M.D.W.; Gooley, P.R. Unravelling the carbohydrate-binding preferences of the carbohydrate-binding modules of AMP-activated protein kinase. *Chembiochem* **2017**. [CrossRef] [PubMed]

35. Mobbs, J.I.; Koay, A.; Di Paolo, A.; Bieri, M.; Petrie, E.J.; Gorman, M.A.; Doughty, L.; Parker, M.W.; Stapleton, D.I.; Griffin, M.D.; et al. Determinants of oligosaccharide specificity of the carbohydrate-binding modules of AMP-activated protein kinase. *Biochem. J.* **2015**, *468*, 245–257. [CrossRef] [PubMed]

36. O'Neill, H.M. AMPK and exercise: Glucose uptake and insulin sensitivity. *Diabetes Metab. J.* **2013**, *37*, 1–21. [CrossRef] [PubMed]

37. Alonso, M.D.; Lomako, J.; Lomako, W.M.; Whelan, W.J. A new look at the biogenesis of glycogen. *FASEB J.* **1995**, *9*, 1126–1137. [CrossRef] [PubMed]

38. Hunter, R.W.; Treebak, J.T.; Wojtaszewski, J.F.; Sakamoto, K. Molecular mechanism by which AMP-activated protein kinase activation promotes glycogen accumulation in muscle. *Diabetes* **2011**, *60*, 766–774. [CrossRef] [PubMed]

39. Roach, P.J.; Depaoli-Roach, A.A.; Hurley, T.D.; Tagliabracci, V.S. Glycogen and its metabolism: Some new developments and old themes. *Biochem. J.* **2012**, *441*, 763–787. [PubMed]

40. Chasiotis, D.; Sahlin, K.; Hultman, E. Regulation of glycogenolysis in human muscle at rest and during exercise. *J. Appl. Physiol. Respir. Environ. Exerc. Physiol.* **1982**, *53*, 708–715. [CrossRef] [PubMed]

41. Richter, E.A.; Ruderman, N.B.; Gavras, H.; Belur, E.R.; Galbo, H. Muscle glycogenolysis during exercise: Dual control by epinephrine and contractions. *Am. J. Physiol.* **1982**, *242*, E25–E32. [CrossRef] [PubMed]

42. Shearer, J.; Graham, T.E. Novel aspects of skeletal muscle glycogen and its regulation during rest and exercise. *Exerc. Sport Sci. Rev.* **2004**, *32*, 120–126. [CrossRef] [PubMed]

43. Prats, C.; Graham, T.E.; Shearer, J. The dynamic life of the glycogen granule. *J. Biol. Chem.* **2018**, *293*, 7089–7098. [CrossRef] [PubMed]

44. Graham, T.E.; Yuan, Z.; Hill, A.K.; Wilson, R.J. The regulation of muscle glycogen: The granule and its proteins. *Acta Physiol.* **2010**, *199*, 489–498. [CrossRef] [PubMed]

45. Nielsen, J.; Ortenblad, N. Physiological aspects of the subcellular localization of glycogen in skeletal muscle. *Appl. Physiol. Nutr. Metab.* **2013**, *38*, 91–99. [CrossRef] [PubMed]

46. Ortenblad, N.; Nielsen, J.; Saltin, B.; Holmberg, H.C. Role of glycogen availability in sarcoplasmic reticulum Ca^{2+} kinetics in human skeletal muscle. *J. Physiol.* **2011**, *589*, 711–725. [CrossRef] [PubMed]

47. Ortenblad, N.; Westerblad, H.; Nielsen, J. Muscle glycogen stores and fatigue. *J. Physiol.* **2013**, *591*, 4405–4413. [CrossRef] [PubMed]

48. Philp, A.; Hargreaves, M.; Baar, K. More than a store: Regulatory roles for glycogen in skeletal muscle adaptation to exercise. *Am. J. Physiol. Endocrinol. Metab.* **2012**, *302*, E1343–E1351. [CrossRef] [PubMed]

49. Nielsen, J.; Holmberg, H.C.; Schroder, H.D.; Saltin, B.; Ortenblad, N. Human skeletal muscle glycogen utilization in exhaustive exercise: Role of subcellular localization and fibre type. *J. Physiol.* **2011**, *589*, 2871–2885. [CrossRef] [PubMed]

50. Yeo, W.K.; Paton, C.D.; Garnham, A.P.; Burke, L.M.; Carey, A.L.; Hawley, J.A. Skeletal muscle adaptation and performance responses to once a day versus twice every second day endurance training regimens. *J. Appl. Physiol.* **2008**, *105*, 1462–1470. [CrossRef] [PubMed]

51. Nielsen, J.; Cheng, A.J.; Ortenblad, N.; Westerblad, H. Subcellular distribution of glycogen and decreased tetanic Ca^{2+} in fatigued single intact mouse muscle fibres. *J. Physiol.* **2014**, *592*, 2003–2012. [CrossRef] [PubMed]

52. Romijn, J.A.; Coyle, E.F.; Sidossis, L.S.; Gastaldelli, A.; Horowitz, J.F.; Endert, E.; Wolfe, R.R. Regulation of endogenous fat and carbohydrate metabolism in relation to exercise intensity and duration. *Am. J. Physiol.* **1993**, *265*, E380–E391. [CrossRef] [PubMed]

53. van Loon, L.J.; Greenhaff, P.L.; Constantin-Teodosiu, D.; Saris, W.H.; Wagenmakers, A.J. The effects of increasing exercise intensity on muscle fuel utilisation in humans. *J. Physiol.* **2001**, *536*, 295–304. [CrossRef] [PubMed]

54. Hughey, C.C.; James, F.D.; Bracy, D.P.; Donahue, E.P.; Young, J.D.; Viollet, B.; Foretz, M.; Wasserman, D.H. Loss of hepatic AMP-activated protein kinase impedes the rate of glycogenolysis but not gluconeogenic fluxes in exercising mice. *J. Biol. Chem.* **2017**. [CrossRef] [PubMed]

55. O'Neill, H.M.; Maarbjerg, S.J.; Crane, J.D.; Jeppesen, J.; Jorgensen, S.B.; Schertzer, J.D.; Shyroka, O.; Kiens, B.; van Denderen, B.J.; Tarnopolsky, M.A.; et al. AMP-activated protein kinase (AMPK) β1β2 muscle null mice reveal an essential role for AMPK in maintaining mitochondrial content and glucose uptake during exercise. *Proc. Natl. Acad. Sci. USA* **2011**, *108*, 16092–16097. [CrossRef] [PubMed]

56. McBride, A.; Ghilagaber, S.; Nikolaev, A.; Hardie, D.G. The glycogen-binding domain on the AMPK β subunit allows the kinase to act as a glycogen sensor. *Cell Metab.* **2009**, *9*, 23–34. [CrossRef] [PubMed]

57. Hudson, E.R.; Pan, D.A.; James, J.; Lucocq, J.M.; Hawley, S.A.; Green, K.A.; Baba, O.; Terashima, T.; Hardie, D.G. A novel domain in AMP-activated protein kinase causes glycogen storage bodies similar to those seen in hereditary cardiac arrhythmias. *Curr. Biol.* **2003**, *13*, 861–866. [CrossRef]

58. Bendayan, M.; Londono, I.; Kemp, B.E.; Hardie, G.D.; Ruderman, N.; Prentki, M. Association of AMP-activated protein kinase subunits with glycogen particles as revealed in situ by immunoelectron microscopy. *J. Histochem. Cytochem.* **2009**, *57*, 963–971. [CrossRef] [PubMed]

59. Scott, J.W.; van Denderen, B.J.; Jorgensen, S.B.; Honeyman, J.E.; Steinberg, G.R.; Oakhill, J.S.; Iseli, T.J.; Koay, A.; Gooley, P.R.; Stapleton, D.; et al. Thienopyridone drugs are selective activators of AMP-activated protein kinase β1-containing complexes. *Chem. Biol.* **2008**, *15*, 1220–1230. [CrossRef] [PubMed]

60. Oligschlaeger, Y.; Miglianico, M.; Chanda, D.; Scholz, R.; Thali, R.F.; Tuerk, R.; Stapleton, D.I.; Gooley, P.R.; Neumann, D. The recruitment of AMP-activated protein kinase to glycogen is regulated by autophosphorylation. *J. Biol. Chem.* **2015**, *290*, 11715–11728. [CrossRef] [PubMed]

61. Xu, H.; Frankenberg, N.T.; Lamb, G.D.; Gooley, P.R.; Stapleton, D.I.; Murphy, R.M. When phosphorylated at Thr[148], the β2-subunit of AMP-activated kinase does not associate with glycogen in skeletal muscle. *Am. J. Physiol. Cell Physiol.* **2016**, *311*, C35–C42. [CrossRef] [PubMed]

62. Carling, D.; Hardie, D.G. The substrate and sequence specificity of the AMP-activated protein kinase. Phosphorylation of glycogen synthase and phosphorylase kinase. *Biochim. Biophys. Acta* **1989**, *1012*, 81–86. [CrossRef]

63. Jorgensen, S.B.; Nielsen, J.N.; Birk, J.B.; Olsen, G.S.; Viollet, B.; Andreelli, F.; Schjerling, P.; Vaulont, S.; Hardie, D.G.; Hansen, B.F.; et al. The α2-5′AMP-activated protein kinase is a site 2 glycogen synthase kinase in skeletal muscle and is responsive to glucose loading. *Diabetes* **2004**, *53*, 3074–3081. [CrossRef] [PubMed]

64. Cokorinos, E.C.; Delmore, J.; Reyes, A.R.; Albuquerque, B.; Kjobsted, R.; Jorgensen, N.O.; Tran, J.L.; Jatkar, A.; Cialdea, K.; Esquejo, R.M.; et al. Activation of skeletal muscle AMPK promotes glucose disposal and glucose lowering in non-human primates and mice. *Cell Metab.* **2017**. [CrossRef] [PubMed]

65. Myers, R.W.; Guan, H.P.; Ehrhart, J.; Petrov, A.; Prahalada, S.; Tozzo, E.; Yang, X.; Kurtz, M.M.; Trujillo, M.; Gonzalez Trotter, D.; et al. Systemic pan-AMPK activator MK-8722 improves glucose homeostasis but induces cardiac hypertrophy. *Science* **2017**, *357*, 507–511. [CrossRef] [PubMed]

66. Fritzen, A.M.; Lundsgaard, A.M.; Jeppesen, J.; Christiansen, M.L.; Bienso, R.; Dyck, J.R.; Pilegaard, H.; Kiens, B. 5′-AMP activated protein kinase α2 controls substrate metabolism during post-exercise recovery via regulation of pyruvate dehydrogenase kinase 4. *J. Physiol.* **2015**, *593*, 4765–4780. [CrossRef] [PubMed]

67. Prats, C.; Helge, J.W.; Nordby, P.; Qvortrup, K.; Ploug, T.; Dela, F.; Wojtaszewski, J.F. Dual regulation of muscle glycogen synthase during exercise by activation and compartmentalization. *J. Biol. Chem.* **2009**, *284*, 15692–15700. [CrossRef] [PubMed]

68. Barre, L.; Richardson, C.; Hirshman, M.F.; Brozinick, J.; Fiering, S.; Kemp, B.E.; Goodyear, L.J.; Witters, L.A. Genetic model for the chronic activation of skeletal muscle AMP-activated protein kinase leads to glycogen accumulation. *Am. J. Physiol. Endocrinol. Metab.* **2007**, *292*, E802–E811. [CrossRef] [PubMed]

69. Prats, C.; Gomez-Cabello, A.; Hansen, A.V. Intracellular compartmentalization of skeletal muscle glycogen metabolism and insulin signalling. *Exp. Physiol.* **2011**, *96*, 385–390. [CrossRef] [PubMed]

70. Young, M.E.; Leighton, B.; Radda, G.K. Glycogen phosphorylase may be activated by AMP-kinase in skeletal muscle. *Biochem. Soc. Trans.* **1996**, *24*, 268S. [CrossRef] [PubMed]

71. Young, M.E.; Radda, G.K.; Leighton, B. Activation of glycogen phosphorylase and glycogenolysis in rat skeletal muscle by AICAR—An activator of AMP-activated protein kinase. *FEBS Lett.* **1996**, *382*, 43–47. [CrossRef]

72. Daignan-Fornier, B.; Pinson, B. 5-Aminoimidazole-4-carboxamide-1-beta-d-ribofuranosyl 5′-monophosphate (AICAR), a highly conserved purine intermediate with multiple effects. *Metabolites* **2012**, *2*, 292–302. [CrossRef] [PubMed]

73. Sakoda, H.; Fujishiro, M.; Fujio, J.; Shojima, N.; Ogihara, T.; Kushiyama, A.; Fukushima, Y.; Anai, M.; Ono, H.; Kikuchi, M.; et al. Glycogen debranching enzyme association with β-Subunit regulates AMP-activated protein kinase activity. *Am. J. Physiol. Endocrinol. Metab.* **2005**, *289*, E474–E481. [CrossRef] [PubMed]

74. Stapleton, D.; Nelson, C.; Parsawar, K.; McClain, D.; Gilbert-Wilson, R.; Barker, E.; Rudd, B.; Brown, K.; Hendrix, W.; O'Donnell, P.; et al. Analysis of hepatic glycogen-associated proteins. *Proteomics* **2010**, *10*, 2320–2329. [CrossRef] [PubMed]

75. Stapleton, D.; Nelson, C.; Parsawar, K.; Flores-Opazo, M.; McClain, D.; Parker, G. The 3T3-L1 adipocyte glycogen proteome. *Proteome Sci.* **2013**. [CrossRef] [PubMed]

76. Steinberg, G.R.; O'Neill, H.M.; Dzamko, N.L.; Galic, S.; Naim, T.; Koopman, R.; Jorgensen, S.B.; Honeyman, J.; Hewitt, K.; Chen, Z.P.; et al. Whole body deletion of AMP-activated protein kinase β2 reduces muscle AMPK activity and exercise capacity. *J. Biol. Chem.* **2010**, *285*, 37198–37209. [CrossRef] [PubMed]

77. Dasgupta, B.; Ju, J.S.; Sasaki, Y.; Liu, X.; Jung, S.R.; Higashida, K.; Lindquist, D.; Milbrandt, J. The AMPK β2 subunit is required for energy homeostasis during metabolic stress. *Mol. Cell. Biol.* **2012**, *32*, 2837–2848. [CrossRef] [PubMed]

78. Hasenour, C.M.; Ridley, D.E.; James, F.D.; Hughey, C.C.; Donahue, E.P.; Viollet, B.; Foretz, M.; Young, J.D.; Wasserman, D.H. Liver AMP-activated protein kinase is unnecessary for gluconeogenesis but protects energy state during nutrient deprivation. *PLoS ONE* **2017**, *12*, e0170382.

79. Hingst, J.R.; Bruhn, L.; Hansen, M.B.; Rosschou, M.F.; Birk, J.B.; Fentz, J.; Foretz, M.; Viollet, B.; Sakamoto, K.; Faergeman, N.J.; et al. Exercise-induced molecular mechanisms promoting glycogen supercompensation in human skeletal muscle. *Mol. Metab.* **2018**. [CrossRef] [PubMed]

80. Wojtaszewski, J.F.; Jorgensen, S.B.; Hellsten, Y.; Hardie, D.G.; Richter, E.A. Glycogen-dependent effects of 5-aminoimidazole-4-carboxamide (AICA)-riboside on AMP-activated protein kinase and glycogen synthase activities in rat skeletal muscle. *Diabetes* **2002**, *51*, 284–292. [CrossRef] [PubMed]

81. Wojtaszewski, J.F.; MacDonald, C.; Nielsen, J.N.; Hellsten, Y.; Hardie, D.G.; Kemp, B.E.; Kiens, B.; Richter, E.A. Regulation of 5′AMP-activated protein kinase activity and substrate utilization in exercising human skeletal muscle. *Am. J. Physiol. Endocrinol. Metab.* **2003**, *284*, E813–E822. [CrossRef] [PubMed]

82. Steinberg, G.R.; Watt, M.J.; McGee, S.L.; Chan, S.; Hargreaves, M.; Febbraio, M.A.; Stapleton, D.; Kemp, B.E. Reduced glycogen availability is associated with increased AMPKα2 activity, nuclear AMPKα2 protein abundance, and GLUT4 mRNA expression in contracting human skeletal muscle. *Appl. Physiol. Nutr. Metab.* **2006**, *31*, 302–312. [CrossRef] [PubMed]

83. Watt, M.J.; Steinberg, G.R.; Chan, S.; Garnham, A.; Kemp, B.E.; Febbraio, M.A. β-Adrenergic stimulation of skeletal muscle HSL can be overridden by AMPK signaling. *FASEB J.* **2004**, *18*, 1445–1446. [CrossRef] [PubMed]

84. Yeo, W.K.; Lessard, S.J.; Chen, Z.P.; Garnham, A.P.; Burke, L.M.; Rivas, D.A.; Kemp, B.E.; Hawley, J.A. Fat adaptation followed by carbohydrate restoration increases AMPK activity in skeletal muscle from trained humans. *J. Appl. Physiol.* **2008**, *105*, 1519–1526. [CrossRef] [PubMed]

85. Yeo, W.K.; McGee, S.L.; Carey, A.L.; Paton, C.D.; Garnham, A.P.; Hargreaves, M.; Hawley, J.A. Acute signalling responses to intense endurance training commenced with low or normal muscle glycogen. *Exp. Physiol.* **2010**, *95*, 351–358. [CrossRef] [PubMed]

86. Lane, S.C.; Camera, D.M.; Lassiter, D.G.; Areta, J.L.; Bird, S.R.; Yeo, W.K.; Jeacocke, N.A.; Krook, A.; Zierath, J.R.; Burke, L.M.; et al. Effects of sleeping with reduced carbohydrate availability on acute training responses. *J. Appl. Physiol.* **2015**, *119*, 643–655. [CrossRef] [PubMed]

87. Philp, A.; MacKenzie, M.G.; Belew, M.Y.; Towler, M.C.; Corstorphine, A.; Papalamprou, A.; Hardie, D.G.; Baar, K. Glycogen content regulates peroxisome proliferator activated receptor-∂ (PPAR-∂) activity in rat skeletal muscle. *PLoS ONE* **2013**, *8*, e77200. [CrossRef] [PubMed]

88. Bartlett, J.D.; Louhelainen, J.; Iqbal, Z.; Cochran, A.J.; Gibala, M.J.; Gregson, W.; Close, G.L.; Drust, B.; Morton, J.P. Reduced carbohydrate availability enhances exercise-induced p53 signaling in human skeletal muscle: Implications for mitochondrial biogenesis. *Am. J. Physiol. Regul. Integr. Comp. Physiol.* **2013**, *304*, R450–R458. [CrossRef] [PubMed]

89. Psilander, N.; Frank, P.; Flockhart, M.; Sahlin, K. Exercise with low glycogen increases PGC-1α gene expression in human skeletal muscle. *Eur. J. Appl. Physiol.* **2013**, *113*, 951–963. [CrossRef] [PubMed]

90. Sriwijitkamol, A.; Coletta, D.K.; Wajcberg, E.; Balbontin, G.B.; Reyna, S.M.; Barrientes, J.; Eagan, P.A.; Jenkinson, C.P.; Cersosimo, E.; DeFronzo, R.A.; et al. Effect of acute exercise on AMPK signaling in skeletal muscle of subjects with type 2 diabetes: A time-course and dose-response study. *Diabetes* **2007**, *56*, 836–848. [CrossRef] [PubMed]

91. Witters, L.A.; Kemp, B.E. Insulin activation of acetyl-CoA carboxylase accompanied by inhibition of the 5'-AMP-activated protein kinase. *J. Biol. Chem.* **1992**, *267*, 2864–2867. [PubMed]

92. Musi, N.; Fujii, N.; Hirshman, M.F.; Ekberg, I.; Froberg, S.; Ljungqvist, O.; Thorell, A.; Goodyear, L.J. AMP-activated protein kinase (AMPK) is activated in muscle of subjects with type 2 diabetes during exercise. *Diabetes* **2001**, *50*, 921–927. [CrossRef] [PubMed]

93. Hojlund, K.; Staehr, P.; Hansen, B.F.; Green, K.A.; Hardie, D.G.; Richter, E.A.; Beck-Nielsen, H.; Wojtaszewski, J.F. Increased phosphorylation of skeletal muscle glycogen synthase at NH2-terminal sites during physiological hyperinsulinemia in type 2 diabetes. *Diabetes* **2003**, *52*, 1393–1402. [CrossRef] [PubMed]

94. Nielsen, J.N.; Wojtaszewski, J.F.; Haller, R.G.; Hardie, D.G.; Kemp, B.E.; Richter, E.A.; Vissing, J. Role of 5'AMP-activated protein kinase in glycogen synthase activity and glucose utilization: Insights from patients with mcardle's disease. *J. Physiol.* **2002**, *541*, 979–989. [CrossRef] [PubMed]

95. Krag, T.O.; Pinos, T.; Nielsen, T.L.; Duran, J.; Garcia-Rocha, M.; Andreu, A.L.; Vissing, J. Differential glucose metabolism in mice and humans affected by mcardle disease. *Am. J. Physiol. Regul. Integr. Comp. Physiol.* **2016**, *311*, R307–R314. [CrossRef] [PubMed]

96. Nielsen, T.L.; Pinos, T.; Brull, A.; Vissing, J.; Krag, T.O. Exercising with blocked muscle glycogenolysis: Adaptation in the mcardle mouse. *Mol. Genet. Metab.* **2018**, *123*, 21–27. [CrossRef] [PubMed]

97. Kollberg, G.; Tulinius, M.; Gilljam, T.; Ostman-Smith, I.; Forsander, G.; Jotorp, P.; Oldfors, A.; Holme, E. Cardiomyopathy and exercise intolerance in muscle glycogen storage disease 0. *N. Engl. J. Med.* **2007**, *357*, 1507–1514. [CrossRef] [PubMed]

98. Pederson, B.A.; Schroeder, J.M.; Parker, G.E.; Smith, M.W.; DePaoli-Roach, A.A.; Roach, P.J. Glucose metabolism in mice lacking muscle glycogen synthase. *Diabetes* **2005**, *54*, 3466–3473. [CrossRef] [PubMed]

99. Pederson, B.A.; Chen, H.; Schroeder, J.M.; Shou, W.; DePaoli-Roach, A.A.; Roach, P.J. Abnormal cardiac development in the absence of heart glycogen. *Mol. Cell. Biol.* **2004**, *24*, 7179–7187. [CrossRef] [PubMed]

100. Xirouchaki, C.E.; Mangiafico, S.P.; Bate, K.; Ruan, Z.; Huang, A.M.; Tedjosiswoyo, B.W.; Lamont, B.; Pong, W.; Favaloro, J.; Blair, A.R.; et al. Impaired glucose metabolism and exercise capacity with muscle-specific glycogen synthase 1 (gys1) deletion in adult mice. *Mol. Metab.* **2016**, *5*, 221–232. [CrossRef] [PubMed]

101. Hoffman, N.J.; Parker, B.L.; Chaudhuri, R.; Fisher-Wellman, K.H.; Kleinert, M.; Humphrey, S.J.; Yang, P.; Holliday, M.; Trefely, S.; Fazakerley, D.J.; et al. Global phosphoproteomic analysis of human skeletal muscle reveals a network of exercise-regulated kinases and AMPK substrates. *Cell Metab.* **2015**, *22*, 922–935. [CrossRef] [PubMed]

102. Schaffer, B.E.; Levin, R.S.; Hertz, N.T.; Maures, T.J.; Schoof, M.L.; Hollstein, P.E.; Benayoun, B.A.; Banko, M.R.; Shaw, R.J.; Shokat, K.M.; et al. Identification of AMPK phosphorylation sites reveals a network of proteins involved in cell invasion and facilitates large-scale substrate prediction. *Cell Metab.* **2015**, *22*, 907–921. [CrossRef] [PubMed]

103. Ducommun, S.; Deak, M.; Sumpton, D.; Ford, R.J.; Nunez Galindo, A.; Kussmann, M.; Viollet, B.; Steinberg, G.R.; Foretz, M.; Dayon, L.; et al. Motif affinity and mass spectrometry proteomic approach for the discovery of cellular AMPK targets: Identification of mitochondrial fission factor as a new AMPK substrate. *Cell. Signal.* **2015**, *27*, 978–988. [CrossRef] [PubMed]

104. Hoffman, N.J. Omics and exercise: Global approaches for mapping exercise biological networks. *Cold Spring Harb. Perspect. Med.* **2017**. [CrossRef] [PubMed]

105. Tsou, P.; Zheng, B.; Hsu, C.H.; Sasaki, A.T.; Cantley, L.C. A fluorescent reporter of AMPK activity and cellular energy stress. *Cell Metab.* **2011**, *13*, 476–486. [CrossRef] [PubMed]

106. Chennell, G.; Willows, R.J.; Warren, S.C.; Carling, D.; French, P.M.; Dunsby, C.; Sardini, A. Imaging of metabolic status in 3D cultures with an improved AMPK fret biosensor for flim. *Sensors* **2016**, *16*, 1312. [CrossRef] [PubMed]

107. Konagaya, Y.; Terai, K.; Hirao, Y.; Takakura, K.; Imajo, M.; Kamioka, Y.; Sasaoka, N.; Kakizuka, A.; Sumiyama, K.; Asano, T.; et al. A highly sensitive fret biosensor for AMPK exhibits heterogeneous AMPK responses among cells and organs. *Cell Rep.* **2017**, *21*, 2628–2638. [CrossRef] [PubMed]

108. Depry, C.; Mehta, S.; Li, R.; Zhang, J. Visualization of compartmentalized kinase activity dynamics using adaptable BimKARs. *Chem. Biol.* **2015**, *22*, 1470–1479. [CrossRef] [PubMed]

109. Yan, Y.; Gu, X.; Xu, H.E.; Melcher, K. A highly sensitive non-radioactive activity assay for AMP-activated protein kinase (AMPK). *Methods Protoc.* **2018**, *1*, 3. [CrossRef] [PubMed]

110. Arad, M.; Benson, D.W.; Perez-Atayde, A.R.; McKenna, W.J.; Sparks, E.A.; Kanter, R.J.; McGarry, K.; Seidman, J.G.; Seidman, C.E. Constitutively active AMP kinase mutations cause glycogen storage disease mimicking hypertrophic cardiomyopathy. *J. Clin. Investig.* **2002**, *109*, 357–362. [CrossRef] [PubMed]

111. Milan, D.; Jeon, J.T.; Looft, C.; Amarger, V.; Robic, A.; Thelander, M.; Rogel-Gaillard, C.; Paul, S.; Iannuccelli, N.; Rask, L.; et al. A mutation in PRKAG3 associated with excess glycogen content in pig skeletal muscle. *Science* **2000**, *288*, 1248–1251. [CrossRef] [PubMed]

112. Garcia-Roves, P.M.; Osler, M.E.; Holmstrom, M.H.; Zierath, J.R. Gain-of-function R225Q mutation in AMP-activated protein kinase γ3 subunit increases mitochondrial biogenesis in glycolytic skeletal muscle. *J. Biol. Chem.* **2008**, *283*, 35724–35734. [CrossRef] [PubMed]

113. Scott, J.W.; Galic, S.; Graham, K.L.; Foitzik, R.; Ling, N.X.; Dite, T.A.; Issa, S.M.; Langendorf, C.G.; Weng, Q.P.; Thomas, H.E.; et al. Inhibition of AMP-activated protein kinase at the allosteric drug-binding site promotes islet insulin release. *Chem. Biol.* **2015**, *22*, 705–711. [CrossRef] [PubMed]

12

AMP-Activated Protein Kinase (AMPK)-Dependent Regulation of Renal Transport

Philipp Glosse [1] and Michael Föller [2,*]

[1]   Institute of Agricultural and Nutritional Sciences, Martin Luther University Halle-Wittenberg,
      D-06120 Halle (Saale), Germany; philipp.glosse@landw.uni-halle.de
[2]   Institute of Physiology, University of Hohenheim, D-70599 Stuttgart, Germany
*   Correspondence: michael.foeller@uni-hohenheim.de

Abstract: AMP-activated kinase (AMPK) is a serine/threonine kinase that is expressed in most cells and activated by a high cellular AMP/ATP ratio (indicating energy deficiency) or by Ca^{2+}. In general, AMPK turns on energy-generating pathways (e.g., glucose uptake, glycolysis, fatty acid oxidation) and stops energy-consuming processes (e.g., lipogenesis, glycogenesis), thereby helping cells survive low energy states. The functional element of the kidney, the nephron, consists of the glomerulus, where the primary urine is filtered, and the proximal tubule, Henle's loop, the distal tubule, and the collecting duct. In the tubular system of the kidney, the composition of primary urine is modified by the reabsorption and secretion of ions and molecules to yield final excreted urine. The underlying membrane transport processes are mainly energy-consuming (active transport) and in some cases passive. Since active transport accounts for a large part of the cell's ATP demands, it is an important target for AMPK. Here, we review the AMPK-dependent regulation of membrane transport along nephron segments and discuss physiological and pathophysiological implications.

Keywords: transporter; carrier; pump; membrane; energy deficiency

---

1. Introduction

The 5′-adenosine monophosphate (AMP)–activated protein kinase (AMPK) is a serine/threonine protein kinase that is evolutionarily conserved and functions as an intracellular energy sensor in mammalian cells [1–5]. It is a central regulator of energy homeostasis and affects many important cellular functions including growth, differentiation, autophagy, and metabolism [1,2,6]. During energy depletion when cellular AMP levels are high relative to the adenosine triphosphate (ATP) concentration, AMPK activates energy-providing pathways including glucose uptake, glycolysis, or fatty acid oxidation [7–10]. Simultaneously, processes consuming ATP (e.g., gluconeogenesis, lipogenesis, or protein synthesis) are inhibited [7–10].

Being expressed in most mammalian cells, AMPK is a heterotrimeric protein consisting of a catalytic α (α1 or α2), scaffolding β (β1 or β2), and a regulatory nucleotide-binding γ (γ1, γ2, or γ3) subunit with the expression pattern differing from cell type to cell type [1,2,11–14]. Induction of AMPK activity involves phosphorylation of the conserved threonine residue Thr172 within the activation loop of the α subunit's kinase domain by various protein kinases including the tumor suppressor liver kinase B1 (LKB1), Ca^{2+}/calmodulin–dependent protein kinase kinase β (CaMKKβ), and transforming growth factor beta-activated kinase 1 [1,15–28]. AMPK activation in cellular energy depletion is primarily mediated by an increase in the AMP/ATP or ADP/ATP ratio [8,29,30]. Thus, AMP or ADP binding to the subunit at cystathionine-beta-synthase repeats results in conformational changes that allows for the phosphorylation at Thr172 by LKB1. This results in an enhancement of AMPK activity by >100-fold [1,8,12,15,31–36]. Moreover, AMP or ADP binding prevents dephosphorylation at Thr172 by protein phosphatases [8,12,37,38]. Additionally, binding of AMP, but not ADP, activates AMPK allosterically [8,11,12,37]. Conversely, ATP binding to the cystathionine-beta-synthase domain results in AMPK dephosphorylation by protein phosphatases [1,8,39].

Besides LKB1-associated regulation of AMPK phosphorylation, an alternative Ca^{2+}-involving activation mechanisms independent of AMP exists [6,12,40,41]. Protein kinase CaMKKβ phosphorylates AMPK at Thr172 in response to elevated intracellular Ca^{2+} levels which may be caused by mediators such as thrombin or ghrelin [6,12,23,40,42,43]. Intracellular Ca^{2+} store depletion detected by the Ca^{2+}-sensing protein stromal interacting molecule-1 leads to store-operated Ca^{2+} entry (SOCE) involving the Ca^{2+} release-activated Ca^{2+} channel Orai1 [44–49]. Orai1-mediated SOCE impacts on many cellular functions including cell proliferation, differentiation, migration, and cytokine production [44,50–55]. SOCE is involved in a sort of feedback mechanism involving AMPK: SOCE activates AMPK through CaMKKβ. AMPK in turn inhibits SOCE [45]. Moreover, AMPK inhibits SOCE by regulating Orai1 membrane abundance (at least in UMR106 cells) [44,56].

AMPK is a major regulator of whole body energy homeostasis [10,12], impacting on a variety of organs including liver [57–61], skeletal [62–66] and cardiac muscle [67–73], kidney [74–77], and bone [78–80]. In the kidney, AMPK regulates epithelial transport, podocyte function, blood pressure, epithelial-to-mesenchymal transition, autophagy as well as nitric oxide synthesis [75,76,81–83]. Not surprisingly, AMPK is highly relevant for renal pathophysiology, including ischemia, diabetic renal hypertrophy, polycystic kidney disease, chronic kidney disease, and hypertension [40,67,74–76]. This review summarizes the contribution of AMPK to the regulation of renal transport and hence to the final composition of excreted urine. Moreover, pathophysiological implications are discussed.

2. AMPK and Renal Tubular Transport

The kidney is particularly relevant for fluid, electrolyte, and acid–base homeostasis. In addition, it is an endocrine organ producing different hormones such as erythropoietin, Klotho, and calcitriol, the active form of vitamin D [84–86]. The kidneys are made up of about 1 million nephrons, their functional elements. A nephron comprises the glomerulus surrounded by the Bowman's capsule, the proximal tubule, Henle's loop, distal tubule, and the collecting duct. The primary urine is filtered in the glomerulus. Its composition is similar to plasma. In general, large molecules and particularly proteins >6000 Dalton are normally filtered to a low extent, if at all. The renal tubular system modifies the primary urine by reabsorbing or secreting ions and molecules, ultimately yielding the final urine [85–87]. Epithelial transport is mainly dependent on ATP-dependent pumps (primary-active), secondary-or tertiary-active transporters, as well as carriers and channels (passive, facilitated diffusion). Since active transport consumes energy by definition, it is not surprising that it is subject to regulation by AMPK. Moreover, even passive transport involving glucose transporter (GLUT) carriers is controlled by AMPK [74,75].

2.1. Na^+/K^+-ATPase

The ubiquitously expressed Na^+/K^+-ATPase is a primary active ATP-driven pump that mediates the basolateral extrusion of $3Na^+$ in exchange of $2K^+$, thereby establishing a transmembrane Na^+ gradient, which is the prerequisite for secondary active Na^+-dependent transport (e.g., through Na^+-dependent glucose cotransporter 1 and 2 (SGLT1/2), Na^+/H^+ exchanger isoform 1 (NHE1), Na^+-coupled phosphate transporter (NaPi-IIa), or Na^+-K^+-$2Cl^-$ cotransporter (NKCC2), as discussed below) [75,88–94]. Almost one-third of the body's energy is consumed by this pump [95]. Therefore, it does make sense that it is regulated by AMPK [74–76,94]: AMPK inhibits Na^+/K^+-ATPase in airway epithelial cells by promoting its endocytosis [96–100]. However, AMPK stimulates Na^+/K^+-ATPase membrane expression in skeletal muscle cells [101] and in renal epithelia [102], thereby counteracting renal ischemia-induced Na^+/K^+-ATPase endocytosis [103]. Interestingly, AMPKβ1 deficiency was found not to alter outcome in an ischemic kidney injury model in mice [104]. Hence, the effect of AMPK on Na^+/K^+-ATPase appears to be highly tissue-specific [74,75].

2.2. Proximal Tubule

A wide variety of luminal Na^+-dependent cotransporters, which are secondary active, are involved in epithelial transport in the proximal tubule. Secondary active transporters utilize the energy of the transmembrane Na^+ gradient generated by the primary active ATP-consuming Na^+/K^+-ATPase to facilitate transport of a substrate against its concentration gradient [105,106]. These transporters and the basolateral Na^+/K^+-ATPase consume substantial amounts of total cellular energy [74,75,107]. Hence, AMPK has been demonstrated to be an important regulator of proximal tubule transport [74,75].

2.2.1. Glucose Transport

Since glucose is freely filtered by the glomerulus, glucose concentration in primary urine is similar to the plasma glucose concentration, whereas excreted urine is usually free of glucose [108–110]. The sugar is reabsorbed in the proximal tubule by the Na^+-dependent glucose cotransporter 1 and 2 (SGLT1 and 2), the different expression patterns and properties of which ensure total glucose reabsorption as long as the plasma glucose concentration is not abnormally high [89,108]. SGLT2 has a high transport capacity but low affinity for glucose and is predominantly expressed in the kidney, while SGLT1 is also expressed in other tissues including the small intestine. SGLT2 contributes to the reabsorption of up to 90% of filtered glucose [108,109,111,112]. On the other hand, AMPK-regulated SGLT1 [7,92,113] has a low transport capacity but high affinity for glucose and reabsorbs the remaining glucose [108–110,114,115]. Glucose leaves the basolateral membrane through passive glucose carriers GLUT1 and GLUT2 [108,116–118]. AMPK activates SGLT1-dependent glucose transport, presumably by stimulating membrane insertion of the cotransporter as observed in colorectal Caco-2 cells [92,119]. In line with this, AMPK activation is associated with increased *SGLT1* expression and glucose uptake in cardiomyocytes [113,120]. Although the AMPK-dependent regulation of SGLT1 in the proximal tubule has not explicitly been addressed, it is tempting to speculate that it is similar to other cell types [92,113,119,120]. The regulation of SGLT by AMPK is a doubled-edged sword: on the one hand, SGLT1-dependent reabsorption of glucose in proximal tubular cells requires energy which is generated by β-oxidation of fatty acids to a large extent [121,122]. On the other hand, it prevents the loss of energy-rich glucose [122,123], thereby maintaining the Na^+/K^+-ATPase-facilitated Na^+ gradient for Na^+-dependent transport and many other cellular processes [75,76]. SGLT1-mediated glucose uptake is linked to the GLUT1-dependent efflux at the basolateral side [108,116]. GLUT1 activity is stimulated by AMPK in various cell types [124–131]. Therefore, it is conceivable that renal GLUT1 might also be regulated by AMPK in order to save energy-providing glucose. In line with this, Baldwin et al. (1997) showed enhanced glucose uptake via GLUT1 in baby hamster kidney cells treated with AMPK activator 5-aminoimidazole-4-carboxamide ribonucleotide (AICAR) [132]. Moreover, Sokolovska et al. (2010) reported that metformin, another pharmacological AMPK activator, increased *GLUT1* gene expression in rat kidneys [133]. Also, AMPK activation was associated with enhanced activity of GLUT2. These studies, however, found reduced SGLT1 membrane abundance upon AMPK activation, at least in the case of murine intestinal tissue [134,135].

2.2.2. Na^+/H^+ Exchanger Isoform 1

The ubiquitous Na^+/H^+ exchanger isoform 1 (NHE1) participates in cell volume and pH regulation by extruding one cytosolic H^+ in exchange for one extracellular Na^+ [136,137]. NHE1 is expressed in all parts of the nephron, including the proximal tubule. However, it cannot be detected in the macula densa and intercalated cells of the distal nephron [136,138,139]. In the proximal tubule, NHE1 is particularly important for HCO_3^- reabsorption [140]. In hypoxia, anaerobic glycolysis is predominant, which results in intracellular accumulation of lactate and H^+ [90]. Acidosis, however, inhibits glycolysis [90,141,142] and would jeopardize cellular energy generation. AMPK-dependent stimulation of NHE1 activity in human embryonic kidney (HEK) cells therefore helps cells keep up anaerobic glycolysis in oxygen deficiency, as demonstrated by Rotte et al. (2010) [90]. Given that NHE1 is needed for proximal tubular HCO_3^- reabsorption [140], AMPK may help retain HCO_3^-, thereby alleviating acidosis in energy deficiency and hypoxia.

2.2.3. Creatine Transporter

In some organs with high metabolic activity, including skeletal muscle, heart, and brain, creatine is used to refuel cellular ATP levels [143–145]. In the proximal tubule, creatine, a small molecule that is freely filtered, is also reabsorbed through secondary active Na^+-dependent creatine transporter (CRT) (SLC6A8) [7,75,143,146]. AMPK has been demonstrated to downregulate CRT activity and apical membrane expression in a polarized mouse S3 proximal tubule cell line, presumably through mammalian target of rapamycin signaling [147]. The AMPK-dependent inhibition of CRT may help reduce unnecessary energy expenditure [75]. Conversely, AMPK stimulates CRT-mediated creatine transport in cardiomyocytes [148,149]. This again demonstrates that AMPK effects are tissue-specific [148].

2.2.4. Na^+-Coupled Phosphate Transporter IIa

Inorganic phosphate is mainly reabsorbed by the secondary active Na^+-coupled phosphate transporter (NaPi-IIa) (SLC34A1) in the proximal tubule [93,150–152]. Employing electrophysiological recordings in *Xenopus* oocytes, it was shown that AMPK inhibits NaPi-IIa [93]. Kinetics analysis revealed that AMPK decreases NaPi-IIa membrane expression rather than changing its properties.

The regulation of phosphate metabolism by AMPK is not restricted to NaPi-IIa: Recently, AMPK was demonstrated to control the formation of bone-derived hormone fibroblast growth factor 23 (FGF23) [56], which induces renal phosphate excretion by extracellular-signal regulated kinases 1/2 (ERK1/2)-mediated degradation of membrane NaPi-IIa [150]. AMPK inhibits FGF23 production in cell culture and in mice [56]. Despite markedly elevated FGF23 serum levels in AMPKα1-deficient mice, renal phosphate excretion was not different from wild-type animals [56]. The same holds true for cellular localization of NaPi-IIa and renal ERK1/2 [56]. Thus, it is possible that AMPK deficiency is paralleled with some FGF23 resistance.

2.3. Loop of Henle

2.3.1. Na^+-K^+-$2Cl^-$ Cotransporter

The Na^+-K^+-$2Cl^-$ cotransporter (NKCC2), expressed in the thick ascending limb (TAL) of the loop of Henle and macula densa, is required for the generation of a hypertonic medullary interstitium, a mechanism needed for concentrating urine [75,76,88,91]. NKCC2 is a direct substrate of AMPK which phosphorylates it at its stimulatory serine residue Ser-126 [153]. Moreover, exposure of murine macula densa-like cells to low salt leads to AMPK activation and increased NKCC2 phosphorylation [154]. In addition, increased subapical expression (and apparent reduced apical expression) of NKCC2 in the medullary TAL of the loop of Henle along with elevated urinary Na^+ excretion in AMPKβ1-deficient mice on a normal salt diet were observed [155]. This is in line with AMPK being an important regulator of NKCC2-mediated salt retention in the medullary TAL of Henle [155]. Efe et al. (2016) recently observed markedly increased outer medullary expression of NKCC2 in rats treated with the AMPK activator metformin [156]. However, according to a recent in vivo study by Udwan et al. (2017), a low salt diet induced upregulation of NKCC2 surface expression in mouse kidneys but left AMPK activity unchanged [157]. Therefore, the exact role of AMPK in stimulating NKCC2 remains to be established.

2.3.2. Renal Outer Medullary K^+ Channel

The apical renal outer medullary K^+ channel (ROMK) is required for NKCC2 to work properly, as it allows the recirculation of K^+ ions taken up by NKCC2 into the lumen [75,88]. AMPK is an inhibitor of ROMK by downregulating both channel activity and membrane abundance of the channel protein in a heterologous expression system using *Xenopus* oocytes [158]. In vivo studies revealed that the AMPK effect on ROMK is relevant for the renal excretion of K^+ after an acute K^+ challenge, as upregulation of renal ROMK1 protein expression and the ability of K^+ elimination were more pronounced in AMPKα1-deficient than in wild-type mice [158].

2.4. Distal Tubule

2.4.1. Cystic Fibrosis Transmembrane Conductance Regulator

The ATP-gated and cyclic AMP (cAMP)-dependent Cl^- channel cystic fibrosis transmembrane conductance regulator (CFTR) participates in Cl^- secretion and is broadly known for its role in cystic fibrosis, the pathophysiology of which is due to channel malfunction [74–76,159]. In the kidney, CFTR contributes to Cl^- secretion in the distal tubule and the principal cells of the cortical and medullary collecting ducts [74,75,160]. AMPK has been demonstrated to inhibit CFTR-dependent Cl^- conductance in Xenopus oocytes [159] and to decrease CFTR channel activity in the lung [161,162] and colon [163]. cAMP-stimulated cell proliferation and CFTR-dependent Cl^- secretion play a decisive role for epithelial cyst enlargement in autosomal dominant polycystic kidney disease (ADPKD) [164]. In line with this, AMPK activation inhibits CFTR in Madin-Darby canine kidney (MDCK) cells [165] as well as decreases cystogenesis in murine models of ADPKD [165,166], suggesting a potential role for pharmacological AMPK activation in the treatment of ADPKD [165,166].

2.4.2. Ca^{2+} Transport

Most Ca^{2+} is reabsorbed by passive paracellular diffusion along with other ions and water through tight junctions in the proximal tubule and the more distal parts of the nephron [88,167]. Conversely, only 5–10% of filtered Ca^{2+} is reabsorbed by transcellular transport involving the apical transient receptor potential vanilloid 5 channel TRPV5 in the distal convoluted tubule [88]: Ca^{2+} enters the cell through TRPV5, whereas basolateral Ca^{2+} efflux is accomplished by the Na^+/Ca^{2+} exchanger (NCX) and the Ca^{2+}-ATPase [88,167,168]. AMPK has been shown to inhibit NCX and decrease Orai1-mediated SOCE in murine dendritic cells [169]. Therefore, it is tempting to speculate that Ca^{2+} reabsorption may be downregulated in the distal tubule in ATP deficiency [169,170]. Indeed, AMPK downregulates Orai1-dependent SOCE in T-lymphocytes [171], endothelial cells [45], and in osteoblast-like cells [56]. Since renal Orai1 activity contributes to kidney fibrosis [172], AMPK-mediated Orai1 downregulation may also be therapeutically desirable.

2.5. Collecting Duct

2.5.1. Epithelial Na^+ Channel

In the collecting duct, fine tuning of Na^+ and K^+ homeostasis is accomplished by epithelial Na^+ channel (ENaC) and ROMK K^+ channel. Both channels are controlled by the renin-angiotensin-aldosterone system [173–175] regulating extracellular volume and hence arterial blood pressure [173–177]. Na^+ reabsorption by ENaC in the late distal convoluted tubule and cortical collecting duct principal cells is a highly energy-demanding process, as it utilizes the electrochemical driving force generated by the basolateral Na^+/K^+-ATPase [74–76,176,178]. AMPK inhibits epithelial Na^+ transport in various tissues, including lung [96,179], colonic [180], and renal cortical collecting duct cells [180–183]. In line with this, AMPKα1-deficient mice exhibit increased renal ENaC expression [180]. In detail, AMPK downregulates ENaC surface expression by inducing the binding of the ubiquitin ligase neural precursor cell expressed developmentally downregulated protein 4-2 (Nedd4-2) to ENaC subunits, resulting in ENaC ubiquitination with subsequent endocytosis and degradation [177,180,184]. In line with this, activation of AMPK enhances the tubuloglomerular feedback and induces urinary diuresis and Na^+ excretion in rats [185]. However, AMPK$\alpha 1^{-/-}$ mice with genetic kidney-specific AMPKα2 deletion exhibit a moderate increase in diuresis and natriuresis, possibly because NKCC2 activity is insufficient despite upregulated ENaC activity [186]. Taken together, AMPK activity limits ENaC-dependent energy-consuming Na^+ reabsorption [177,180,181,185].

2.5.2. Voltage-Gated K^+ Channel

The voltage-gated K^+ channel (KCNQ1) is important for the cardiovascular system as well as for electrolyte and fluid homeostasis and is expressed in the distal nephron including the collecting duct [170,187–189]. Its exact role is ill-defined, although a contribution to cell volume regulation is postulated [75,187]. Similar to ENaC, AMPK inhibits KCNQ1 via Nedd4-2, as demonstrated in collecting duct principal cells of rat ex vivo kidney slices [187], MDCK cells [190], and *Xenopus* oocytes [191].

2.5.3. Vacuolar H^+-ATPase

The primary active vacuolar H^+-ATPase (V-ATPase) is located at the apical membrane of proximal tubule cells and collecting duct type A intercalated cells. It contributes to the regulation of acid–base homeostasis by secreting H^+ ions into the tubular lumen [76,192,193]. AMPK inhibits the protein kinase A (PKA)-dependent membrane expression of V-ATPase in collecting duct intercalated cells of rat ex vivo kidney slices [193]. Moreover, epididymal proton-secreting clear cells, developmentally related to intercalated cells, exhibit reduced apical membrane abundance of V-ATPase after in vivo perfusion with the AMPK activator 5-aminoimidazole-4-carboxamide-1-beta-D-ribofuranoside (AICAR) into rats [194]. It appears to be likely that energy deficiency limits highly energy-consuming primary active H^+ excretion in the proximal tubule, whereas secondary active NHE1-dependent H^+ secretion is maintained, thereby keeping up at least anaerobic glycolysis [192]. The opposing effects of AMPK and PKA on V-ATPase expression and activity in kidney intercalated cells can be explained by different phosphorylation sites, as AMPK and PKA phosphorylate the A subunit at Ser-384 and Ser-175, respectively [195,196]. McGuire and Forgac (2018) further demonstrated that AMPK increases lysosomal V-ATPase assembly and activity in HEK293T cells under conditions of energy depletion [197]. In cells depleted of energy, acidification of autophagic intracellular compartments by V-ATPases enables the lysosomal degradation of proteins and lipids to generate energy substrates for ATP production [197,198]. Thus, it appears to be likely that AMPK-regulated V-ATPase activity depends on its concrete cellular localization and function [197].

2.5.4. Water and Urea Handling

AMPK also regulates renal urea and water handling [76,199]. In the inner medullary collecting duct, osmotic gradients are generated by NKCC2 and urea transporter UT-A1 and water is reabsorbed through aquaporin 2 (AQP2) [76,156,199,200]. The concentration of urine requires the antidiuretic hormone vasopressin, which binds to vasopressin type 2 receptors of collecting duct principal cells, resulting in cAMP-mediated activation of PKA and subsequent phosphorylation and apical membrane insertion of AQP2 and UT-A1 [76,156,199]. Congenital nephrogenic diabetes insipidus (NDI) is a disease primarily caused by mutations of vasopressin type 2 receptors that is characterized by renal resistance to vasopressin and limited urine concentrating capacity [156,201]. According to two in vivo studies using rodent models of congenital NDI, the metformin-stimulated AMPK activation ameliorates the ability of the kidney to concentrate urine by increasing the phosphorylation and apical membrane expression of inner medullary AQP2 and UT-A1 [156,202]. In contrast, an ex vivo treatment of rat kidney slices with AICAR led to reduced apical membrane insertion of AQP2 [203]. Moreover, AMPK antagonizes the desmopressin-induced AQP2 phosphorylation in vitro, thus also suggesting an inhibitory function of AMPK on AQP2 regulation [203]. It appears likely that AMPK-independent effects of the pharmacological AMPK agonists contribute to this discrepancy [156,202,203]. Thus, further studies are clearly required.

3. Conclusions and Perspectives

A growing list of studies indicates the pivotal role of AMPK as a metabolic-sensing regulator of a multitude of transport processes in the kidney [7,74–76,170]. Particularly, AMPK activation under conditions of energy deficiency is expected to differentially modulate renal epithelial ion transport in order to preserve cellular energy homeostasis (Figure 1) [7,74–76,94,170]. Alongside the above discussed function of AMPK in kidney tubular transport, a variety of other transport proteins, which are expressed in the kidney as well, are regulated by AMPK in extrarenal tissues [7,94,170,204] that are reviewed elsewhere [170] and [7] and summarized in Table 1. Future studies are required to focus on the therapeutic value of pharmacological AMPK manipulation to combat kidney disease [74–76,205,206].

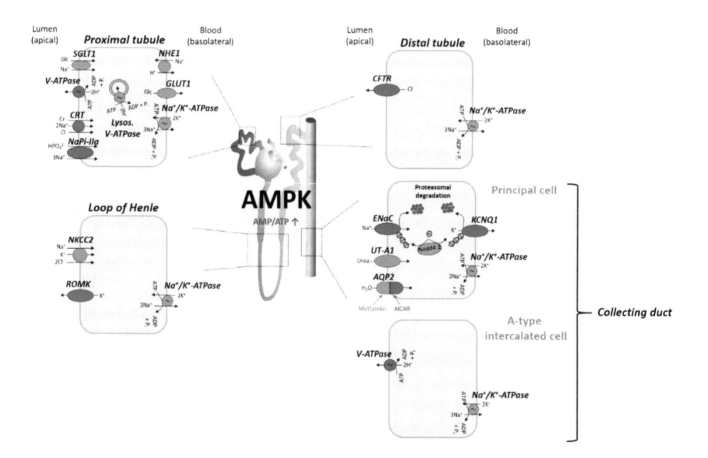

Figure 1. Tentative model illustrating AMPK-dependent effects on renal transport along the nephron. Cellular energy depletion (e.g., during hypoxia) leads to an elevated AMP/ATP ratio and subsequent AMPK activation. AMPK in turn regulates a multitude of active and passive epithelial transport processes along the renal tubular system in order to maintain cellular energy homeostasis. Ion channels, transport proteins, and ATPases that are activated upon AMPK stimulation are depicted as green icons, whereas red coloring indicates AMPK-dependent inhibition (see text for details). AMP, 5'-adenosine monophosphate; AMPK, AMP-activated protein kinase; SGLT1, Na^+-dependent glucose cotransporter 1; V-ATPase, vacuolar H^+-ATPase; CRT, creatine transporter; NaPi-IIa, Na^+-coupled phosphate transporter IIa; NHE1, Na^+/H^+ exchanger isoform 1; GLUT1, glucose transporter 1; NKCC2, Na^+-K^+-$2Cl^-$ cotransporter; ROMK, renal outer medullary K^+ channel; CFTR, cystic fibrosis transmembrane conductance regulator; ENaC, epithelial Na^+ channel; KCNQ1, voltage-gated K^+ channel; Nedd4-2, neural precursor cell expressed developmentally downregulated protein 4-2; UT-A1, urea transporter A1; AQP2, aquaporin 2.

Table 1. Overview of transport proteins regulated by AMPK in extrarenal tissues and evidence for renal expression.

Ion Channel/Transporter and Method of Modifying AMPK Activity	AMPK Effect	Cell Type of Studied AMPK Effect/Ref.	Evidence for Renal Expression/Ref.
Heterologous expression systems			
Kir2.1	Reduction of channel activity and membrane abundance via Nedd4-2 mediated endocytosis	Xenopus oocytes [207]	Human proximal tubular cells [208]
Kv1.5	Reduction of channel activity and membrane abundance via Nedd4-2 mediated endocytosis	Xenopus oocytes [209]	Human kidney biopsies [210]
Kv11.1 (hERG)	Reduction of channel activity and membrane abundance via Nedd4-2 mediated endocytosis	Xenopus oocytes [211]	Human proximal and distal convoluted tubule [212]
SMIT	Reduction of channel activity	Xenopus oocytes [213]	Rat kidney medulla [214]
BGT1	Reduction of channel activity	Xenopus oocytes [213]	Human kidney inner medulla [215] and mouse kidney medulla (basolateral membranes of collecting ducts and TAL of Henle) [216]
EAAT3	Reduction of channel activity and membrane abundance	Xenopus oocytes [217]	Mouse renal proximal tubule [218]
NCX	Reduction of channel activity and membrane abundance	Xenopus oocytes [169]	Rat distal convoluted tubule [219]
K$_{2P}$10.1 (TREK-2)	Inhibition of channel activity via phosphorylation at Ser-326 and Ser-359	HEK293 cells [220]	Human proximal tubule [221]
K$_{Ca}$1.1	Increase in channel activity and membrane abundance	Xenopus oocytes [222]	Human clear cell renal cell carcinoma (ccRCC) and healthy kidney cortex [223]
Pharmacological Manipulation			
K$_{Ca}$1.1	Inhibition of channel activity	Rat carotid body type I cells [224]	
Kir6.2	Upregulation of channel activity / Up- or down-regulation of channel activity	Rat cardiomyocytes [225] / Rat pancreatic beta-cells [226,227]	Rat renal tubular epithelial cells [228]
KCa3.1	Reduction of channel activity	Human airway epithelial cells [229]	Human proximal tubular cells [230]
MCT1 and MCT4	Upregulation of mRNA expression	Rat skeletal muscle [231]	MCT1: basolateral membrane of mouse proximal tubular epithelial cells [232] MCT4: human ccRCC [233]
PepT1	Downregulation of channel activity and brush-border membrane abundance	Caco-2 cells [234]	Rat renal proximal tubule [235]
Orai1	Downregulation of cell membrane abundance and SOCE	Rat UMR106 osteoblast-like cells [56]	Rat glomerular mesangial cells [236]
Genetically Modified Mouse Models			
Orai1		Mouse T-lymphocytes [171] Mouse dendritic cells [169]	

Author Contributions: P.G. and M.F. wrote this review.

Abbreviations

ADP	Adenosine diphosphate
ADPKD	Autosomal dominant polycystic kidney disease
AMPK	5′-adenosine monophosphate (AMP)–activated protein kinase
AQP2	Aquaporin 2
ATP	Adenosine triphosphate
BGT1	Betaine/γ-aminobutyric acid (GABA) transporter 1
CaMKKβ	Ca^{2+}/calmodulin–dependent protein kinase kinase β
cAMP	Cyclic adenosine monophosphate
ccRCC	Clear cell renal cell carcinoma
CFTR	Cystic fibrosis transmembrane conductance regulator
CRT	Creatine transporter
EAAT3	Excitatory amino acid transporter 3
ENaC	Epithelial Na^+ channel
ERK1/2	Extracellular-signal regulated kinases 1/2
FGF23	Fibroblast growth factor 23
GLUT	Glucose transporter
HEK	Human embryonic kidney cells
hERG	Human ether-a-go-go-related gene
Kca	Ca^{2+} activated K^+ channels
KCNQ1	Voltage-gated K^+ channel
Kir	Inwardly rectifying K^+ channels
Kv	Voltage gated K^+ channels
LKB1	Liver kinase B1
MCT	Monocarboxylate transporters
MDCK	Madin-Darby canine kidney cells
NaPi-IIa	Na^+-coupled phosphate transporter
NCX	Na^+/Ca^{2+} exchanger
NDI	Nephrogenic diabetes insipidus
Nedd4-2	Neural precursor cell expressed developmentally down-regulated protein 4-2
NHE1	Na^+/H^+ exchanger isoform 1
NKCC2	Na^+-K^+-2Cl$^-$ cotransporter
PepT1	H^+-coupled di- and tripeptide transporter 1
PKA	Protein kinase A
ROMK	Renal outer medullary K^+ channel
SGLT	Na^+-dependent glucose cotransporter
SMIT	Na^+ coupled myoinositol transporter
SOCE	Store-operated Ca^{2+} entry
TAL	Thick ascending limb
TREK-2	Tandem pore domain K^+ channel 2
TRPV5	Transient receptor potential vanilloid 5 channel
UT	Urea transporter
V-ATPase	Vacuolar H^+-ATPase

References

1. Ramesh, M.; Vepuri, S.B.; Oosthuizen, F.; Soliman, M.E. Adenosine Monophosphate-Activated Protein Kinase (AMPK) as a Diverse Therapeutic Target: A Computational Perspective. *Appl. Biochem. Biotechnol.* **2016**, *178*, 810–830. [CrossRef] [PubMed]

2. Mihaylova, M.M.; Shaw, R.J. The AMPK signalling pathway coordinates cell growth, autophagy and metabolism. *Nat. Cell Biol.* **2011**, *13*, 1016–1023. [CrossRef] [PubMed]

3. Hardie, D.G. The AMP-activated protein kinase pathway—New players upstream and downstream. *J. Cell Sci.* **2004**, *117*, 5479–5487. [CrossRef] [PubMed]

4. Viollet, B. AMPK: Lessons from transgenic and knockout animals. *Front. Biosci.* **2009**, *14*, 19–44. [CrossRef]

5. Viollet, B.; Andreelli, F.; Jørgensen, S.B.; Perrin, C.; Flamez, D.; Mu, J.; Wojtaszewski, J.F.P.; Schuit, F.C.; Birnbaum, M.; Richter, E.; et al. Physiological role of AMP-activated protein kinase (AMPK): Insights from knockout mouse models. *Biochem. Soc. Trans.* **2003**, *31*, 216–219. [CrossRef] [PubMed]

6. Hardie, D.G.; Schaffer, B.E.; Brunet, A. AMPK: An Energy-Sensing Pathway with Multiple Inputs and Outputs. *Trends Cell Biol.* **2016**, *26*, 190–201. [CrossRef] [PubMed]

7. Dërmaku-Sopjani, M.; Abazi, S.; Faggio, C.; Kolgeci, J.; Sopjani, M. AMPK-sensitive cellular transport. *J. Biochem.* **2014**, *155*, 147–158. [CrossRef] [PubMed]

8. Hardie, D.G. AMPK—Sensing energy while talking to other signalling pathways. *Cell Metab.* **2014**, *20*, 939–952. [CrossRef] [PubMed]

9. Hardie, D.G.; Carling, D.; Gamblin, S.J. AMP-activated protein kinase: Also regulated by ADP? *Trends Biochem. Sci.* **2011**, *36*, 470–477. [CrossRef] [PubMed]

10. Hardie, D.G.; Ross, F.A.; Hawley, S.A. AMPK: A nutrient and energy sensor that maintains energy homeostasis. *Nat. Rev. Mol. Cell Biol.* **2012**, *13*, 251–262. [CrossRef] [PubMed]

11. Ross, F.A.; Jensen, T.E.; Hardie, D.G. Differential regulation by AMP and ADP of AMPK complexes containing different γ subunit isoforms. *Biochem. J.* **2016**, *473*, 189–199. [CrossRef] [PubMed]

12. Hardie, D.G.; Lin, S.-C. AMP-activated protein kinase—Not just an energy sensor. *F1000Research* **2017**, *6*, 1724. [CrossRef] [PubMed]

13. Thornton, C.; Snowden, M.A.; Carling, D. Identification of a novel AMP-activated protein kinase β subunit isoform that is highly expressed in skeletal muscle. *J. Biol. Chem.* **1998**, *273*, 12443–12450. [CrossRef] [PubMed]

14. Viollet, B.; Andreelli, F.; Jørgensen, S.B.; Perrin, C.; Geloen, A.; Flamez, D.; Mu, J.; Lenzner, C.; Baud, O.; Bennoun, M.; et al. The AMP-activated protein kinase α2 catalytic subunit controls whole-body insulin sensitivity. *J. Clin. Investig.* **2003**, *111*, 91–98. [CrossRef] [PubMed]

15. Hawley, S.A.; Davison, M.; Woods, A.; Davies, S.P.; Beri, R.K.; Carling, D.; Hardie, D.G. Characterization of the AMP-activated Protein Kinase Kinase from Rat Liver and Identification of Threonine 172 as the Major Site at Which It Phosphorylates AMP-activated Protein Kinase. *J. Biol. Chem.* **1996**, *271*, 27879–27887. [CrossRef] [PubMed]

16. Hong, S.-P.; Leiper, F.C.; Woods, A.; Carling, D.; Carlson, M. Activation of yeast Snf1 and mammalian AMP-activated protein kinase by upstream kinases. *Proc. Natl. Acad. Sci. USA* **2003**, *100*, 8839–8843. [CrossRef] [PubMed]

17. Hawley, S.A.; Boudeau, J.; Reid, J.L.; Mustard, K.J.; Udd, L.; Mäkelä, T.P.; Alessi, D.R.; Hardie, D.G. Complexes between the LKB1 tumor suppressor, STRAD α/β and MO25 α/β are upstream kinases in the AMP-activated protein kinase cascade. *J. Biol.* **2003**, *2*, 28. [CrossRef] [PubMed]

18. Woods, A.; Johnstone, S.R.; Dickerson, K.; Leiper, F.C.; Fryer, L.G.D.; Neumann, D.; Schlattner, U.; Wallimann, T.; Carlson, M.; Carling, D. LKB1 Is the Upstream Kinase in the AMP-Activated Protein Kinase Cascade. *Curr. Biol.* **2003**, *13*, 2004–2008. [CrossRef] [PubMed]

19. Shaw, R.J.; Kosmatka, M.; Bardeesy, N.; Hurley, R.L.; Witters, L.A.; DePinho, R.A.; Cantley, L.C. The tumor suppressor LKB1 kinase directly activates AMP-activated kinase and regulates apoptosis in response to energy stress. *Proc. Natl. Acad. Sci. USA* **2004**, *101*, 3329–3335. [CrossRef] [PubMed]

20. Herrero-Martín, G.; Høyer-Hansen, M.; García-García, C.; Fumarola, C.; Farkas, T.; López-Rivas, A.; Jäättelä, M. TAK1 activates AMPK-dependent cytoprotective autophagy in TRAIL-treated epithelial cells. *EMBO J.* **2009**, *28*, 677–685. [CrossRef] [PubMed]

21. Momcilovic, M.; Hong, S.-P.; Carlson, M. Mammalian TAK1 activates Snf1 protein kinase in yeast and phosphorylates AMP-activated protein kinase in vitro. *J. Biol. Chem.* **2006**, *281*, 25336–25343. [CrossRef] [PubMed]

22. Fujiwara, Y.; Kawaguchi, Y.; Fujimoto, T.; Kanayama, N.; Magari, M.; Tokumitsu, H. Differential AMP-activated Protein Kinase (AMPK) Recognition Mechanism of Ca^{2+}/Calmodulin-dependent Protein Kinase Kinase Isoforms. *J. Biol. Chem.* **2016**, *291*, 13802–13808. [CrossRef] [PubMed]

23. Hawley, S.A.; Pan, D.A.; Mustard, K.J.; Ross, L.; Bain, J.; Edelman, A.M.; Frenguelli, B.G.; Hardie, D.G. Calmodulin-dependent protein kinase kinase-β is an alternative upstream kinase for AMP-activated protein kinase. *Cell Metab.* **2005**, *2*, 9–19. [CrossRef] [PubMed]

24. Hurley, R.L.; Anderson, K.A.; Franzone, J.M.; Kemp, B.E.; Means, A.R.; Witters, L.A. The Ca^{2+}/calmodulin-dependent protein kinase kinases are AMP-activated protein kinase kinases. *J. Biol. Chem.* **2005**, *280*, 29060–29066. [CrossRef] [PubMed]

25. Burkewitz, K.; Zhang, Y.; Mair, W.B. AMPK at the nexus of energetics and aging. *Cell Metab.* **2014**, *20*, 10–25. [CrossRef] [PubMed]

26. Neumann, D. Is TAK1 a Direct Upstream Kinase of AMPK? *Int. J. Mol. Sci.* **2018**, *19*, 2412. [CrossRef] [PubMed]

27. Zhu, X.; Dahlmans, V.; Thali, R.; Preisinger, C.; Viollet, B.; Voncken, J.W.; Neumann, D. AMP-activated Protein Kinase Up-regulates Mitogen-activated Protein (MAP) Kinase-interacting Serine/Threonine Kinase 1a-dependent Phosphorylation of Eukaryotic Translation Initiation Factor 4E. *J. Biol. Chem.* **2016**, *291*, 17020–17027. [CrossRef] [PubMed]

28. Viollet, B.; Foretz, M. Revisiting the mechanisms of metformin action in the liver. *Ann. Endocrinol.* **2013**, *74*, 123–129. [CrossRef] [PubMed]

29. Sakamoto, K.; Göransson, O.; Hardie, D.G.; Alessi, D.R. Activity of LKB1 and AMPK-related kinases in skeletal muscle: Effects of contraction, phenformin, and AICAR. *Am. J. Physiol. Endocrinol. Metab.* **2004**, *287*, E310–E317. [CrossRef] [PubMed]

30. Sakamoto, K.; McCarthy, A.; Smith, D.; Green, K.A.; Grahame Hardie, D.; Ashworth, A.; Alessi, D.R. Deficiency of LKB1 in skeletal muscle prevents AMPK activation and glucose uptake during contraction. *EMBO J.* **2005**, *24*, 1810–1820. [CrossRef] [PubMed]

31. Cheung, P.C.F.; Salt, I.P.; Davies, S.P.; Hardie, D.G.; Carling, D. Characterization of AMP-activated protein kinase γ-subunit isoforms and their role in AMP binding. *Biochem. J.* **2000**, *346*, 659–669. [CrossRef] [PubMed]

32. Sanders, M.J.; Grondin, P.O.; Hegarty, B.D.; Snowden, M.A.; Carling, D. Investigating the mechanism for AMP activation of the AMP-activated protein kinase cascade. *Biochem. J.* **2007**, *403*, 139–148. [CrossRef] [PubMed]

33. Xiao, B.; Sanders, M.J.; Underwood, E.; Heath, R.; Mayer, F.V.; Carmena, D.; Jing, C.; Walker, P.A.; Eccleston, J.F.; Haire, L.F.; et al. Structure of mammalian AMPK and its regulation by ADP. *Nature* **2011**, *472*, 230–233. [CrossRef] [PubMed]

34. Oakhill, J.S.; Chen, Z.-P.; Scott, J.W.; Steel, R.; Castelli, L.A.; Ling, N.; Macaulay, S.L.; Kemp, B.E. β-Subunit myristoylation is the gatekeeper for initiating metabolic stress sensing by AMP-activated protein kinase (AMPK). *Proc. Natl. Acad. Sci. USA* **2010**, *107*, 19237–19241. [CrossRef] [PubMed]

35. Oakhill, J.S.; Steel, R.; Chen, Z.-P.; Scott, J.W.; Ling, N.; Tam, S.; Kemp, B.E. AMPK is a direct adenylate charge-regulated protein kinase. *Science* **2011**, *332*, 1433–1435. [CrossRef] [PubMed]

36. Viollet, B.; Mounier, R.; Leclerc, J.; Yazigi, A.; Foretz, M.; Andreelli, F. Targeting AMP-activated protein kinase as a novel therapeutic approach for the treatment of metabolic disorders. *Diabetes Metab.* **2007**, *33*, 395–402. [CrossRef] [PubMed]

37. Gowans, G.J.; Hawley, S.A.; Ross, F.A.; Hardie, D.G. AMP Is a True Physiological Regulator of AMP-Activated Protein Kinase by Both Allosteric Activation and Enhancing Net Phosphorylation. *Cell Metab.* **2013**, *18*, 556–566. [CrossRef] [PubMed]

38. Davies, S.P.; Helps, N.R.; Cohen, P.T.; Hardie, D.G. 5′-AMP inhibits dephosphorylation, as well as promoting phosphorylation, of the AMP-activated protein kinase. Studies using bacterially expressed human protein phosphatase-2C α and native bovine protein phosphatase-2A c. *FEBS Lett.* **1995**, *377*, 421–425. [PubMed]

39. Chen, L.; Wang, J.; Zhang, Y.-Y.; Yan, S.F.; Neumann, D.; Schlattner, U.; Wang, Z.-X.; Wu, J.-W. AMP-activated protein kinase undergoes nucleotide-dependent conformational changes. *Nat. Struct. Mol. Biol.* **2012**, *19*, 716–718. [CrossRef] [PubMed]

40. Garcia, D.; Shaw, R.J. AMPK: Mechanisms of Cellular Energy Sensing and Restoration of Metabolic Balance. *Mol. Cell.* **2017**, *66*, 789–800. [CrossRef] [PubMed]

41. Woods, A.; Dickerson, K.; Heath, R.; Hong, S.-P.; Momcilovic, M.; Johnstone, S.R.; Carlson, M.; Carling, D. Ca^{2+}/calmodulin-dependent protein kinase kinase-β acts upstream of AMP-activated protein kinase in mammalian cells. *Cell Metab.* **2005**, *2*, 21–33. [CrossRef] [PubMed]

42.	Stahmann, N.; Woods, A.; Carling, D.; Heller, R. Thrombin activates AMP-activated protein kinase in endothelial cells via a pathway involving Ca^{2+}/calmodulin-dependent protein kinase kinase β. *Mol. Cell. Biol.* **2006**, *26*, 5933–5945. [CrossRef] [PubMed]
43.	Yang, Y.; Atasoy, D.; Su, H.H.; Sternson, S.M. Hunger states switch a flip-flop memory circuit via a synaptic AMPK-dependent positive feedback loop. *Cell* **2011**, *146*, 992–1003. [CrossRef] [PubMed]
44.	Lang, F.; Eylenstein, A.; Shumilina, E. Regulation of Orai1/STIM1 by the kinases SGK1 and AMPK. *Cell Calcium* **2012**, *52*, 347–354. [CrossRef] [PubMed]
45.	Sundivakkam, P.C.; Natarajan, V.; Malik, A.B.; Tiruppathi, C. Store-operated Ca^{2+} entry (SOCE) induced by protease-activated receptor-1 mediates STIM1 protein phosphorylation to inhibit SOCE in endothelial cells through AMP-activated protein kinase and p38β mitogen-activated protein kinase. *J. Biol. Chem.* **2013**, *288*, 17030–17041. [CrossRef] [PubMed]
46.	Zhang, B.; Yan, J.; Umbach, A.T.; Fakhri, H.; Fajol, A.; Schmidt, S.; Salker, M.S.; Chen, H.; Alexander, D.; Spichtig, D.; et al. NFκB-sensitive Orai1 expression in the regulation of FGF23 release. *J. Mol. Med.* **2016**, *94*, 557–566. [CrossRef] [PubMed]
47.	Prakriya, M.; Feske, S.; Gwack, Y.; Srikanth, S.; Rao, A.; Hogan, P.G. Orai1 is an essential pore subunit of the CRAC channel. *Nature* **2006**, *443*, 230–233. [CrossRef] [PubMed]
48.	Zhang, S.L.; Kozak, J.A.; Jiang, W.; Yeromin, A.V.; Chen, J.; Yu, Y.; Penna, A.; Shen, W.; Chi, V.; Cahalan, M.D. Store-dependent and -independent modes regulating Ca^{2+} release-activated Ca^{2+} channel activity of human Orai1 and Orai3. *J. Biol. Chem.* **2008**, *283*, 17662–17671. [CrossRef] [PubMed]
49.	Tiruppathi, C.; Ahmmed, G.U.; Vogel, S.M.; Malik, A.B. Ca^{2+} signaling, TRP channels, and endothelial permeability. *Microcirculation* **2006**, *13*, 693–708. [CrossRef] [PubMed]
50.	Yu, F.; Sun, L.; Machaca, K. Constitutive recycling of the store-operated Ca^{2+} channel Orai1 and its internalization during meiosis. *J. Cell Biol.* **2010**, *191*, 523–535. [CrossRef] [PubMed]
51.	Baryshnikov, S.G.; Pulina, M.V.; Zulian, A.; Linde, C.I.; Golovina, V.A. Orai1, a critical component of store-operated Ca^{2+} entry, is functionally associated with Na^+/Ca^{2+} exchanger and plasma membrane Ca^{2+} pump in proliferating human arterial myocytes. *Am. J. Physiol. Cell Physiol.* **2009**, *297*, C1103–C1112. [CrossRef] [PubMed]
52.	Johnstone, L.S.; Graham, S.J.L.; Dziadek, M.A. STIM proteins: Integrators of signalling pathways in development, differentiation and disease. *J. Cell. Mol. Med.* **2010**, *14*, 1890–1903. [CrossRef] [PubMed]
53.	Yang, S.; Zhang, J.J.; Huang, X.-Y. Orai1 and STIM1 are critical for breast tumor cell migration and metastasis. *Cancer Cell* **2009**, *15*, 124–134. [CrossRef] [PubMed]
54.	Stathopulos, P.B.; Ikura, M. Store operated calcium entry: From concept to structural mechanisms. *Cell Calcium* **2017**, *63*, 3–7. [CrossRef] [PubMed]
55.	Ambudkar, I.S.; de Souza, L.B.; Ong, H.L. TRPC1, Orai1, and STIM1 in SOCE: Friends in tight spaces. *Cell Calcium* **2017**, *63*, 33–39. [CrossRef] [PubMed]
56.	Glosse, P.; Feger, M.; Mutig, K.; Chen, H.; Hirche, F.; Hasan, A.A.; Gaballa, M.M.S.; Hocher, B.; Lang, F.; Foller, M. AMP-activated kinase is a regulator of fibroblast growth factor 23 production. *Kidney Int.* **2018**, *94*, 491–501. [CrossRef] [PubMed]
57.	Hasenour, C.M.; Berglund, E.D.; Wasserman, D.H. Emerging role of AMP-activated protein kinase in endocrine control of metabolism in the liver. *Mol. Cell. Endocrinol.* **2013**, *366*, 152–162. [CrossRef] [PubMed]
58.	Li, Y.; Xu, S.; Mihaylova, M.M.; Zheng, B.; Hou, X.; Jiang, B.; Park, O.; Luo, Z.; Lefai, E.; Shyy, J.Y.-J.; et al. AMPK phosphorylates and inhibits SREBP activity to attenuate hepatic steatosis and atherosclerosis in diet-induced insulin-resistant mice. *Cell Metab.* **2011**, *13*, 376–388. [CrossRef] [PubMed]
59.	Foretz, M.; Viollet, B. Activation of AMPK for a Break in Hepatic Lipid Accumulation and Circulating Cholesterol. *EBioMedicine* **2018**, *31*, 15–16. [CrossRef] [PubMed]
60.	Merlen, G.; Gentric, G.; Celton-Morizur, S.; Foretz, M.; Guidotti, J.-E.; Fauveau, V.; Leclerc, J.; Viollet, B.; Desdouets, C. AMPKα1 controls hepatocyte proliferation independently of energy balance by regulating Cyclin A2 expression. *J. Hepatol.* **2014**, *60*, 152–159. [CrossRef] [PubMed]
61.	Foretz, M.; Viollet, B. Regulation of hepatic metabolism by AMPK. *J. Hepatol.* **2011**, *54*, 827–829. [CrossRef] [PubMed]
62.	Kjøbsted, R.; Hingst, J.R.; Fentz, J.; Foretz, M.; Sanz, M.-N.; Pehmøller, C.; Shum, M.; Marette, A.; Mounier, R.; Treebak, J.T.; et al. AMPK in skeletal muscle function and metabolism. *FASEB J.* **2018**, *32*, 1741–1777. [CrossRef] [PubMed]

63. Mounier, R.; Théret, M.; Lantier, L.; Foretz, M.; Viollet, B. Expanding roles for AMPK in skeletal muscle plasticity. *Trends Endocrinol. Metab.* **2015**, *26*, 275–286. [CrossRef] [PubMed]

64. Kjøbsted, R.; Munk-Hansen, N.; Birk, J.B.; Foretz, M.; Viollet, B.; Björnholm, M.; Zierath, J.R.; Treebak, J.T.; Wojtaszewski, J.F.P. Enhanced Muscle Insulin Sensitivity After Contraction/Exercise Is Mediated by AMPK. *Diabetes* **2017**, *66*, 598–612. [CrossRef] [PubMed]

65. Cokorinos, E.C.; Delmore, J.; Reyes, A.R.; Albuquerque, B.; Kjøbsted, R.; Jørgensen, N.O.; Tran, J.-L.; Jatkar, A.; Cialdea, K.; Esquejo, R.M.; et al. Activation of Skeletal Muscle AMPK Promotes Glucose Disposal and Glucose Lowering in Non-human Primates and Mice. *Cell Metab.* **2017**, *25*, 1147–1159. [CrossRef] [PubMed]

66. Fentz, J.; Kjøbsted, R.; Birk, J.B.; Jordy, A.B.; Jeppesen, J.; Thorsen, K.; Schjerling, P.; Kiens, B.; Jessen, N.; Viollet, B.; et al. AMPKα is critical for enhancing skeletal muscle fatty acid utilization during in vivo exercise in mice. *FASEB J.* **2015**, *29*, 1725–1738. [CrossRef] [PubMed]

67. Arad, M.; Seidman, C.E.; Seidman, J.G. AMP-Activated Protein Kinase in the Heart: Role during Health and Disease. *Circ. Res.* **2007**, *100*, 474–488. [CrossRef] [PubMed]

68. Voelkl, J.; Alesutan, I.; Primessnig, U.; Feger, M.; Mia, S.; Jungmann, A.; Castor, T.; Viereck, R.; Stöckigt, F.; Borst, O.; et al. AMP-activated protein kinase α1-sensitive activation of AP-1 in cardiomyocytes. *J. Mol. Cell. Cardiol.* **2016**, *97*, 36–43. [CrossRef] [PubMed]

69. Liao, Y.; Takashima, S.; Maeda, N.; Ouchi, N.; Komamura, K.; Shimomura, I.; Hori, M.; Matsuzawa, Y.; Funahashi, T.; Kitakaze, M. Exacerbation of heart failure in adiponectin-deficient mice due to impaired regulation of AMPK and glucose metabolism. *Cardiovasc. Res.* **2005**, *67*, 705–713. [CrossRef] [PubMed]

70. Russell, R.R.; Li, J.; Coven, D.L.; Pypaert, M.; Zechner, C.; Palmeri, M.; Giordano, F.J.; Mu, J.; Birnbaum, M.J.; Young, L.H. AMP-activated protein kinase mediates ischemic glucose uptake and prevents postischemic cardiac dysfunction, apoptosis, and injury. *J. Clin. Investig.* **2004**, *114*, 495–503. [CrossRef] [PubMed]

71. Gélinas, R.; Mailleux, F.; Dontaine, J.; Bultot, L.; Demeulder, B.; Ginion, A.; Daskalopoulos, E.P.; Esfahani, H.; Dubois-Deruy, E.; Lauzier, B.; et al. AMPK activation counteracts cardiac hypertrophy by reducing O-GlcNAcylation. *Nat. Commun.* **2018**, *9*, 374. [CrossRef] [PubMed]

72. Chen, K.; Kobayashi, S.; Xu, X.; Viollet, B.; Liang, Q. AMP activated protein kinase is indispensable for myocardial adaptation to caloric restriction in mice. *PLoS ONE* **2013**, *8*, e59682. [CrossRef] [PubMed]

73. Zhang, P.; Hu, X.; Xu, X.; Fassett, J.; Zhu, G.; Viollet, B.; Xu, W.; Wiczer, B.; Bernlohr, D.A.; Bache, R.J.; et al. AMP activated protein kinase-α2 deficiency exacerbates pressure-overload-induced left ventricular hypertrophy and dysfunction in mice. *Hypertension* **2008**, *52*, 918–924. [CrossRef] [PubMed]

74. Hallows, K.R.; Mount, P.F.; Pastor-Soler, N.M.; Power, D.A. Role of the energy sensor AMP-activated protein kinase in renal physiology and disease. *Am. J. Physiol. Renal. Physiol.* **2010**, *298*, F1067–F1077. [CrossRef] [PubMed]

75. Pastor-Soler, N.M.; Hallows, K.R. AMP-activated protein kinase regulation of kidney tubular transport. *Curr. Opin. Nephrol. Hypertens.* **2012**, *21*, 523–533. [CrossRef] [PubMed]

76. Rajani, R.; Pastor-Soler, N.M.; Hallows, K.R. Role of AMP-activated protein kinase in kidney tubular transport, metabolism, and disease. *Curr. Opin. Nephrol. Hypertens.* **2017**, *26*, 375–383. [CrossRef] [PubMed]

77. Lee, M.-J.; Feliers, D.; Mariappan, M.M.; Sataranatarajan, K.; Mahimainathan, L.; Musi, N.; Foretz, M.; Viollet, B.; Weinberg, J.M.; Choudhury, G.G.; et al. A role for AMP-activated protein kinase in diabetes-induced renal hypertrophy. *Am. J. Physiol. Renal. Physiol.* **2007**, *292*, F617–F627. [CrossRef] [PubMed]

78. Jeyabalan, J.; Shah, M.; Viollet, B.; Chenu, C. AMP-activated protein kinase pathway and bone metabolism. *J. Endocrinol.* **2012**, *212*, 277–290. [CrossRef] [PubMed]

79. McCarthy, A.D.; Cortizo, A.M.; Sedlinsky, C. Metformin revisited: Does this regulator of AMP-activated protein kinase secondarily affect bone metabolism and prevent diabetic osteopathy. *World J. Diabetes* **2016**, *7*, 122–133. [CrossRef] [PubMed]

80. Kanazawa, I. Interaction between bone and glucose metabolism. *Endocr. J.* **2017**, *64*, 1043–1053. [CrossRef] [PubMed]

81. Tain, Y.-L.; Hsu, C.-N. AMP-Activated Protein Kinase as a Reprogramming Strategy for Hypertension and Kidney Disease of Developmental Origin. *Int. J. Mol. Sci.* **2018**, *19*, 1744. [CrossRef] [PubMed]

82. Tsai, C.-M.; Kuo, H.-C.; Hsu, C.-N.; Huang, L.-T.; Tain, Y.-L. Metformin reduces asymmetric dimethylarginine and prevents hypertension in spontaneously hypertensive rats. *Transl. Res.* **2014**, *164*, 452–459. [CrossRef] [PubMed]

83. Allouch, S.; Munusamy, S. AMP-activated Protein Kinase as a Drug Target in Chronic Kidney Disease. *Curr. Drug Targets* **2018**, *19*, 709–720. [CrossRef] [PubMed]

84. Curthoys, N.P.; Moe, O.W. Proximal tubule function and response to acidosis. *Clin. J. Am. Soc. Nephrol.* **2014**, *9*, 1627–1638. [CrossRef] [PubMed]

85. Wallace, M.A. Anatomy and Physiology of the Kidney. *AORN J.* **1998**, *68*, 799–820. [CrossRef]

86. Mount, D.B. Thick ascending limb of the loop of Henle. *Clin. J. Am. Soc. Nephrol.* **2014**, *9*, 1974–1986. [CrossRef] [PubMed]

87. Zhang, J.L.; Rusinek, H.; Chandarana, H.; Lee, V.S. Functional MRI of the kidneys. *J. Magn. Reson. Imaging* **2013**, *37*, 282–293. [CrossRef] [PubMed]

88. Blaine, J.; Chonchol, M.; Levi, M. Renal control of calcium, phosphate, and magnesium homeostasis. *Clin. J. Am. Soc. Nephrol.* **2015**, *10*, 1257–1272. [CrossRef] [PubMed]

89. Lee, Y.J.; Han, H.J. Regulatory mechanisms of Na^+/glucose cotransporters in renal proximal tubule cells. *Kidney Int. Suppl.* **2007**, S27–S35. [CrossRef] [PubMed]

90. Rotte, A.; Pasham, V.; Eichenmüller, M.; Bhandaru, M.; Föller, M.; Lang, F. Upregulation of Na^+/H^+ exchanger by the AMP-activated protein kinase. *Biochem. Biophys. Res. Commun.* **2010**, *398*, 677–682. [CrossRef] [PubMed]

91. Palmer, L.G.; Schnermann, J. Integrated control of Na transport along the nephron. *Clin. J. Am. Soc. Nephrol.* **2015**, *10*, 676–687. [CrossRef] [PubMed]

92. Sopjani, M.; Bhavsar, S.K.; Fraser, S.; Kemp, B.E.; Föller, M.; Lang, F. Regulation of Na^+-coupled glucose carrier SGLT1 by AMP-activated protein kinase. *Mol. Membr. Biol.* **2010**, *27*, 137–144. [CrossRef] [PubMed]

93. Dërmaku-Sopjani, M.; Almilaji, A.; Pakladok, T.; Munoz, C.; Hosseinzadeh, Z.; Blecua, M.; Sopjani, M.; Lang, F. Down-regulation of the Na^+-coupled phosphate transporter NaPi-IIa by AMP-activated protein kinase. *Kidney Blood Press. Res.* **2013**, *37*, 547–556. [CrossRef] [PubMed]

94. Hallows, K.R. Emerging role of AMP-activated protein kinase in coupling membrane transport to cellular metabolism. *Curr. Opin. Nephrol. Hypertens.* **2005**, *14*, 464–471. [CrossRef] [PubMed]

95. Noske, R.; Cornelius, F.; Clarke, R.J. Investigation of the enzymatic activity of the Na^+, K^+-ATPase via isothermal titration microcalorimetry. *Biochim. Biophys. Acta* **2010**, *1797*, 1540–1545. [CrossRef] [PubMed]

96. Woollhead, A.M.; Scott, J.W.; Hardie, D.G.; Baines, D.L. Phenformin and 5-aminoimidazole-4-carboxamide-1-β-D-ribofuranoside (AICAR) activation of AMP-activated protein kinase inhibits transepithelial Na^+ transport across H441 lung cells. *J. Physiol.* **2005**, *566*, 781–792. [CrossRef] [PubMed]

97. Woollhead, A.M.; Sivagnanasundaram, J.; Kalsi, K.K.; Pucovsky, V.; Pellatt, L.J.; Scott, J.W.; Mustard, K.J.; Hardie, D.G.; Baines, D.L. Pharmacological activators of AMP-activated protein kinase have different effects on Na^+ transport processes across human lung epithelial cells. *Br. J. Pharmacol.* **2007**, *151*, 1204–1215. [CrossRef] [PubMed]

98. Vadász, I.; Dada, L.A.; Briva, A.; Trejo, H.E.; Welch, L.C.; Chen, J.; Tóth, P.T.; Lecuona, E.; Witters, L.A.; Schumacker, P.T.; et al. AMP-activated protein kinase regulates CO_2-induced alveolar epithelial dysfunction in rats and human cells by promoting Na, K-ATPase endocytosis. *J. Clin. Investig.* **2008**, *118*, 752–762. [CrossRef] [PubMed]

99. Gusarova, G.A.; Dada, L.A.; Kelly, A.M.; Brodie, C.; Witters, L.A.; Chandel, N.S.; Sznajder, J.I. α1-AMP-activated protein kinase regulates hypoxia-induced Na, K-ATPase endocytosis via direct phosphorylation of protein kinase C zeta. *Mol. Cell. Biol.* **2009**, *29*, 3455–3464. [CrossRef] [PubMed]

100. Gusarova, G.A.; Trejo, H.E.; Dada, L.A.; Briva, A.; Welch, L.C.; Hamanaka, R.B.; Mutlu, G.M.; Chandel, N.S.; Prakriya, M.; Sznajder, J.I. Hypoxia leads to Na, K-ATPase downregulation via Ca^{2+} release-activated Ca(2+) channels and AMPK activation. *Mol. Cell. Biol.* **2011**, *31*, 3546–3556. [CrossRef] [PubMed]

101. Benziane, B.; Björnholm, M.; Pirkmajer, S.; Austin, R.L.; Kotova, O.; Viollet, B.; Zierath, J.R.; Chibalin, A.V. Activation of AMP-activated protein kinase stimulates Na^+, K^+-ATPase activity in skeletal muscle cells. *J. Biol. Chem.* **2012**, *287*, 23451–23463. [CrossRef] [PubMed]

102. Alves, D.S.; Farr, G.A.; Seo-Mayer, P.; Caplan, M.J. AS160 associates with the Na^+, K^+-ATPase and mediates the adenosine monophosphate-stimulated protein kinase-dependent regulation of sodium pump surface expression. *Mol. Biol. Cell* **2010**, *21*, 4400–4408. [CrossRef] [PubMed]

103. Seo-Mayer, P.W.; Thulin, G.; Zhang, L.; Alves, D.S.; Ardito, T.; Kashgarian, M.; Caplan, M.J. Preactivation of AMPK by metformin may ameliorate the epithelial cell damage caused by renal ischemia. *Am. J. Physiol. Renal. Physiol.* **2011**, *301*, F1346–F1357. [CrossRef] [PubMed]

104. Mount, P.F.; Gleich, K.; Tam, S.; Fraser, S.A.; Choy, S.-W.; Dwyer, K.M.; Lu, B.; van Denderen, B.; Fingerle-Rowson, G.; Bucala, R.; et al. The outcome of renal ischemia-reperfusion injury is unchanged in AMPK-β1 deficient mice. *PLoS ONE* **2012**, *7*, e29887. [CrossRef] [PubMed]

105. Fitzgerald, G.A.; Mulligan, C.; Mindell, J.A. A general method for determining secondary active transporter substrate stoichiometry. *eLife* **2017**, *6*. [CrossRef] [PubMed]

106. Forrest, L.R.; Krämer, R.; Ziegler, C. The structural basis of secondary active transport mechanisms. *Biochim. Biophys. Acta* **2011**, *1807*, 167–188. [CrossRef] [PubMed]

107. Mandel, L.J.; Balaban, R.S. Stoichiometry and coupling of active transport to oxidative metabolism in epithelial tissues. *Am. J. Physiol.* **1981**, *240*, F357–F371. [CrossRef] [PubMed]

108. Mather, A.; Pollock, C. Glucose handling by the kidney. *Kidney Int. Suppl.* **2011**, *79*, S1–S6. [CrossRef] [PubMed]

109. Bakris, G.L.; Fonseca, V.A.; Sharma, K.; Wright, E.M. Renal sodium-glucose transport: Role in diabetes mellitus and potential clinical implications. *Kidney Int.* **2009**, *75*, 1272–1277. [CrossRef] [PubMed]

110. Wright, E.M.; Hirayama, B.A.; Loo, D.F. Active sugar transport in health and disease. *J. Intern. Med.* **2007**, *261*, 32–43. [CrossRef] [PubMed]

111. Hawley, S.A.; Ford, R.J.; Smith, B.K.; Gowans, G.J.; Mancini, S.J.; Pitt, R.D.; Day, E.A.; Salt, I.P.; Steinberg, G.R.; Hardie, D.G. The Na$^+$/Glucose Cotransporter Inhibitor Canagliflozin Activates AMPK by Inhibiting Mitochondrial Function and Increasing Cellular AMP Levels. *Diabetes* **2016**, *65*, 2784–2794. [CrossRef] [PubMed]

112. You, G.; Lee, W.-S.; Barros, E.J.G.; Kanai, Y.; Huo, T.-L.; Khawaja, S.; Wells, R.G.; Nigam, S.K.; Hediger, M.A. Molecular Characteristics of Na$^+$-coupled Glucose Transporters in Adult and Embryonic Rat Kidney. *J. Biol. Chem.* **1995**, *270*, 29365–29371. [CrossRef] [PubMed]

113. Banerjee, S.K.; Wang, D.W.; Alzamora, R.; Huang, X.N.; Pastor-Soler, N.M.; Hallows, K.R.; McGaffin, K.R.; Ahmad, F. SGLT1, a novel cardiac glucose transporter, mediates increased glucose uptake in PRKAG2 cardiomyopathy. *J. Mol. Cell. Cardiol.* **2010**, *49*, 683–692. [CrossRef] [PubMed]

114. Wright, E.M. Renal Na$^+$-glucose cotransporters. *Am. J. Physiol. Renal. Physiol.* **2001**, *280*, F10–F18. [CrossRef] [PubMed]

115. Pajor, A.M.; Wright, E.M. Cloning and functional expression of a mammalian Na$^+$/nucleoside cotransporter. A member of the SGLT family. *J. Biol. Chem.* **1992**, *267*, 3557–3560. [PubMed]

116. Linden, K.C.; DeHaan, C.L.; Zhang, Y.; Glowacka, S.; Cox, A.J.; Kelly, D.J.; Rogers, S. Renal expression and localization of the facilitative glucose transporters GLUT1 and GLUT12 in animal models of hypertension and diabetic nephropathy. *Am. J. Physiol. Renal. Physiol.* **2006**, *290*, F205–F213. [CrossRef] [PubMed]

117. Dominguez, J.H.; Camp, K.; Maianu, L.; Garvey, W.T. Glucose transporters of rat proximal tubule: Differential expression and subcellular distribution. *Am. J. Physiol.* **1992**, *262*, F807–F812. [CrossRef] [PubMed]

118. Thorens, B.; Lodish, H.F.; Brown, D. Differential localization of two glucose transporter isoforms in rat kidney. *Am. J. Physiol.* **1990**, *259*, C286–C294. [CrossRef] [PubMed]

119. Castilla-Madrigal, R.; Barrenetxe, J.; Moreno-Aliaga, M.J.; Lostao, M.P. EPA blocks TNF-α-induced inhibition of sugar uptake in Caco-2 cells via GPR120 and AMPK. *J. Cell. Physiol.* **2018**, *233*, 2426–2433. [CrossRef] [PubMed]

120. Di Franco, A.; Cantini, G.; Tani, A.; Coppini, R.; Zecchi-Orlandini, S.; Raimondi, L.; Luconi, M.; Mannucci, E. Sodium-dependent glucose transporters (SGLT) in human ischemic heart: A new potential pharmacological target. *Int. J. Cardiol.* **2017**, *243*, 86–90. [CrossRef] [PubMed]

121. Portilla, D. Energy metabolism and cytotoxicity. *Semin. Nephrol.* **2003**, *23*, 432–438. [CrossRef]

122. Le Hir, M.; Dubach, U.C. Peroxisomal and mitochondrial β-oxidation in the rat kidney: Distribution of fatty acyl-coenzyme A oxidase and 3-hydroxyacyl-coenzyme A dehydrogenase activities along the nephron. *J. Histochem. Cytochem.* **1982**, *30*, 441–444. [CrossRef] [PubMed]

123. Uchida, S.; Endou, H. Substrate specificity to maintain cellular ATP along the mouse nephron. *Am. J. Physiol.* **1988**, *255*, F977–F983. [CrossRef] [PubMed]

124. Fryer, L.G.D.; Foufelle, F.; Barnes, K.; Baldwin, S.A.; Woods, A.; Carling, D. Characterization of the role of the AMP-activated protein kinase in the stimulation of glucose transport in skeletal muscle cells. *Biochem. J.* **2002**, *363*, 167–174. [CrossRef] [PubMed]

125. Al-Bayati, A.; Lukka, D.; Brown, A.E.; Walker, M. Effects of thrombin on insulin signalling and glucose uptake in cultured human myotubes. *J. Diabetes Complicat.* **2016**, *30*, 1209–1216. [CrossRef] [PubMed]

126. Andrade, B.M.; Cazarin, J.; Zancan, P.; Carvalho, D.P. AMP-activated protein kinase upregulates glucose uptake in thyroid PCCL3 cells independent of thyrotropin. *Thyroid* **2012**, *22*, 1063–1068. [CrossRef] [PubMed]

127. Takeno, A.; Kanazawa, I.; Notsu, M.; Tanaka, K.-I.; Sugimoto, T. Glucose uptake inhibition decreases expressions of receptor activator of nuclear factor-kappa B ligand (RANKL) and osteocalcin in osteocytic MLO-Y4-A2 cells. *Am. J. Physiol. Endocrinol. Metab.* **2018**, *314*, E115–E123. [CrossRef] [PubMed]

128. Wang, Y.; Zhang, Y.; Wang, Y.; Peng, H.; Rui, J.; Zhang, Z.; Wang, S.; Li, Z. WSF-P-1, a novel AMPK activator, promotes adiponectin multimerization in 3T3-L1 adipocytes. *Biosci. Biotechnol. Biochem.* **2017**, *81*, 1529–1535. [CrossRef] [PubMed]

129. Yamada, S.; Kotake, Y.; Sekino, Y.; Kanda, Y. AMP-activated protein kinase-mediated glucose transport as a novel target of tributyltin in human embryonic carcinoma cells. *Metallomics* **2013**, *5*, 484–491. [CrossRef] [PubMed]

130. Yu, H.; Zhang, H.; Dong, M.; Wu, Z.; Shen, Z.; Xie, Y.; Kong, Z.; Dai, X.; Xu, B. Metabolic reprogramming and AMPKα1 pathway activation by caulerpin in colorectal cancer cells. *Int. J. Oncol.* **2017**, *50*, 161–172. [CrossRef] [PubMed]

131. Abbud, W.; Habinowski, S.; Zhang, J.Z.; Kendrew, J.; Elkairi, F.S.; Kemp, B.E.; Witters, L.A.; Ismail-Beigi, F. Stimulation of AMP-activated protein kinase (AMPK) is associated with enhancement of Glut1-mediated glucose transport. *Arch. Biochem. Biophys.* **2000**, *380*, 347–352. [CrossRef] [PubMed]

132. Baldwin, S.A.; Barros, L.F.; Griffiths, M.; Ingram, J.; Robbins, E.C.; Streets, A.J.; Saklatvala, J. Regulation of GLUTI in response to cellular stress. *Biochem. Soc. Trans.* **1997**, *25*, 954–958. [CrossRef] [PubMed]

133. Sokolovska, J.; Isajevs, S.; Sugoka, O.; Sharipova, J.; Lauberte, L.; Svirina, D.; Rostoka, E.; Sjakste, T.; Kalvinsh, I.; Sjakste, N. Influence of metformin on GLUT1 gene and protein expression in rat streptozotocin diabetes mellitus model. *Arch. Physiol. Biochem.* **2010**, *116*, 137–145. [CrossRef] [PubMed]

134. Walker, J.; Jijon, H.B.; Diaz, H.; Salehi, P.; Churchill, T.; Madsen, K.L. 5-aminoimidazole-4-carboxamide riboside (AICAR) enhances GLUT2-dependent jejunal glucose transport: A possible role for AMPK. *Biochem. J.* **2005**, *385*, 485–491. [CrossRef] [PubMed]

135. Sakar, Y.; Meddah, B.; Faouzi, M.A.; Cherrah, Y.; Bado, A.; Ducroc, R. Metformin-induced regulation of the intestinal D-glucose transporters. *J. Physiol. Pharmacol.* **2010**, *61*, 301–307. [PubMed]

136. Parker, M.D.; Myers, E.J.; Schelling, J.R. Na$^+$–H$^+$ exchanger-1 (NHE1) regulation in kidney proximal tubule. *Cell. Mol. Life Sci.* **2015**, *72*, 2061–2074. [CrossRef] [PubMed]

137. Odunewu, A.; Fliegel, L. Acidosis-mediated regulation of the NHE1 isoform of the Na$^+$/H$^+$ exchanger in renal cells. *Am. J. Physiol. Renal. Physiol.* **2013**, *305*, F370–F381. [CrossRef] [PubMed]

138. Biemesderfer, D.; Reilly, R.F.; Exner, M.; Igarashi, P.; Aronson, P.S. Immunocytochemical characterization of Na$^+$-H$^+$ exchanger isoform NHE-1 in rabbit kidney. *Am. J. Physiol.* **1992**, *263*, F833–F840. [CrossRef] [PubMed]

139. Peti-Peterdi, J.; Chambrey, R.; Bebok, Z.; Biemesderfer, D.; St John, P.L.; Abrahamson, D.R.; Warnock, D.G.; Bell, P.D. Macula densa Na$^+$/H$^+$ exchange activities mediated by apical NHE2 and basolateral NHE4 isoforms. *Am. J. Physiol. Renal. Physiol.* **2000**, *278*, F452–F463. [CrossRef] [PubMed]

140. Baum, M.; Moe, O.W.; Gentry, D.L.; Alpern, R.J. Effect of glucocorticoids on renal cortical NHE-3 and NHE-1 mRNA. *Am. J. Physiol.* **1994**, *267*, F437–F442. [CrossRef] [PubMed]

141. Hue, L.; Beauloye, C.; Marsin, A.-S.; Bertrand, L.; Horman, S.; Rider, M.H. Insulin and Ischemia Stimulate Glycolysis by Acting on the Same Targets Through Different and Opposing Signaling Pathways. *J. Mol. Cell. Cardiol.* **2002**, *34*, 1091–1097. [CrossRef] [PubMed]

142. Marsin, A.-S.; Bouzin, C.; Bertrand, L.; Hue, L. The stimulation of glycolysis by hypoxia in activated monocytes is mediated by AMP-activated protein kinase and inducible 6-phosphofructo-2-kinase. *J. Biol. Chem.* **2002**, *277*, 30778–30783. [CrossRef] [PubMed]

143. Wyss, M.; Kaddurah-Daouk, R. Creatine and creatinine metabolism. *Physiol. Rev.* **2000**, *80*, 1107–1213. [CrossRef] [PubMed]

144. García-Delgado, M.; Peral, M.J.; Cano, M.; Calonge, M.L.; Ilundáin, A.A. Creatine transport in brush-border membrane vesicles isolated from rat kidney cortex. *J. Am. Soc. Nephrol.* **2001**, *12*, 1819–1825. [PubMed]

145. Wallimann, T.; Wyss, M.; Brdiczka, D.; Nicolay, K.; Eppenberger, H.M. Intracellular compartmentation, structure and function of creatine kinase isoenzymes in tissues with high and fluctuating energy demands: The 'phosphocreatine circuit' for cellular energy homeostasis. *Biochem. J.* **1992**, *281*, 21–40. [CrossRef] [PubMed]

146. Neumann, D.; Schlattner, U.; Wallimann, T. A molecular approach to the concerted action of kinases involved in energy homoeostasis. *Biochem. Soc. Trans.* **2003**, *31*, 169–174. [CrossRef] [PubMed]

147. Li, H.; Thali, R.F.; Smolak, C.; Gong, F.; Alzamora, R.; Wallimann, T.; Scholz, R.; Pastor-Soler, N.M.; Neumann, D.; Hallows, K.R. Regulation of the creatine transporter by AMP-activated protein kinase in kidney epithelial cells. *Am. J. Physiol. Renal. Physiol.* **2010**, *299*, F167–F177. [CrossRef] [PubMed]

148. Darrabie, M.D.; Arciniegas, A.J.L.; Mishra, R.; Bowles, D.E.; Jacobs, D.O.; Santacruz, L. AMPK and substrate availability regulate creatine transport in cultured cardiomyocytes. *Am. J. Physiol. Endocrinol. Metab.* **2011**, *300*, E870–E876. [CrossRef] [PubMed]

149. Santacruz, L.; Arciniegas, A.J.L.; Darrabie, M.; Mantilla, J.G.; Baron, R.M.; Bowles, D.E.; Mishra, R.; Jacobs, D.O. Hypoxia decreases creatine uptake in cardiomyocytes, while creatine supplementation enhances HIF activation. *Physiol. Rep.* **2017**, *5*. [CrossRef] [PubMed]

150. Erben, R.G.; Andrukhova, O. FGF23-Klotho signaling axis in the kidney. *Bone* **2017**, *100*, 62–68. [CrossRef] [PubMed]

151. Biber, J.; Hernando, N.; Forster, I.; Murer, H. Regulation of phosphate transport in proximal tubules. *Pflugers Arch.* **2009**, *458*, 39–52. [CrossRef] [PubMed]

152. Murer, H.; Forster, I.; Biber, J. The sodium phosphate cotransporter family SLC34. *Pflugers Arch.* **2004**, *447*, 763–767. [CrossRef] [PubMed]

153. Fraser, S.A.; Gimenez, I.; Cook, N.; Jennings, I.; Katerelos, M.; Katsis, F.; Levidiotis, V.; Kemp, B.E.; Power, D.A. Regulation of the renal-specific Na$^+$-K$^+$-2Cl$^-$ co-transporter NKCC2 by AMP-activated protein kinase (AMPK). *Biochem. J.* **2007**, *405*, 85–93. [CrossRef] [PubMed]

154. Cook, N.; Fraser, S.A.; Katerelos, M.; Katsis, F.; Gleich, K.; Mount, P.F.; Steinberg, G.R.; Levidiotis, V.; Kemp, B.E.; Power, D.A. Low salt concentrations activate AMP-activated protein kinase in mouse macula densa cells. *Am. J. Physiol. Renal. Physiol.* **2009**, *296*, F801–F809. [CrossRef] [PubMed]

155. Fraser, S.A.; Choy, S.-W.; Pastor-Soler, N.M.; Li, H.; Davies, M.R.P.; Cook, N.; Katerelos, M.; Mount, P.F.; Gleich, K.; McRae, J.L.; et al. AMPK couples plasma renin to cellular metabolism by phosphorylation of ACC1. *Am. J. Physiol. Renal. Physiol.* **2013**, *305*, F679–F690. [CrossRef] [PubMed]

156. Efe, O.; Klein, J.D.; LaRocque, L.M.; Ren, H.; Sands, J.M. Metformin improves urine concentration in rodents with nephrogenic diabetes insipidus. *JCI Insight* **2016**, *1*. [CrossRef] [PubMed]

157. Udwan, K.; Abed, A.; Roth, I.; Dizin, E.; Maillard, M.; Bettoni, C.; Loffing, J.; Wagner, C.A.; Edwards, A.; Feraille, E. Dietary sodium induces a redistribution of the tubular metabolic workload. *J. Physiol.* **2017**, *595*, 6905–6922. [CrossRef] [PubMed]

158. Siraskar, B.; Huang, D.Y.; Pakladok, T.; Siraskar, G.; Sopjani, M.; Alesutan, I.; Kucherenko, Y.; Almilaji, A.; Devanathan, V.; Shumilina, E.; et al. Downregulation of the renal outer medullary K$^+$ channel ROMK by the AMP-activated protein kinase. *Pflugers Arch.* **2013**, *465*, 233–245. [CrossRef] [PubMed]

159. Hallows, K.R.; Raghuram, V.; Kemp, B.E.; Witters, L.A.; Foskett, J.K. Inhibition of cystic fibrosis transmembrane conductance regulator by novel interaction with the metabolic sensor AMP-activated protein kinase. *J. Clin. Investig.* **2000**, *105*, 1711–1721. [CrossRef] [PubMed]

160. Morales, M.M.; Falkenstein, D.; Lopes, A.G. The Cystic Fibrosis Transmembrane Regulator (CFTR) in the kidney. *An. Acad. Bras. Ciênc.* **2000**, *72*, 399–406. [CrossRef] [PubMed]

161. Hallows, K.R.; McCane, J.E.; Kemp, B.E.; Witters, L.A.; Foskett, J.K. Regulation of channel gating by AMP-activated protein kinase modulates cystic fibrosis transmembrane conductance regulator activity in lung submucosal cells. *J. Biol. Chem.* **2003**, *278*, 998–1004. [CrossRef] [PubMed]

162. King, J.D.; Fitch, A.C.; Lee, J.K.; McCane, J.E.; Mak, D.-O.D.; Foskett, J.K.; Hallows, K.R. AMP-activated protein kinase phosphorylation of the R domain inhibits PKA stimulation of CFTR. *Am. J. Physiol. Cell Physiol.* **2009**, *297*, C94–C101. [CrossRef] [PubMed]

163. Kongsuphol, P.; Hieke, B.; Ousingsawat, J.; Almaca, J.; Viollet, B.; Schreiber, R.; Kunzelmann, K. Regulation of Cl$^-$ secretion by AMPK in vivo. *Pflugers Arch.* **2009**, *457*, 1071–1078. [CrossRef] [PubMed]

164. Li, H.; Findlay, I.A.; Sheppard, D.N. The relationship between cell proliferation, Cl-secretion, and renal cyst growth: A study using CFTR inhibitors. *Kidney Int.* **2004**, *66*, 1926–1938. [CrossRef] [PubMed]

165. Takiar, V.; Nishio, S.; Seo-Mayer, P.; King, J.D.; Li, H.; Zhang, L.; Karihaloo, A.; Hallows, K.R.; Somlo, S.; Caplan, M.J. Activating AMP-activated protein kinase (AMPK) slows renal cystogenesis. *Proc. Natl. Acad. Sci. USA* **2011**, *108*, 2462–2467. [CrossRef] [PubMed]

166. Yuajit, C.; Muanprasat, C.; Gallagher, A.-R.; Fedeles, S.V.; Kittayaruksakul, S.; Homvisasevongsa, S.; Somlo, S.; Chatsudthipong, V. Steviol retards renal cyst growth through reduction of CFTR expression and inhibition of epithelial cell proliferation in a mouse model of polycystic kidney disease. *Biochem. Pharmacol.* **2014**, *88*, 412–421. [CrossRef] [PubMed]

167. Jeon, U.S. Kidney and calcium homeostasis. *Electrolyte Blood Press.* **2008**, *6*, 68–76. [CrossRef] [PubMed]

168. Na, T.; Peng, J.-B. TRPV5: A Ca^{2+} channel for the fine-tuning of Ca^{2+} reabsorption. *Handb. Exp. Pharmacol.* **2014**, *222*, 321–357. [PubMed]

169. Nurbaeva, M.K.; Schmid, E.; Szteyn, K.; Yang, W.; Viollet, B.; Shumilina, E.; Lang, F. Enhanced Ca^{2+} entry and Na^+/Ca^{2+} exchanger activity in dendritic cells from AMP-activated protein kinase-deficient mice. *FASEB J.* **2012**, *26*, 3049–3058. [CrossRef] [PubMed]

170. Lang, F.; Föller, M. Regulation of ion channels and transporters by AMP-activated kinase (AMPK). *Channels (Austin)* **2014**, *8*, 20–28. [CrossRef] [PubMed]

171. Bhavsar, S.K.; Schmidt, S.; Bobbala, D.; Nurbaeva, M.K.; Hosseinzadeh, Z.; Merches, K.; Fajol, A.; Wilmes, J.; Lang, F. AMPKα1-sensitivity of Orai1 and Ca^{2+} entry in T-lymphocytes. *Cell. Physiol. Biochem.* **2013**, *32*, 687–698. [CrossRef] [PubMed]

172. Mai, X.; Shang, J.; Liang, S.; Yu, B.; Yuan, J.; Lin, Y.; Luo, R.; Zhang, F.; Liu, Y.; Lv, X.; et al. Blockade of Orai1 Store-Operated Calcium Entry Protects against Renal Fibrosis. *J. Am. Soc. Nephrol.* **2016**, *27*, 3063–3078. [CrossRef] [PubMed]

173. Shigaev, A.; Asher, C.; Latter, H.; Garty, H.; Reuveny, E. Regulation of sgk by aldosterone and its effects on the epithelial Na^+ channel. *Am. J. Physiol. Renal. Physiol.* **2000**, *278*, F613–F619. [CrossRef] [PubMed]

174. Zaika, O.; Mamenko, M.; Staruschenko, A.; Pochynyuk, O. Direct activation of ENaC by angiotensin II: Recent advances and new insights. *Curr. Hypertens. Rep.* **2013**, *15*, 17–24. [CrossRef] [PubMed]

175. Staruschenko, A. Regulation of transport in the connecting tubule and cortical collecting duct. *Compr. Physiol.* **2012**, *2*, 1541–1584. [PubMed]

176. Bhalla, V.; Hallows, K.R. Mechanisms of ENaC regulation and clinical implications. *J. Am. Soc. Nephrol.* **2008**, *19*, 1845–1854. [CrossRef] [PubMed]

177. Bhalla, V.; Oyster, N.M.; Fitch, A.C.; Wijngaarden, M.A.; Neumann, D.; Schlattner, U.; Pearce, D.; Hallows, K.R. AMP-activated kinase inhibits the epithelial Na^+ channel through functional regulation of the ubiquitin ligase Nedd4-2. *J. Biol. Chem.* **2006**, *281*, 26159–26169. [CrossRef] [PubMed]

178. Hager, H.; Kwon, T.H.; Vinnikova, A.K.; Masilamani, S.; Brooks, H.L.; Frøkiaer, J.; Knepper, M.A.; Nielsen, S. Immunocytochemical and immunoelectron microscopic localization of α-, β-, and γ-ENaC in rat kidney. *Am. J. Physiol. Renal. Physiol.* **2001**, *280*, F1093–F1106. [CrossRef] [PubMed]

179. Myerburg, M.M.; King, J.D.; Oyster, N.M.; Fitch, A.C.; Magill, A.; Baty, C.J.; Watkins, S.C.; Kolls, J.K.; Pilewski, J.M.; Hallows, K.R. AMPK agonists ameliorate sodium and fluid transport and inflammation in cystic fibrosis airway epithelial cells. *Am. J. Respir. Cell Mol. Biol.* **2010**, *42*, 676–684. [CrossRef] [PubMed]

180. Almaça, J.; Kongsuphol, P.; Hieke, B.; Ousingsawat, J.; Viollet, B.; Schreiber, R.; Amaral, M.D.; Kunzelmann, K. AMPK controls epithelial Na^+ channels through Nedd4-2 and causes an epithelial phenotype when mutated. *Pflugers Arch.* **2009**, *458*, 713–721. [CrossRef] [PubMed]

181. Carattino, M.D.; Edinger, R.S.; Grieser, H.J.; Wise, R.; Neumann, D.; Schlattner, U.; Johnson, J.P.; Kleyman, T.R.; Hallows, K.R. Epithelial sodium channel inhibition by AMP-activated protein kinase in oocytes and polarized renal epithelial cells. *J. Biol. Chem.* **2005**, *280*, 17608–17616. [CrossRef] [PubMed]

182. Yu, H.; Yang, T.; Gao, P.; Wei, X.; Zhang, H.; Xiong, S.; Lu, Z.; Li, L.; Wei, X.; Chen, J.; et al. Caffeine intake antagonizes salt sensitive hypertension through improvement of renal sodium handling. *Sci. Rep.* **2016**, *6*, 25746. [CrossRef] [PubMed]

183. Weixel, K.M.; Marciszyn, A.; Alzamora, R.; Li, H.; Fischer, O.; Edinger, R.S.; Hallows, K.R.; Johnson, J.P. Resveratrol inhibits the epithelial sodium channel via phopshoinositides and AMP-activated protein kinase in kidney collecting duct cells. *PLoS ONE* **2013**, *8*, e78019. [CrossRef] [PubMed]

184. Ho, P.-Y.; Li, H.; Pavlov, T.S.; Tuerk, R.D.; Tabares, D.; Brunisholz, R.; Neumann, D.; Staruschenko, A.; Hallows, K.R. β1Pix exchange factor stabilizes the ubiquitin ligase Nedd4-2 and plays a critical role in ENaC regulation by AMPK in kidney epithelial cells. *J. Biol. Chem.* **2018**, *293*, 11612–11624. [CrossRef] [PubMed]

185. Huang, D.Y.; Gao, H.; Boini, K.M.; Osswald, H.; Nürnberg, B.; Lang, F. In vivo stimulation of AMP-activated protein kinase enhanced tubuloglomerular feedback but reduced tubular sodium transport during high dietary NaCl intake. *Pflugers Arch.* **2010**, *460*, 187–196. [CrossRef] [PubMed]

186. Lazo-Fernández, Y.; Baile, G.; Meade, P.; Torcal, P.; Martínez, L.; Ibañez, C.; Bernal, M.L.; Viollet, B.; Giménez, I. Kidney-specific genetic deletion of both AMPK α-subunits causes salt and water wasting. *Am J. Physiol. Renal. Physiol.* **2017**, *312*, F352–F365. [CrossRef] [PubMed]

187. Alzamora, R.; Gong, F.; Rondanino, C.; Lee, J.K.; Smolak, C.; Pastor-Soler, N.M.; Hallows, K.R. AMP-activated protein kinase inhibits KCNQ1 channels through regulation of the ubiquitin ligase Nedd4-2 in renal epithelial cells. *Am. J. Physiol. Renal. Physiol.* **2010**, *299*, F1308–F1319. [CrossRef] [PubMed]

188. Vallon, V.; Grahammer, F.; Richter, K.; Bleich, M.; Lang, F.; Barhanin, J.; Völkl, H.; Warth, R. Role of KCNE1-dependent K⁺ fluxes in mouse proximal tubule. *J. Am. Soc. Nephrol.* **2001**, *12*, 2003–2011. [PubMed]

189. Vallon, V.; Grahammer, F.; Volkl, H.; Sandu, C.D.; Richter, K.; Rexhepaj, R.; Gerlach, U.; Rong, Q.; Pfeifer, K.; Lang, F. KCNQ1-dependent transport in renal and gastrointestinal epithelia. *Proc. Natl. Acad. Sci. USA* **2005**, *102*, 17864–17869. [CrossRef] [PubMed]

190. Andersen, M.N.; Krzystanek, K.; Jespersen, T.; Olesen, S.-P.; Rasmussen, H.B. AMP-activated protein kinase downregulates Kv7.1 cell surface expression. *Traffic* **2012**, *13*, 143–156. [CrossRef] [PubMed]

191. Alesutan, I.; Föller, M.; Sopjani, M.; Dërmaku-Sopjani, M.; Zelenak, C.; Fröhlich, H.; Velic, A.; Fraser, S.; Kemp, B.E.; Seebohm, G.; et al. Inhibition of the heterotetrameric K⁺ channel KCNQ1/KCNE1 by the AMP-activated protein kinase. *Mol. Membr. Biol.* **2011**, *28*, 79–89. [CrossRef] [PubMed]

192. Al-Bataineh, M.M.; Gong, F.; Marciszyn, A.L.; Myerburg, M.M.; Pastor-Soler, N.M. Regulation of proximal tubule vacuolar H⁺-ATPase by PKA and AMP-activated protein kinase. *Am. J. Physiol. Renal. Physiol.* **2014**, *306*, F981–F995. [CrossRef] [PubMed]

193. Gong, F.; Alzamora, R.; Smolak, C.; Li, H.; Naveed, S.; Neumann, D.; Hallows, K.R.; Pastor-Soler, N.M. Vacuolar H⁺-ATPase apical accumulation in kidney intercalated cells is regulated by PKA and AMP-activated protein kinase. *Am. J. Physiol. Renal. Physiol.* **2010**, *298*, F1162–F1169. [CrossRef] [PubMed]

194. Hallows, K.R.; Alzamora, R.; Li, H.; Gong, F.; Smolak, C.; Neumann, D.; Pastor-Soler, N.M. AMP-activated protein kinase inhibits alkaline pH- and PKA-induced apical vacuolar H⁺-ATPase accumulation in epididymal clear cells. *Am. J. Physiol. Cell Physiol.* **2009**, *296*, C672–C681. [CrossRef] [PubMed]

195. Alzamora, R.; Al-bataineh, M.M.; Liu, W.; Gong, F.; Li, H.; Thali, R.F.; Joho-Auchli, Y.; Brunisholz, R.A.; Satlin, L.M.; Neumann, D.; et al. AMP-activated protein kinase regulates the vacuolar H⁺-ATPase via direct phosphorylation of the A subunit (ATP6V1A) in the kidney. *Am. J. Physiol. Renal. Physiol.* **2013**, *305*, F943–F956. [CrossRef] [PubMed]

196. Alzamora, R.; Thali, R.F.; Gong, F.; Smolak, C.; Li, H.; Baty, C.J.; Bertrand, C.A.; Auchli, Y.; Brunisholz, R.A.; Neumann, D.; et al. PKA regulates vacuolar H⁺-ATPase localization and activity via direct phosphorylation of the a subunit in kidney cells. *J. Biol. Chem.* **2010**, *285*, 24676–24685. [CrossRef] [PubMed]

197. McGuire, C.M.; Forgac, M. Glucose starvation increases V-ATPase assembly and activity in mammalian cells through AMP kinase and phosphatidylinositide 3-kinase/Akt signaling. *J. Biol. Chem.* **2018**, *293*, 9113–9123. [CrossRef] [PubMed]

198. Collins, M.P.; Forgac, M. Regulation of V-ATPase Assembly in Nutrient Sensing and Function of V-ATPases in Breast Cancer Metastasis. *Front. Physiol* **2018**, *9*, 902. [CrossRef] [PubMed]

199. Sands, J.M.; Klein, J.D. Physiological insights into novel therapies for nephrogenic diabetes insipidus. *Am. J. Physiol. Renal. Physiol.* **2016**, *311*, F1149–F1152. [CrossRef] [PubMed]

200. Denton, J.S.; Pao, A.C.; Maduke, M. Novel diuretic targets. *Am. J. Physiol. Renal. Physiol.* **2013**, *305*, F931–F942. [CrossRef] [PubMed]

201. Bech, A.P.; Wetzels, J.F.M.; Nijenhuis, T. Effects of sildenafil, metformin, and simvastatin on ADH-independent urine concentration in healthy volunteers. *Physiol. Rep.* **2018**, *6*, e13665. [CrossRef] [PubMed]

202. Klein, J.D.; Wang, Y.; Blount, M.A.; Molina, P.A.; LaRocque, L.M.; Ruiz, J.A.; Sands, J.M. Metformin, an AMPK activator, stimulates the phosphorylation of aquaporin 2 and urea transporter A1 in inner medullary collecting ducts. *Am. J. Physiol. Renal. Physiol.* **2016**, *310*, F1008–F1012. [CrossRef] [PubMed]

203. Al-bataineh, M.M.; Li, H.; Ohmi, K.; Gong, F.; Marciszyn, A.L.; Naveed, S.; Zhu, X.; Neumann, D.; Wu, Q.; Cheng, L.; et al. Activation of the metabolic sensor AMP-activated protein kinase inhibits aquaporin-2 function in kidney principal cells. *Am. J. Physiol. Renal. Physiol.* **2016**, *311*, F890–F900. [CrossRef] [PubMed]

204. Andersen, M.N.; Rasmussen, H.B. AMPK: A regulator of ion channels. *Commun. Integr. Biol.* **2012**, *5*, 480–484. [CrossRef] [PubMed]

205. Nickolas, T.L.; Jamal, S.A. Bone kidney interactions. *Rev. Endocr. Metab. Disord.* **2015**, *16*, 157–163. [CrossRef] [PubMed]

206. Graciolli, F.G.; Neves, K.R.; Barreto, F.; Barreto, D.V.; Dos Reis, L.M.; Canziani, M.E.; Sabbagh, Y.; Carvalho, A.B.; Jorgetti, V.; Elias, R.M.; et al. The complexity of chronic kidney disease-mineral and bone disorder across stages of chronic kidney disease. *Kidney Int.* **2017**, *91*, 1436–1446. [CrossRef] [PubMed]

207. Alesutan, I.; Munoz, C.; Sopjani, M.; Dërmaku-Sopjani, M.; Michael, D.; Fraser, S.; Kemp, B.E.; Seebohm, G.; Föller, M.; Lang, F. Inhibition of Kir2.1 (KCNJ2) by the AMP-activated protein kinase. *Biochem. Biophys. Res. Commun.* **2011**, *408*, 505–510. [CrossRef] [PubMed]

208. Derst, C.; Karschin, C.; Wischmeyer, E.; Hirsch, J.R.; Preisig-Müller, R.; Rajan, S.; Engel, H.; Grzeschik, K.-H.; Daut, J.; Karschin, A. Genetic and functional linkage of Kir5.1 and Kir2.1 channel subunits. *FEBS Lett.* **2001**, *491*, 305–311. [CrossRef]

209. Mia, S.; Munoz, C.; Pakladok, T.; Siraskar, G.; Voelkl, J.; Alesutan, I.; Lang, F. Downregulation of Kv1.5 K channels by the AMP-activated protein kinase. *Cell. Physiol. Biochem.* **2012**, *30*, 1039–1050. [CrossRef] [PubMed]

210. Bielanska, J.; Hernandez-Losa, J.; Perez-Verdaguer, M.; Moline, T.; Somoza, R.; Cajal, S.; Condom, E.; Ferreres, J.; Felipe, A. Voltage-Dependent Potassium Channels Kv1.3 and Kv1.5 in Human Cancer. *Curr. Cancer Drug Targets* **2009**, *9*, 904–914. [CrossRef] [PubMed]

211. Almilaji, A.; Munoz, C.; Elvira, B.; Fajol, A.; Pakladok, T.; Honisch, S.; Shumilina, E.; Lang, F.; Föller, M. AMP-activated protein kinase regulates hERG potassium channel. *Pflugers Arch.* **2013**, *465*, 1573–1582. [CrossRef] [PubMed]

212. Wadhwa, S.; Wadhwa, P.; Dinda, A.K.; Gupta, N.P. Differential expression of potassium ion channels in human renal cell carcinoma. *Int. Urol. Nephrol.* **2009**, *41*, 251–257. [CrossRef] [PubMed]

213. Munoz, C.; Sopjani, M.; Dërmaku-Sopjani, M.; Almilaji, A.; Föller, M.; Lang, F. Downregulation of the osmolyte transporters SMIT and BGT1 by AMP-activated protein kinase. *Biochem. Biophys. Res. Commun.* **2012**, *422*, 358–362. [CrossRef] [PubMed]

214. Yamauchi, A.; Nakanishi, T.; Takamitsu, Y.; Sugita, M.; Imai, E.; Noguchi, T.; Fujiwara, Y.; Kamada, T.; Ueda, N. In vivo osmoregulation of Na/myo-inositol cotransporter mRNA in rat kidney medulla. *J. Am. Soc. Nephrol.* **1994**, *5*, 62–67. [PubMed]

215. Rasola, A.; Galietta, L.J.; Barone, V.; Romeo, G.; Bagnasco, S. Molecular cloning and functional characterization of a GABA/betaine transporter from human kidney. *FEBS Lett.* **1995**, *373*, 229–233. [CrossRef]

216. Zhou, Y.; Holmseth, S.; Hua, R.; Lehre, A.C.; Olofsson, A.M.; Poblete-Naredo, I.; Kempson, S.A.; Danbolt, N.C. The betaine-GABA transporter (BGT1, slc6a12) is predominantly expressed in the liver and at lower levels in the kidneys and at the brain surface. *Am. J. Physiol. Renal. Physiol.* **2012**, *302*, F316–F328. [CrossRef] [PubMed]

217. Sopjani, M.; Alesutan, I.; Dërmaku-Sopjani, M.; Fraser, S.; Kemp, B.E.; Föller, M.; Lang, F. Down-regulation of Na$^+$-coupled glutamate transporter EAAT3 and EAAT4 by AMP-activated protein kinase. *J. Neurochem.* **2010**, *113*, 1426–1435. [CrossRef] [PubMed]

218. Hu, Q.X.; Ottestad-Hansen, S.; Holmseth, S.; Hassel, B.; Danbolt, N.C.; Zhou, Y. Expression of Glutamate Transporters in Mouse Liver, Kidney, and Intestine. *J. Histochem. Cytochem.* **2018**, *66*, 189–202. [CrossRef] [PubMed]

219. Schmitt, R.; Ellison, D.H.; Farman, N.; Rossier, B.C.; Reilly, R.F.; Reeves, W.B.; Oberbäumer, I.; Tapp, R.; Bachmann, S. Developmental expression of sodium entry pathways in rat nephron. *Am. J. Physiol. Renal. Physiol.* **1999**, *276*, F367–F381. [CrossRef]

220. Kréneisz, O.; Benoit, J.P.; Bayliss, D.A.; Mulkey, D.K. AMP-activated protein kinase inhibits TREK channels. *J. Physiol.* **2009**, *587*, 5819–5830. [CrossRef] [PubMed]

221. Gu, W.; Schlichthörl, G.; Hirsch, J.R.; Engels, H.; Karschin, C.; Karschin, A.; Derst, C.; Steinlein, O.K.; Daut, J. Expression pattern and functional characteristics of two novel splice variants of the two-pore-domain potassium channel TREK-2. *J. Physiol.* **2002**, *539*, 657–668. [CrossRef] [PubMed]

222. Föller, M.; Jaumann, M.; Dettling, J.; Saxena, A.; Pakladok, T.; Munoz, C.; Ruth, P.; Sopjani, M.; Seebohm, G.; Rüttiger, L.; et al. AMP-activated protein kinase in BK-channel regulation and protection against hearing loss following acoustic overstimulation. *FASEB J.* **2012**, *26*, 4243–4253. [CrossRef] [PubMed]

223. Rabjerg, M.; Oliván-Viguera, A.; Hansen, L.K.; Jensen, L.; Sevelsted-Møller, L.; Walter, S.; Jensen, B.L.; Marcussen, N.; Köhler, R. High expression of KCa3.1 in patients with clear cell renal carcinoma predicts high metastatic risk and poor survival. *PLoS ONE* **2015**, *10*, e0122992. [CrossRef] [PubMed]

224. Wyatt, C.N.; Mustard, K.J.; Pearson, S.A.; Dallas, M.L.; Atkinson, L.; Kumar, P.; Peers, C.; Hardie, D.G.; Evans, A.M. AMP-activated protein kinase mediates carotid body excitation by hypoxia. *J. Biol. Chem.* **2007**, *282*, 8092–8098. [CrossRef] [PubMed]

225. Yoshida, H.; Bao, L.; Kefaloyianni, E.; Taskin, E.; Okorie, U.; Hong, M.; Dhar-Chowdhury, P.; Kaneko, M.; Coetzee, W.A. AMP-activated protein kinase connects cellular energy metabolism to KATP channel function. *J. Mol. Cell. Cardiol.* **2012**, *52*, 410–418. [CrossRef] [PubMed]

226. Lim, A.; Park, S.-H.; Sohn, J.-W.; Jeon, J.-H.; Park, J.-H.; Song, D.-K.; Lee, S.-H.; Ho, W.-K. Glucose deprivation regulates KATP channel trafficking via AMP-activated protein kinase in pancreatic β-cells. *Diabetes* **2009**, *58*, 2813–2819. [CrossRef] [PubMed]

227. Chang, T.-J.; Chen, W.-P.; Yang, C.; Lu, P.-H.; Liang, Y.-C.; Su, M.-J.; Lee, S.-C.; Chuang, L.-M. Serine-385 phosphorylation of inwardly rectifying K⁺ channel subunit (Kir6.2) by AMP-dependent protein kinase plays a key role in rosiglitazone-induced closure of the K(ATP) channel and insulin secretion in rats. *Diabetologia* **2009**, *52*, 1112–1121. [CrossRef] [PubMed]

228. Tan, X.-H.; Zheng, X.-M.; Yu, L.-X.; He, J.; Zhu, H.-M.; Ge, X.-P.; Ren, X.-L.; Ye, F.-Q.; Bellusci, S.; Xiao, J.; et al. Fibroblast growth factor 2 protects against renal ischaemia/reperfusion injury by attenuating mitochondrial damage and proinflammatory signalling. *J. Cell. Mol. Med.* **2017**, *21*, 2909–2925. [CrossRef] [PubMed]

229. Klein, H.; Garneau, L.; Trinh, N.T.N.; Privé, A.; Dionne, F.; Goupil, E.; Thuringer, D.; Parent, L.; Brochiero, E.; Sauvé, R. Inhibition of the KCa3.1 channels by AMP-activated protein kinase in human airway epithelial cells. *Am. J Physiol. Cell Physiol.* **2009**, *296*, C285–C295. [CrossRef] [PubMed]

230. Huang, C.; Shen, S.; Ma, Q.; Chen, J.; Gill, A.; Pollock, C.A.; Chen, X.-M. Blockade of KCa3.1 ameliorates renal fibrosis through the TGF-β1/Smad pathway in diabetic mice. *Diabetes* **2013**, *62*, 2923–2934. [CrossRef] [PubMed]

231. Takimoto, M.; Takeyama, M.; Hamada, T. Possible involvement of AMPK in acute exercise-induced expression of monocarboxylate transporters MCT1 and MCT4 mRNA in fast-twitch skeletal muscle. *Metab. Clin. Exp.* **2013**, *62*, 1633–1640. [CrossRef] [PubMed]

232. Becker, H.M.; Mohebbi, N.; Perna, A.; Ganapathy, V.; Capasso, G.; Wagner, C.A. Localization of members of MCT monocarboxylate transporter family Slc16 in the kidney and regulation during metabolic acidosis. *Am. J. Physiol. Renal. Physiol.* **2010**, *299*, F141–F154. [CrossRef] [PubMed]

233. Fisel, P.; Kruck, S.; Winter, S.; Bedke, J.; Hennenlotter, J.; Nies, A.T.; Scharpf, M.; Fend, F.; Stenzl, A.; Schwab, M.; et al. DNA methylation of the SLC16A3 promoter regulates expression of the human lactate transporter MCT4 in renal cancer with consequences for clinical outcome. *Clin. Cancer Res.* **2013**, *19*, 5170–5181. [CrossRef] [PubMed]

234. Pieri, M.; Christian, H.C.; Wilkins, R.J.; Boyd, C.A.R.; Meredith, D. The apical (hPepT1) and basolateral peptide transport systems of Caco-2 cells are regulated by AMP-activated protein kinase. *Am. J. Physiol. Gastrointest. Liver Physiol.* **2010**, *299*, G136–G143. [CrossRef] [PubMed]

235. Shen, H.; Smith, D.E.; Yang, T.; Huang, Y.G.; Schnermann, J.B.; Brosius, F.C. Localization of PEPT1 and PEPT2 proton-coupled oligopeptide transporter mRNA and protein in rat kidney. *Am. J. Physiol. Renal. Physiol.* **1999**, *276*, F658–F665. [CrossRef]

236. Shen, B.; Zhu, J.; Zhang, J.; Jiang, F.; Wang, Z.; Zhang, Y.; Li, J.; Huang, D.; Ke, D.; Ma, R.; et al. Attenuated mesangial cell proliferation related to store-operated Ca²⁺ entry in aged rat: The role of STIM 1 and Orai 1. *Age* **2013**, *35*, 2193–2202. [CrossRef] [PubMed]

13

Reciprocal Regulation of AMPK/SNF1 and Protein Acetylation

Ales Vancura *, Shreya Nagar, Pritpal Kaur, Pengli Bu, Madhura Bhagwat and Ivana Vancurova

Department of Biological Sciences, St. John's University, New York, NY 11439, USA;
shreya.nagar17@my.stjohns.edu (S.N.); pritpal.kaur17@my.stjohns.edu (P.K.); bup@stjohns.edu (P.B.);
madhura.bhagwat16@my.stjohns.edu (M.B.); vancuroi@stjohns.edu (I.V.)
* Correspondence: vancuraa@stjohns.edu

Abstract: Adenosine monophosphate (AMP)-activated protein kinase (AMPK) serves as an energy sensor and master regulator of metabolism. In general, AMPK inhibits anabolism to minimize energy consumption and activates catabolism to increase ATP production. One of the mechanisms employed by AMPK to regulate metabolism is protein acetylation. AMPK regulates protein acetylation by at least five distinct mechanisms. First, AMPK phosphorylates and inhibits acetyl-CoA carboxylase (ACC) and thus regulates acetyl-CoA homeostasis. Since acetyl-CoA is a substrate for all lysine acetyltransferases (KATs), AMPK affects the activity of KATs by regulating the cellular level of acetyl-CoA. Second, AMPK activates histone deacetylases (HDACs) sirtuins by increasing the cellular concentration of NAD^+, a cofactor of sirtuins. Third, AMPK inhibits class I and II HDACs by upregulating hepatic synthesis of α-hydroxybutyrate, a natural inhibitor of HDACs. Fourth, AMPK induces translocation of HDACs 4 and 5 from the nucleus to the cytoplasm and thus increases histone acetylation in the nucleus. Fifth, AMPK directly phosphorylates and downregulates p300 KAT. On the other hand, protein acetylation regulates AMPK activity. Sirtuin SIRT1-mediated deacetylation of liver kinase B1 (LKB1), an upstream kinase of AMPK, activates LKB1 and AMPK. AMPK phosphorylates and inactivates ACC, thus increasing acetyl-CoA level and promoting LKB1 acetylation and inhibition. In yeast cells, acetylation of Sip2p, one of the regulatory β-subunits of the SNF1 complex, results in inhibition of SNF1. This results in activation of ACC and reduced cellular level of acetyl-CoA, which promotes deacetylation of Sip2p and activation of SNF1. Thus, in both yeast and mammalian cells, AMPK/SNF1 regulate protein acetylation and are themselves regulated by protein acetylation.

Keywords: AMP-activated protein kinase; epigenetics; protein acetylation; KATs; HDACs; acetyl-CoA; NAD^+

1. AMPK Links Metabolism and Signaling with Protein Acetylation, Epigenetics, and Transcriptional Regulation

AMP-activated protein kinase (AMPK) is highly conserved across eukaryotes and serves as an energy sensor and master regulator of metabolism, functioning as a fuel gauge monitoring systemic and cellular energy status [1–3]. Activation of AMPK occurs when the intracellular AMP/ATP ratio increases. In general, AMPK inhibits anabolism to minimize energy consumption and activates catabolism to increase ATP production.

AMPK is a heterotrimeric complex composed of subunit and two regulatory subunits, α and γ. The human genome contains two genes encoding two distinct subunits, 1 and 2, two α subunits, $\beta1$ and $\beta2$, and three γ subunits, $\gamma1$, $\gamma2$, and $\gamma3$ [1–3]. Different combinations of α, β, and γ subunits can produce 12 distinct AMPK complexes; however, it is not known whether these complexes differ in substrate specificities, subcellular localization, or other aspects of regulation. The subunit features the catalytic protein kinase domain, the α subunit contains a carbohydrate-binding domain that allows AMPK to interact with glycogen [4], and the γ subunit contains domains that bind AMP and thus impart AMPK regulation by cellular energy state [5–7].

The AMPK complex is activated more than 100-fold by phosphorylation on Thr172 of the catalytic α subunit [8]. The major upstream kinase targeting this site is the tumor suppressor liver kinase B1 (LKB1) [9–11]. LKB1 is responsible for most of AMPK activation under low energy conditions in the majority of tissues, including liver and muscle [2,12–14]. LKB1 is also responsible for AMPK activation in response to mitochondrial insults [15].

AMPK targets a number of metabolic enzymes and transporters, such as glucose transporter (GLUT) 1 and GLUT4, glycogen synthase (GS), acetyl-CoA carboxylase (ACC), and hydroxymethylglutaryl-CoA reductase (HMGCR) [16–18]. AMPK also regulates metabolism at the transcriptional level by phosphorylating sterol regulatory element-binding protein 1 (SREBP1), carbohydrate-responsive element-binding protein (ChREBP), transcriptional coactivator peroxisome proliferator-activated receptor gamma coactivator 1-alpha (PGC1) and transcriptional factor forkhead box O3 (FOXO3) [19–22].

One of the most important targets of AMPK is mechanistic target of rapamycin complex 1 (mTORC1). mTOR is a conserved serine/threonine protein kinase from the phosphatidylinositol-3-kinase (PI3K) family. mTOR is found in all eukaryotes and forms the catalytic subunit of mTORC1 and mTORC2. mTORC1 is regulated by nutrients and growth factors, and functions as a master regulator of cell growth and metabolism by phosphorylating a host of targets [23]. The AMPK-dependent mechanisms of mTORC1 inhibition are mediated by phosphorylation of the tuberous sclerosis complex (TSC) and raptor subunit of mTORC1 [24,25]. TSC functions as a GTPase activating protein (GAP) for the small GTPase Rheb, which directly binds and activates mTORC1. Thus, by downregulating Rheb, TSC inhibits mTORC1 and downregulation of TSC, which therefore leads to activation of mTORC1. In addition to integrating signals from several growth factor pathways, TSC is also regulated by AMPK. Activated AMPK directly phosphorylates TSC2 on serine residues that are distinct from those regulated by growth factor pathways, resulting in TSC activation and mTORC1 inhibition. In addition to TSC2, AMPK also phosphorylates mTORC1 subunit Raptor, leading again to mTORC1 inhibition [23,25]. Under low energy conditions, AMPK assembles in a complex with v-ATPase, Ragulator, scaffold protein Axin, and LKB1 on the lysosome surface, resulting in AMPK activation. At the same time, mTORC1 dissociates from the Ragulator and lysosome, resulting in mTORC1 inhibition [26,27]. These results further illustrate that AMPK and mTORC1 are inversely regulated and represent a molecular switch between catabolism and anabolism.

This review focuses on the previously little explored role of AMPK in regulation of acetyl-CoA and NAD$^+$ homeostasis and on reciprocal regulation of AMPK and protein acetylation, which places AMPK at the interface between metabolism and other essential cellular functions, including transcription, replication, DNA repair, and aging [14,28–30].

2. Protein Acetylation

Protein acetylation is a posttranslational protein modification in which the acetyl group from acetyl-CoA is transferred onto ε-amino group of lysine residues. Histones were the first proteins known to be acetylated. More recently, genomic and proteomic approaches in bacteria, yeast, and higher eukaryotes identified many non-histone proteins that are acetylated, suggesting that acetylation extends beyond histones. A number of proteomic studies show that acetylation occurs at thousands of sites throughout eukaryotic cells and that the human proteome contains at least ~2500 acetylated proteins [31–34]. In comparison, similar analyses of human and mouse proteins identified ~2200 phosphoproteins [35,36]. Thus, it appears that protein acetylation is as widespread as phosphorylation [37]. In human cells, acetylated proteins are involved in the regulation of diverse cellular processes, including chromatin remodeling, the cell cycle, RNA metabolism, cytoskeleton dynamics, membrane trafficking, and key metabolic pathways, such as glycolysis, gluconeogenesis, and the citric acid cycle [32,34]. In general, protein acetylation can both activate and inhibit enzymatic activity of proteins as well as interactions between proteins [28,29,38,39]. Acetylation of histones affects the chromatin structure and transcriptional regulation by two mechanisms. It neutralizes positive charges of lysines and thus diminishes interaction of histone tails with DNA. By forming acetyllysines, histone acetylation creates sites that are recognized and bound by proteins and protein complexes that contain bromodomains. Many of these bromodomain-containing protein complexes covalently or noncovalently modify chromatin structure and thus regulate transcription [40,41].

2.1. KATs and HDACs

The enzymes that catalyze protein acetylation were originally called histone acetyltransferases (HATs) [40]. With the realization that histones are not the only substrates, these enzymes are now more commonly referred to as lysine acetyltransferases (KATs) [41–45]. The human genome contains 22 genes that encode proteins currently known to possess protein acetyltransferase activity [39]. The KATs can be classified into three major groups: The GNAT, MYST, and p300/CBP families. Most KATs are catalytic subunits of multiprotein complexes; the noncatalytic subunits of these complexes are typically responsible for substrate recognition, regulation, and subcellular localization.

Protein acetylation is a dynamic modification. The acetyl groups are removed from proteins by histone deacetylases (HDACs), sometimes also called lysine deacetylases (KDACs) to indicate that acetylated histones are not the only substrates [39]. However, the name HDACs is still more commonly used. HDACs hydrolyze the amide linkage between the acetyl group and amino group of lysine residues, yielding acetate. HDACs are grouped into four classes. Class I, II, and IV are Zn^{2+}-dependent amidohydrolases, while class III uses NAD^+ as a cosubstrate [46]. Class III HDACs are known as the sirtuins [47].

The dynamic balance between protein acetylation and deacetylation, mediated by the activities of KATs and HDACs, is well regulated in healthy cells, but is often dysregulated in cancer and other pathologic conditions. For example, change in the acetylation status of chromatin histones alters the structure of chromatin and expression pattern of genes in cancer cells [43].

2.2. Nonenzymatic Acetylation of Mitochondrial Proteins

The reactivity of metabolites depends on the presence of nucleophilic or electrophilic groups. The carbonyl group is electrophilic and can be further enzymatically activated by adding electronegative groups, such as thiols or phosphates. Compounds containing a reactive thioester group in the form of CoA are common in many metabolic pathways and reactions, involving fatty acid synthesis, tricarboxylic acid cycle, amino acid metabolism, and protein acetylation [48]. Protein acetylation by KATs employs a common catalytic mechanism which involves the formation of a ternary complex of KAT-acetyl-CoA-histone and the deprotonation of the ε-amino group of lysine by a glutamate or aspartate residue within the active site of a KAT, followed by a nucleophilic attack on the carbonyl group of acetyl-CoA [42].

High concentration of acetyl-CoA coupled with high pH, conditions that exist in the mitochondrial matrix, create a permissive environment for non-enzymatic acetylation of proteins [49,50]. In *Saccharomyces cerevisiae*, about 4000 lysine acetylation sites were identified, many of them on mitochondrial proteins [49,51]. The acetylation of mitochondrial proteins correlates with acetyl-CoA levels in mitochondria, as demonstrated by the fact that acetylation of mitochondrial proteins is dependent on *PDA1*, encoding a subunit of the pyruvate dehydrogenase (PDH) complex [49]. The acetylation of mitochondrial proteins was also elevated by introducing the *cit1Δ* mutation. *CIT1* encodes mitochondrial citrate synthase; *cit1Δ* mutants are not able to utilize acetyl-CoA for citrate synthesis and probably have an elevated level of mitochondrial acetyl-CoA. These results suggest that most of the mitochondrial acetyl-CoA in exponentially growing cells is derived from glycolytically-produced pyruvate that was translocated into mitochondria and converted to acetyl-CoA by the PDH complex. Inactivation of the PDH complex results in about a 30% decrease in cellular acetyl-CoA; this indicates that mitochondrial acetyl-CoA represents about 30% of the cellular pool. However, since mitochondria occupy only 1–2% of the cellular volume in *S. cerevisiae* [52], the mitochondrial acetyl-CoA concentration is about 20–30-fold higher than the concentration in the nucleocytosolic compartment and is probably within the millimolar range [49,50,53,54]. Due to the extrusion of protons across the inner mitochondrial membrane, the pH of the mitochondrial matrix is higher than the pH in the cytosol or nucleus, about 8.0 [50,55]. The high pH coupled with the high concentration of acetyl-CoA in the mitochondrial matrix create a permissive environment for non-enzymatic acetylation of mitochondrial proteins [49,50]. However, these considerations do not exclude the possibility that at least some protein acetylation in the mitochondria is catalyzed by KATs. In addition, some acyl-CoAs, such as 3-hydroxy-3-methylglutaryl-CoA, and glutaryl-CoA, are sufficiently reactive under the in vivo conditions and are able to non-enzymatically modify proteins [56].

3. AMPK Regulation of Protein Acetylation

AMPK regulates protein acetylation by at least five distinct mechanisms (Figure 1). First, AMPK phosphorylates and inhibits ACC and thus regulates acetyl-CoA homeostasis. Second, AMPK activates sirtuin SIRT1 by increasing the cellular concentration of NAD^+, a cofactor of sirtuins. Third, AMPK inhibits class I and II histone deacetylases (HDACs) by upregulating hepatic synthesis of α-hydroxybutyrate, a natural inhibitor of HDACs. Fourth, AMPK induces translocation of HDACs 4 and 5 from the nucleus to the cytoplasm and thus increases histone acetylation in the nucleus. Fifth, AMPK directly phosphorylates and downregulates p300 KAT.

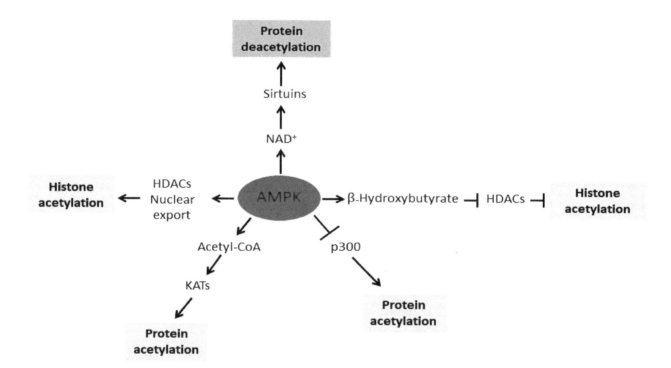

Figure 1. AMP-activated protein kinase (AMPK) regulates protein acetylation by several different mechanisms: (i) AMPK phosphorylates and inhibits acetyl-CoA carboxylase (ACC) and thus elevates acetyl-CoA level and activity of lysine acetyltransferases (KATs); (ii) AMPK increases the cellular concentration of NAD^+ and thus activates sirtuins; (iii) AMPK upregulates hepatic synthesis of α-hydroxybutyrate, and thus inhibits histone deacetylases (HDACs) and promotes histone acetylation; (iv) AMPK increases histone acetylation in the nucleus by inducing nuclear export of HDACs; (v) AMPK directly phosphorylates and downregulates p300 KAT. Arrows denote activation and t-bars denote inhibition.

3.1. Acetyl-CoA Level Regulates Protein Acetylation

Acetyl-CoA is the donor of acetyl groups for protein acetylation and KATs depend on intermediary metabolism for supplying acetyl-CoA in the nucleocytosolic compartment (Figure 1). Acetyl-CoA is thus a key metabolite that links metabolism with signaling, chromatin structure, and transcription [29,30,45,53,57–59]. Changing metabolic conditions drive fluctuations of the cellular level of acetyl-CoA to the extent that the activity of KATs is regulated by the availability of acetyl-CoA, resulting in dynamic protein acetylations that regulate a variety of cell functions, including transcription, replication, DNA repair, cell cycle progression, and aging. Acetyl-CoA can freely diffuse through the nuclear pore complex and changes in the pool of available acetyl-CoA in the cytoplasm cause changes in protein acetylation in both the nucleus and cytoplasm. However, the mitochondrial pool of acetyl-CoA is biochemically isolated and cannot be used for histone acetylation in the nucleocytosolic compartment [60]. In mammalian cells, glycolytically produced pyruvate is translocated from the cytosol into mitochondria, where pyruvate dehydrogenase converts it into acetyl-CoA. Acetyl-CoA then enters the tricarboxylic acid (TCA) cycle and condenses with oxaloacetate, producing citrate. Citrate can be subsequently exported from the mitochondrial matrix into the cytosol, where ATP-citrate lyase (ACL) converts it into acetyl-CoA and oxaloacetate. This acetyl-CoA is then used by KATs for protein acetylation in the nucleocytosolic compartment [61], in addition to being a precursor of several anabolic pathways, including de novo synthesis of fatty acids. Since ACL generates acetyl-CoA from glucose-derived citrate, glucose availability affects histone acetylation in an ACL-dependent manner, and when the synthesis of acetyl-CoA is compromised, rapid histone deacetylation ensues [60,61].

Translocation of pyruvate dehydrogenase complex (PDH) from the mitochondria to the nucleus provides an alternative mechanism for synthesis of acetyl-CoA in the nucleus. PDH translocated to the nucleus in a cell-cycle-dependent manner and in response to serum, epidermal growth factor, or mitochondrial stress. Inhibition of nuclear PDH decreased acetylation of specific lysine residues in histones and transcription of genes important for G1-S phase progression [62]. In addition to ACL and nuclear PDH, direct de novo synthesis of acetate from pyruvate for acetyl-CoA production occurs under conditions of nutritional excess [63]. The conversion of pyruvate to acetate takes place either by coupling to reactive oxygen species (ROS) or by the activity of keto acid dehydrogenases, which function under certain conditions as pyruvate decarboxylase [63].

Since nucleocytosolic acetyl-CoA is also used for de novo synthesis of fatty acids, histone acetylation and synthesis of fatty acids compete for the same acetyl-CoA pool. ACC catalyzes the carboxylation of acetyl-CoA to malonyl-CoA, the first and rate-limiting reaction in the de novo synthesis of fatty acids. The ACC activity affects the concentration of nucleocytosolic acetyl-CoA. Attenuated expression of yeast ACC encoded by the ACC1 gene, increases global acetylation of chromatin histones as well as non-histone proteins, and alters transcriptional regulation [64]. Direct pharmacological inhibition of ACC in human cancer cells also induces histone acetylation [65,66]. ACC is phosphorylated and inhibited by AMPK. In yeast, inactivation of the SNF1 complex, the budding yeast ortholog of mammalian AMPK [67–69], results in increased Acc1p activity, reduced pool of cellular acetyl-CoA, and globally decreased histone acetylation [70]. Activation of AMPK with metformin or with the AMP mimetic 5-aminoimidazole-4-carboxamide ribonucleotide (AICAR) increases the inhibitory phosphorylation of ACC, and decreases the conversion of acetyl-CoA to malonyl-CoA, leading to increased protein acetylation and altered gene expression in prostate and ovarian cancer cells [65].

3.2. NAD+ Synthesis Regulates Protein Acetylation

NAD+ is a cofactor used by many oxidoreductases to carry electrons in redox reactions. In addition, NAD+ is used as a cosubstrate by a group of HDACs called sirtuins, named after the budding yeast protein Sir2. In mammals, there are seven NAD+-dependent sirtuins. Similar to the dependence of KATs on acetyl-CoA level, the activity of sirtuins is regulated by metabolically driven changes in the cellular level of NAD$^+$ [71,72]. AMPK activation induces expression of nicotinamide phosphoribosyltransferase (NAMPT), the rate-limiting enzyme in the NAD$^+$ salvage pathway that converts nicotinamide to nicotinamide mononucleotide to enable NAD$^+$ biosynthesis. AMPK activation thus increases NAD$^+$ level, elevating SIRT1 activity [73,74]. SIRT1-mediated protein deacetylation subsequently activates downstream targets, including peroxisome proliferator-activated receptor gamma coactivator 1-α (PGC-1α) and forkhead box protein O1 (FOXO1) [74]. In addition, AMPK directly phosphorylates SIRT1 at T344, which results in dissociation of SIRT1 from the inhibitory bladder cancer protein 1 (DBC1), and SIRT1-mediated deacetylation of p53 and inhibition of its transcriptional activity [75]. Yet another mechanism for SIRT1 activation by AMPK involves phosphorylation of glyceraldehyde 3-phosphate dehydrogenase (GAPDH) by AMPK, leading to nuclear translocation of GAPDH and GAPDH-dependent dissociation of SIRT1 from DBC1 [76].

3.3. α-Hydroxybutyrate Synthesis Regulates Protein Acetylation

The ketone body α-hydroxybutyrate is structurally similar to butyrate, an effective inhibitor of class I and II HDACs [77]. Ketone bodies, including α-hydroxybutyrate, are produced during starvation or prolonged exercise, when liver switches the metabolic mode from catabolism of glucose to catabolism of triacylglycerols and fatty acids. This metabolic switch is partly orchestrated by AMPK and activation of AMPK increases fatty acid oxidation, leading to production of ketone bodies, including α-hydroxybutyrate [77]. Administration of exogenous α-hydroxybutyrate or inducing catabolism of fatty acids by calorie restriction increased global histone acetylation in mouse tissues [77]. Acetylation of histones in the promoters of genes required for protection against oxidative stress was also increased, leading to increased expression of the corresponding genes and elevated protection against oxidative stress [77]. These results indicate that AMPK activation induced by calorie restriction or prolonged exercise leads to α-hydroxybutyrate-mediated inhibition of HDACs and globally increased histone acetylation, and may represent one of the health-promoting mechanisms of calorie restriction.

3.4. AMPK Induces Nuclear Export of HDACs

Type II HDACs belong into two subgroups, IIa and IIb. HDACs of the IIa subgroup, HDAC4, HDAC5, HDAC7, and HDAC9, are able to shuttle between the nucleus and cytoplasm [78–81]. AMPK phosphorylates HDAC5 at Ser259 and Ser498, which promotes export of HDAC5 from the nucleus to the cytoplasm. This removal of HDAC5 from the nucleus results in decreased occupancy of HDAC5 and increased histone acetylation at promoters of glucose transporter member 4 (GLUT4), myogenin, and α-catenin genes, leading to increased transcription of the corresponding genes [82–86]. It appears that AMPK regulates expression of host of genes involved in differentiation or development by phosphorylating HDAC5 and HDAC4 and promoting their export from the nucleus. AMPK also mediates nuclear accumulation of transcription factor hypoxia-inducible factor 1α (HIF-1α) by a mechanism that involves HDAC5. Activation of nuclear AMPK promotes nuclear export of HDAC5, presumably by directly phosphorylating HDAC5. Cytosolic HDAC then deacetylates heat shock protein 70 (HSP70), triggering dissociation of HSP70 from HIF-1 and nuclear transport of HIF-1α [84]. Depending on their phosphorylation level, also HDAC7 and HDAC9 shuttle between nucleus and cytoplasm. However, it is not known whether they are AMPK substrates and whether AMPK regulates their nucleocytosolic shuttling.

3.5. AMPK Phosphorylates p300 KAT and Histone H3

AMPK directly phosphorylates transcriptional coactivator p300 KAT on Ser89, which inhibits the interaction of p300 with peroxisome proliferator-activated receptor γ (PPAR-γ) and retinoid acid receptor [19]. The AMPK-mediated phosphorylation of p300 also results in decreased acetylation and reduced activity of transcription factors nuclear factor kappa B (NFκB) and SMAD3 [87,88]. AMPK also promotes histone acetylation indirectly through phosphorylation of histone H2B at Ser36, particularly at promoters occupied by p53. During metabolic or genotoxic stress, AMPK translocates to the nucleus, binds to chromatin, and phosphorylates histone H2B at Ser36. This phosphorylation leads to increased assembly and recruitment of KATs to specific promoters, associated with increased transcription [89]. An analogous situation was described also in yeast. SNF1, the yeast AMPK ortholog, phosphorylates histone H3 at Ser10, which results in increased acetylation of Lys14 of histone H3 by KAT Gcn5 in the promoter of the *INO1* gene [90]. It appears that SNF1-mediated phosphorylation of Ser10 of histone H3 does not represent a general mechanism of histone acetylation. Rather, inactivation of SNF1 results in decreased nucleocytosolic level of acetyl-CoA by increasing conversion of acetyl-CoA into malonyl-CoA, which results in globally reduced acetylation of histone and non-histone proteins [70].

4. AMPK Is Regulated by Protein Acetylation

AMPK activity is regulated by an upstream kinase LKB1, which activates AMPK by phosphorylating it on Thr172. LKB1 is a low energy sensor that regulates tumorigenesis and apoptosis by regulating AMPK and mechanistic target of rapamycin (mTOR) pathways [10]. LKB1 is acetylated and the acetylation reduces its ability to activate AMPK (Figure 2). Deacetylation of LKB1 by SIRT1 activates LKB1 and AMPK and increases inhibitory phosphorylation of ACC [91]. Inhibition of ACC increases acetyl-CoA level and protein acetylation [65] and presumably should also promote LKB1 acetylation and reduced activation of AMPK (Figure 2). It is tempting to speculate that LKB1 acetylation and diminished activation of AMPK form a regulatory loop with ACC, which contributes to regulation of acetyl-CoA homeostasis and protein acetylation [70]. Increased acetyl-CoA level would promote LKB1 acetylation and diminished activation of AMPK. Decreased activity of AMPK would result in lower AMPK-mediated phosphorylation and inhibition of ACC, increased conversion of acetyl-CoA to malonyl-CoA, decreased acetyl-CoA level and decreased protein acetylation. Decreased acetyl-CoA level would also result in hypoacetylation of LKB1 and increased activation of AMPK. This, in turn, would lead to increased phosphorylation and inhibition of ACC, decreased conversion of acetyl-CoA to malonyl-CoA, increased acetyl-CoA level, and increased protein acetylation. In addition, activation of AMPK would promote NAD^+ synthesis, leading to elevated SIRT1 activity (Figure 2). This homeostatic mechanism would contribute to the regulation of the nucleocytosolic level of acetyl-CoA and NAD^+ within certain limits and would prevent gross hypoacetylation or hyperacetylation of proteins, a condition that might alter regulation of many essential processes. It also appears that SIRT1 and AMPK mediate the positive effect of some dietary compounds, such as resveratrol and other polyphenols. Resveratrol increases SIRT1 activity, which activates AMPK signaling, presumably by deacetylating LKB1. Activated AMPK then suppresses lipid accumulation in hepatocytes [92].

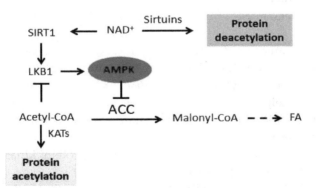

Figure 2. Model of the feedback regulation of AMPK by protein acetylation. AMPK phosphorylates and inhibits ACC, thus increasing acetyl-CoA cellular level and promoting KAT-mediated protein acetylation. Acetylation of liver kinase B1 (LKB1) inhibits the ability of LKB1 to activate AMPK. AMPK also promotes synthesis of NAD^+, thus activating SIRT1 and other sirtuins and promoting protein deacetylation. SIRT1 deacetylates and activates LKB1, resulting in AMPK activation. Arrows denote activation, t-bars denote inhibition, and dashed arrow indicates multistep pathway.

An analogous regulatory loop seems to operate in yeast. The yeast SNF1 complex consists of the catalytic α subunit Snf1p, one of three different regulatory α subunits, Sip1p, Sip2p, or Gal83p, and the stimulatory γ subunit Snf4p [93]. Acetylation of Sip2p, one of the regulatory α-subunits of the SNF1 complex, results in inhibition of SNF1. The level of Sip2p acetylation depends on the nucleocytosolic level of acetyl-CoA and is increased when *ACC1* transcription is repressed [64] and decreased in *snf1Δ* cells [70]. The acetylation of Sip2p increases its interaction and inhibition of Snf1p [94]. This results in activation of the *ACC1* gene and reduced cellular level of acetyl-CoA, which promotes deacetylation of Sip2p and activation of SNF1. Thus, in both yeast and mammalian cells, AMPK/SNF1 regulate protein acetylation and are themselves regulated by protein acetylation. Since SNF1 also phosphorylates Sch9, a yeast ortholog of the Akt kinase, inhibition of SNF1 by Sip2p results in reduced phosphorylation of Sch9, ultimately leading to extended life span. Acetylation of Sip2p thus promotes life span extension and Sip2p acetylation mimetics are more resistant to oxidative stress [94].

5. Conclusions

AMPK is an energy sensor and master regulator of metabolism, functioning as a fuel gauge. AMPK phosphorylates and regulates a number of metabolic enzymes and transporters, as well as transcription factors. In this review article, we have focused on the role of AMPK in regulation of protein acetylation and on regulation of AMPK by protein acetylation. Taken together, AMPK regulates protein acetylation by regulating synthesis of acetyl-CoA, NAD^+, and α-hydroxybutyrate, as well as by directly phosphorylating and regulating KATs and HDACs. Using these two general mechanisms, AMPK contributes to the global regulation of protein acetylation and connects epigenetic chromatin modifications with the cellular metabolic state. Since protein acetylation appears to be as widespread as protein phosphorylation, it endows AMPK with yet another mechanism of regulation of cellular and organismal physiology.

References

1. Herzig, S.; Shaw, R. AMPK: Guardian of metabolism and mitochondrial homeostasis. *Nat. Rev. Mol. Cell. Biol.* **2018**, *19*, 121–135. [CrossRef] [PubMed]
2. Hardie, D.G. Keeping the home fires burning: AMP-activated protein kinase. *J. R. Soc. Interface* **2018**, *15*, 20170774. [CrossRef] [PubMed]
3. Kjobsted, R. AMPK in skeletal muscle function and metabolism. *FASEB J.* **2018**, *32*, 1741–1777. [CrossRef] [PubMed]
4. Hudson, E.R.; Pan, D.A.; James, J.; Lucocq, J.M.; Hawley, S.A.; Green, K.A.; Baba, O.; Terashima, T.; Hardie, D.G. A novel domain in AMP-activated protein kinase causes glycogen storage bodies similar to those seen in hereditary cardiac arrhythmias. *Curr. Biol.* **2003**, *13*, 861–866. [CrossRef]
5. Xiao, B.; Heath, R.; Saiu, P.; Leiper, F.C.; Leone, P.; Jing, C.; Walker, P.A.; Haire, L.; Eccleston, J.F.; Davis, C.T.; et al. Structural basis for AMP binding to mammalian AMP-activated protein kinase. *Nature* **2007**, *449*, 496–500. [CrossRef] [PubMed]
6. Hardie, D.G.; Carling, D.; Gamblin, S.J. AMP-activated protein kinase: Also regulated by ADP? *Trends Biochem Sci.* **2011**, *36*, 470–477. [CrossRef] [PubMed]
7. Gowans, G.J.; Hawley, S.A.; Ross, F.A.; Hardie, D.G. AMP is a true physiological regulator of AMP-activated protein kinase by both allosteric activation and enhancing net phosphorylation. *Cell Metab.* **2013**, *18*, 556–566. [CrossRef] [PubMed]
8. Hawley, S.A.; Davison, M.; Woods, A.; Davies, S.P.; Beri, R.K.; Carling, D.; Hardie, D.G. Characterization of the AMP-activated protein kinase kinase from rat liver and identification of threonine 172 as the major site at which it phosphorylates AMP-activated protein kinase. *J. Biol. Chem.* **1996**, *271*, 27879–27887. [CrossRef] [PubMed]

9.   Hawley, S.A.; Boudeau, J.; Reid, J.L.; Mustard, K.J.; Udd, L.; Mäkelä, T.P.; Alessi, D.R.; Hardie, D.G. Complexes between the LKB1 tumor suppressor, STRAD α/β and MO25 α/β are upstream kinases in the AMP-activated protein kinase cascade. *J. Biol.* **2003**, *2*, 28. [CrossRef] [PubMed]

10. Shaw, R.J.; Kosmatka, M.; Bardeesy, N.; Hurley, R.L.; Witters, L.A.; de Pinho, R.A.; Cantley, L.C. The tumor suppressor LKB1 kinase directly activates AMP-activated kinase and regulates apoptosis in response to energy stress. *Proc. Natl. Acad. Sci. USA* **2004**, *101*, 3329–3335. [CrossRef] [PubMed]

11. Woods, A.; Johnstone, S.R.; Dickerson, K.; Leiper, F.C.; Fryer, L.G.; Neumann, D.; Schlattner, U.; Wallimann, T.; Carlson, M.; Carling, D. LKB1 is the upstream kinase in the AMP-activated protein kinase cascade. *Curr. Biol.* **2003**, *13*, 2004–2008. [CrossRef] [PubMed]

12. Shaw, R.J.; Lamia, K.A.; Vasquez, D.; Koo, S.H.; Bardeesy, N.; Depinho, R.A.; Montminy, M.; Cantley, L.C. The kinase LKB1 mediates glucose homeostasis in liver and therapeutic effects of metformin. *Science* **2005**, *310*, 1642–1646. [CrossRef] [PubMed]

13. Hardie, D.G. AMPK-Sensing energy while talking to other signaling pathways. *Cell Metab.* **2014**, *20*, 939–952. [CrossRef] [PubMed]

14. Burkewitz, K.; Zhang, Y.; Mair, W.B. AMPK at the nexus of energetics and aging. *Cell Metab.* **2014**, *20*, 10–25. [CrossRef] [PubMed]

15. Shackelford, D.B.; Shaw, R.J. The LKB1-AMPK pathway: Metabolism and growth control in tumour suppression. *Nat Rev Cancer.* **2009**, *9*, 563–575. [CrossRef] [PubMed]

16. Carling, D.; Zammit, V.A.; Hardie, D.G. A common bicyclic protein kinase cascade inactivates the regulatory enzymes of fatty acid and cholesterol biosynthesis. *FEBS Lett.* **1987**, *223*, 217–222. [CrossRef]

17. Munday, M.R.; Campbell, D.G.; Carling, D.; Hardie, D.G. Identification by amino acid sequencing of three major regulatory phosphorylation sites on rat acetyl-CoA carboxylase. *Eur. J. Biochem.* **1988**, *175*, 331–338. [CrossRef] [PubMed]

18. Wu, N.; Zheng, B.; Shaywitz, A.; Dagon, Y.; Tower, C.; Bellinger, G.; Shen, C.H.; Wen, J.; Asara, J.; McGraw, T.E.; et al. AMPK-dependent degradation of TXNIP upon energy stress leads to enhanced glucose uptake via GLUT1. *Mol. Cell.* **2013**, *49*, 1167–1175. [CrossRef] [PubMed]

19. Yang, W.; Hong, Y.H.; Shen, X.Q.; Frankowski, C.; Camp, H.S.; Leff, T. Regulation of transcription by AMP-activated protein kinase: Phosphorylation of p300 blocks its interaction with nuclear receptors. *J. Biol. Chem.* **2001**, *276*, 38341–38344. [CrossRef] [PubMed]

20. Koo, S.H.; Flechner, L.; Qi, L.; Zhang, X.; Screaton, R.A.; Jeffries, S.; Hedrick, S.; Xu, W.; Boussouar, F.; Brindle, P.; et al. The CREB coactivator TORC2 is a key regulator of fasting glucose metabolism. *Nature* **2005**, *437*, 1109–1111. [CrossRef] [PubMed]

21. Li, Y.; Xu, S.; Mihaylova, M.M.; Zheng, B.; Hou, X.; Jiang, B.; Park, O.; Luo, Z.; Lefai, E.; Shyy, J.Y.; et al. AMPK phosphorylates and inhibits SREBP activity to attenuate hepatic steatosis and atherosclerosis in diet-induced insulin-resistant mice. *Cell Metab.* **2011**, *13*, 376–388. [CrossRef] [PubMed]

22. Mihaylova, M.M.; Vasquez, D.S.; Ravnskjaer, K.; Denechaud, P.D.; Yu, R.T.; Alvarez, J.G.; Downes, M.; Evans, R.M.; Montminy, M.; Shaw, R.J. Class IIa histone deacetylases are hormone-activated regulators of FOXO and mammalian glucose homeostasis. *Cell* **2011**, *145*, 607–621. [CrossRef] [PubMed]

23. Saxton, R.A.; Sabatini, D.M. mTOR Signaling in Growth, Metabolism, and Disease. *Cell* **2017**, *169*, 361–371. [CrossRef] [PubMed]

24. Inoki, K.; Zhu, T.; Guan, K.L. TSC2 mediates cellular energy response to control cell growth and survival. *Cell* **2003**, *115*, 577–590. [CrossRef]

25. Gwinn, D.M.; Shackelford, D.; Egan, D.F.; Mihaylova, M.M.; Mery, A.; Vasquez, D.S.; Turk, B.E.; Shaw, R.J. AMPK phosphorylation of raptor mediates a metabolic checkpoint. *Mol. Cell* **2008**, *30*, 214–226. [CrossRef] [PubMed]

26. Zhang, Y.L.; Guo, H.; Zhang, C.S.; Lin, S.Y.; Yin, Z.; Peng, Y.; Luo, H.; Shi, Y.; Lian, G.; Zhang, C.; et al. AMP as a low-energy charge signal autonomously initiates assembly of AXIN-AMPK-LKB1 complex for AMPK activation. *Cell Metab.* **2013**, *18*, 546–555. [CrossRef] [PubMed]

27. Zhang, C.S.; Jiang, B.; Li, M.; Zhu, M.; Peng, Y.; Zhang, Y.L.; Wu, Y.Q.; Li, T.Y.; Liang, Y.; Lu, Z.; et al. The lysosomal v-ATPase-Ragulator complex is a common activator for AMPK and mTORC1, acting as a switch between catabolism and anabolism. *Cell Metab.* **2014**, *20*, 526–540. [CrossRef] [PubMed]

28. Salminen, A.; Kauppinen, A.; Kaarniranta, K. AMPK/Snf1 signaling regulates histone acetylation: Impact on gene expression and epigenetic functions. *Cell Signal.* **2016**, *28*, 887–895. [CrossRef] [PubMed]

29. Guarente, L. The logic linking protein acetylation and metabolism. *Cell Metab.* **2011**, *14*, 151–153. [CrossRef] [PubMed]

30. Pietrocola, F.; Galluzzi, L.; Bravo-San Pedro, J.M.; Madeo, F.; Kroemer, G. Acetyl coenzyme A: A central metabolite and second messenger. *Cell Metab.* **2015**, *21*, 805–821. [CrossRef] [PubMed]

31. Kim, S.C.; Sprung, R.; Chen, Y.; Xu, Y.; Ball, H.; Pei, J.; Cheng, T.; Kho, Y.; Xiao, H.; Xiao, L.; et al. Substrate and functional diversity of lysine acetylation revealed by a proteomics survey. *Mol. Cell* **2006**, *23*, 607–618. [CrossRef] [PubMed]

32. Choudhary, C.; Kumar, C.; Nielsen, M.L.; Rehman, M.; Walther, T.C.; Olsen, J.V.; Mann, M. Lysine acetylation targets protein complexes and co-regulates major cellular functions. *Science* **2009**, *325*, 834–840. [CrossRef] [PubMed]

33. Wang, Q.; Zhang, Y.; Yang, C.; Xiong, H.; Lin, Y.; Yao, J.; Li, H.; Xie, L.; Zhao, W.; Yao, Y.; et al. Acetylation of metabolic enzymes coordinates carbon source utilization and metabolic flux. *Science* **2010**, *327*, 1004–1007. [CrossRef] [PubMed]

34. Zhao, S.; Xu, W.; Jiang, W.; Yu, W.; Lin, Y.; Zhang, T.; Yao, J.; Zhou, L.; Zeng, Y.; Li, H.; et al. Regulation of cellular metabolism by protein lysine acetylation. *Science* **2010**, *327*, 1000–1004. [CrossRef] [PubMed]

35. Olsen, J.V.; Blagoev, B.; Gnad, F.; Macek, B.; Mortensen, P.; Mann, M. Global, in vivo, and site-specific phosphorylation dynamics in signaling networks. *Cell* **2006**, *127*, 635–648. [CrossRef] [PubMed]

36. Villen, J.; Beausoleil, S.A.; Gerber, S.A.; Gygi, S.P. Large scale phosphorylation analysis of mouse liver. *Proc Natl. Acad. Sci. USA* **2007**, *104*, 1488–1493. [CrossRef] [PubMed]

37. Kim, G.W.; Yang, X.J. Comprehensive lysine acetylomes emerging from bacteria to humans. *Trens. Biochem. Sci.* **2011**, *36*, 211–220. [CrossRef] [PubMed]

38. Galdieri, L.; Zhang, T.; Rogerson, D.; Lleshi, R.; Vancura, A. Protein acetylation and acetyl coenzyme A metabolism in budding yeast. *Eukaryot. Cell* **2014**, *13*, 1472–1483. [CrossRef] [PubMed]

39. Drazic, A.; Myklebust, L.M.; Ree, R.; Arnesen, T. The world of protein acetylation. *Biochim. Biophys. Acta* **2016**, *1864*, 1372–1401. [CrossRef] [PubMed]

40. Roth, S.Y.; Denu, J.M.; Allis, C.D. Histone acetyltransferases. *Annu. Rev. Biochem.* **2001**, *70*, 81–120. [CrossRef] [PubMed]

41. Kouzarides, T. Chromatin modifications and their function. *Cell* **2007**, *128*, 693–705. [CrossRef] [PubMed]

42. Albaugh, B.N.; Arnold, K.M.; Denu, J.M. KAT(ching) metabolism by the tail: Insight into the links between lysine acetyltransferases and metabolism. *ChemBioChem* **2011**, *12*, 290–298. [CrossRef] [PubMed]

43. Farria, A.; Li, W.; Dent, S.Y.R. KATs in cancer: Function and therapies. *Oncogene* **2015**, *34*, 4901–4913. [CrossRef] [PubMed]

44. Fan, J.; Krautkramer, K.A.; Feldman, J.L.; Denu, J.M. Metabolic regulation of histone post-translational modifications. *ACS Chem. Biol.* **2015**, *10*, 95–108. [CrossRef] [PubMed]

45. Janke, R.; Dodson, A.E.; Rine, J. Metabolism and epigenetics. *Annu. Rev. Cell Dev. Biol.* **2015**, *31*, 473–496. [CrossRef] [PubMed]

46. Haberland, M.; Montgomery, R.L.; Olson, E.N. The many roles of histone deacetylases in development and physiology: Implications for disease and therapy. *Nat. Rev. Genet.* **2009**, *10*, 32–42. [CrossRef] [PubMed]

47. Chalkiadaki, A.; Guarente, L. The multifaceted functions of sirtuins in cancer. *Nat. Rev. Cancer* **2015**, *15*, 608–624. [CrossRef] [PubMed]

48. Wagner, G.R.; Hirschey, M.D. Nonezymatic protein acylation as a carbon stress regulated by sirtuin deacylases. *Mol. Cell* **2014**, *54*, 5–16. [CrossRef] [PubMed]

49. Weinert, B.T.; Iesmantavicius, V.; Moustafa, T.; Scholz, C.; Wagner, S.A.; Magnes, C.; Zechner, R.; Choudhary, C. Acetylation dynamics and stoichiometry in *Saccharomyces cerevisiae*. *Mol. Systems Biol.* **2014**, *10*, 716. [CrossRef] [PubMed]

50. Wagner, G.R.; Payne, R.M. Widespread and enzyme-independent *N*-acetylation and *N*-succinylation of proteins in the chemical conditions of the mitochondrial matrix. *J. Biol. Chem.* **2013**, *288*, 29036–29045. [CrossRef] [PubMed]

51. Henriksen, P.; Wagner, S.A.; Weinert, B.T.; Sharma, S.; Bacinskaja, G.; Rehman, M.; Juffer, A.H.; Walther, T.C.; Lisby, M.; Choudhary, C. Proteome-wide analysis of lysine acetylation suggests its broad regulatory scope in Saccharomyces cerevisiae. *Mol. Cell. Proteomics* **2012**, *11*, 1510–1522. [CrossRef] [PubMed]

52. Uchida, M.; Sun, Y.; Mcdermott, G.; Knoechel, C.; le Gros, M.A.; Parkinson, D.; Drubin, D.G.; Larabell, C.A. Quantitative analysis of yeast internal architecture using soft X-ray tomography. *Yeast* **2011**, *28*, 227–236. [CrossRef] [PubMed]

53. Cai, L.; Tu, B.P. Acetyl-CoA drives the transcriptional growth program in yeast. *Cell Cycle* **2011**, *10*, 3045–3046. [CrossRef] [PubMed]

54. Cai, L.; Sutter, B.M.; Li, B.; Tu, B.P. Acetyl-CoA induces cell growth and proliferation by promoting the acetylation of histones at growth genes. *Mol. Cell* **2011**, *42*, 426–437. [CrossRef] [PubMed]

55. Casey, J.R.; Grinstein, S.; Orlowski, J. Sensors and regulators of intracellular pH. *Nat. Rev. Mol. Cell Biol.* **2010**, *11*, 50–61. [CrossRef] [PubMed]

56. Wagner, G.R.; Bhatt, D.P.; O'Connell, T.M.; Thompson, J.W.; Dubois, L.G.; Backos, D.S.; Yang, H.; Mitchell, G.A.; Ilkayeva, O.R.; Stevens, R.D.; et al. A Class of Reactive Acyl-CoA Species Reveals the Non-enzymatic Origins of Protein Acylation. *Cell Metab.* **2017**, *25*, 823–837. [CrossRef] [PubMed]

57. Cai, L.; Tu, B.P. On acetyl-CoA as a gauge of cellular metabolic state. *Cold Spring Harbor Symp. Quant. Biol.* **2011**, *76*, 195–202. [CrossRef] [PubMed]

58. Shi, L.; Tu, B.P. Acetyl-CoA and the regulation of metabolism: Mechanisms and consequences. *Curr. Opin. Cell Biol.* **2015**, *33*, 125–131. [CrossRef] [PubMed]

59. Sivanand, S.; Viney, I.; Wellen, K.E. Spatiotemporal Control of Acetyl-CoA Metabolism in Chromatin Regulation. *Trends Biochem Sci.* **2018**, *43*, 61–74. [CrossRef] [PubMed]

60. Takahashi, H.; MacCaffery, J.M.; Irizarry, R.A.; Boeke, J.D. Nucleocytosolic acetyl-coenzyme A synthetase is required for histone acetylation and global transcription. *Mol. Cell* **2006**, *23*, 207–217. [CrossRef] [PubMed]

61. Wellen, K.E.; Hatzivassiliou, G.; Sachdeva, U.M.; Bui, T.V.; Cross, J.R.; Thompson, C.B. ATP-citrate lyase links cellular metabolism to histone acetylation. *Science* **2009**, *324*, 1076–1080. [CrossRef] [PubMed]

62. Sutendra, G.; Kinnaird, A.; Dromparis, P.; Paulin, R.; Stenson, T.H.; Haromy, A.; Hashimoto, K.; Zhang, N.; Flaim, E.; Michelakis, E.D. A nuclear pyruvate dehydrogenase complex is important for the generation of acetyl-CoA and histone acetylation. *Cell* **2014**, *158*, 84–97. [CrossRef] [PubMed]

63. Liu, X.; Cooper, D.E.; Cluntun, A.A.; Warmoes, M.O.; Zhao, S.; Reid, M.A.; Liu, J.; Lund, P.J.; Lopes, M.; Garcia, B.A.; et al. Acetate production from glucose and coupling to mitochondrial metabolism in mammals. *Cell* **2018**, *175*, 502–513. [CrossRef] [PubMed]

64. Galdieri, L.; Vancura, A. Acetyl-CoA carboxylase regulates global histone acetylation. *J. Biol. Chem.* **2012**, *28*, 23865–23876. [CrossRef] [PubMed]

65. Galdieri, L.; Gatla, H.; Vancurova, I.; Vancura, A. Activation of AMP-activated protein kinase by metformin induces protein acetylation in prostate and ovarian cancer cells. *J. Biol. Chem.* **2016**, *291*, 25154–25166. [CrossRef] [PubMed]

66. Vancura, A.; Vancurova, I. Metformin induces protein acetylation in cancer cells. *Oncotarget* **2017**, *8*, 39939–39940. [CrossRef] [PubMed]

67. Hardie, D.G.; Carling, D.; Carlson, M. The AMP-activated/SNF1 protein kinase subfamily: Metabolic sensors of the eukaryotic cell? *Annu. Rev. Biochem.* **1998**, *67*, 821–855. [CrossRef] [PubMed]

68. Hardie, D.G. AMP-activated/SNF1 protein kinases: Conserved guardians of cellular energy. *Nat. Rev. Mol. Cell Biol.* **2007**, *8*, 774–785. [CrossRef] [PubMed]

69. Hedbacker, K.; Carlson, M. SNF1/AMPK pathways in yeast. *Front. Biosci.* **2009**, *13*, 2408–2420. [CrossRef]

70. Zhang, M.; Galdieri, L.; Vancura, A. The yeast AMPK homolog SNF1 regulates acetyl coenzyme A homeostasis and histone acetylation. *Mol. Cell. Biol.* **2013**, *33*, 4701–4717. [CrossRef] [PubMed]

71. Haigis, M.C.; Guarente, L.P. Mammalian sirtuins-emerging roles in physipology, aging, and calorie restriction. *Genes Dev.* **2006**, *20*, 2913–2921. [CrossRef] [PubMed]

72. Houtkooper, R.H.; Pirinen, E.; Auwerx, J. Sirtuins as regulators of metabolism and healthspan. *Nat. Rev. Mol. Cell Biol.* **2012**, *13*, 225–238. [CrossRef] [PubMed]

73. Fulco, M.; Ce, Y.; Zhao, P.; Hoffman, E.P.; McBurney, M.W.; Sauve, A.A.; DSartorelli, V. Glucose restriction inhibits skeletal myoblast differentiation by activating SIRT1 through AMPK-mediated regulation of Nampt. *Dev. Cell* **2008**, *14*, 661–673. [CrossRef] [PubMed]

74. Canto, C.; Gerhart-Hines, Z.; Feige, J.N.; Lagouge, M.; Noriega, L.; Milne, J.C.; Elliot, P.J.; Puigserver, P.; Auwerx, J. AMPK regulates energy expenditure by modulating NAD$^+$ metabolism and SIRT1 activity. *Nature* **2009**, *458*, 1056–1060. [CrossRef] [PubMed]

75. Lau, A.W.; Liu, P.; Inuzuka, H.; Gao, D. SIRT1 phosphorylation by AMP-activated protein kinase regulates p53 acetylation. *Am. J. Cancer Res.* **2014**, *4*, 245–255. [PubMed]

76. Chang, C.; Su, H.; Zhang, D.; Wang, Y.; Shen, Q.; Liu, B.; Huang, R.; Zhou, T.; Peng, C.; Wong, C.C.; et al. AMPK-dependent phosphorylation of GAPDH triggers Sirt1 activation and is necessary for autophagy upon glucose starvation. *Mol. Cell* **2015**, *60*, 930–940. [CrossRef] [PubMed]

77. Shimazu, T.; Hirschey, M.D.; Newman, J.; He, W.; Shirakawa, K.; Le Moan, N.; Grueter, C.A.; Lim, H.; Saunders, L.R.; Stevens, R.D.; et al. Suppression of oxidative stress by -hydroxybutyrate, an endogenous histone deacetylase inhibitor. *Science* **2013**, *339*, 211–214. [CrossRef] [PubMed]

78. Verdin, E.; Dequiedt, F.; Kasler, H.G. Class II histone deacetylase: Versatile regulators. *Trends Genet.* **2003**, *19*, 286–293. [CrossRef]

79. Yang, X.J.; Gregoire, S. Class II histone deacetylases: From sequence to function, regulation, and clinical implication. *Mol. Cell. Biol.* **2005**, *25*, 2873–2884. [CrossRef] [PubMed]

80. Martin, M.; Kettmann, R.; Dequiedt, F. Class IIa histone deacetylases: Regulating the regulators. *Oncogene* **2007**, *26*, 5450–5467. [CrossRef] [PubMed]

81. Parra, M. Class IIa HDACs-new insights into their function in physiology and pathology. *FEBS J.* **2015**, *282*, 1736–1744. [CrossRef] [PubMed]

82. McGee, S.L.; van Denderen, B.J.; Howlett, K.F.; Mollica, J.; Schertzer, J.D.; Kemp, B.E.; Hargreaves, M. AMP-actoivated protein kinase regulates GLUT4 transcription by phosphorylating histone deacetylase 5. *Diabetes* **2008**, *57*, 860–867. [CrossRef] [PubMed]

83. Zhao, J.X.; Yue, W.F.; Zhu, M.J.; Du, M. AMP-activated protein kinase regulates -catenin transcription via histone deacetylase 5. *J. Biol. Chem.* **2011**, *286*, 16426–16434. [CrossRef] [PubMed]

84. Chen, S.; Yin, C.; Lao, T.; Liang, D.; He, D.; Wang, C.; Sang, N. AMPK-HDAC5 pathway facilitates nuclear accumulation of HIF-1 and functional activation of HIF-1 by deacetylating HSP70 in the cytosol. *Cell Cycle* **2015**, *14*, 2520–2536. [CrossRef] [PubMed]

85. Fu, X.; Zhao, J.X.; Liang, J.; Zhu, M.J.; Foretz, M.; Viollet, B.; Du, M. AMP-activated protein kinase mediates myogenin expression and myogenesis via histone deacetylase 5. *Am. J. Phys. Cell Physiol.* **2013**, *305*, C887–C895. [CrossRef] [PubMed]

86. Fu, X.; Zhao, J.X.; Zhu, M.J.; Foretz, M.; Viollet, B.; Dodson, M.V.; Du, M. AMP-activated protein kinase a1 but not a2 catalytic subunit potentiates myogenin expression and myogenesis. *Mol. Cell. Biol.* **2013**, *33*, 4517–4525. [CrossRef] [PubMed]

87. Zhang, Y.; Qiu, J.; Wang, X.; Zhang, Y.; Xia, M. AMP-activated protein kinase suppresses endothelial cell inflammation through phosphorylation of transcriptional coactivator p300. *Arterioscler. Thromb. Vasc. Biol.* **2011**, *31*, 2897–2908. [CrossRef] [PubMed]

88. Lim, J.Y.; Oh, M.A.; Kim, W.H.; Sohn, H.Y.; Park, S.I. AMP-activated protein kinase inhibits TGF-B-induced fibrogenic responses of hepatic stellate cells by targeting transcriptional coactivator p300. *J. Cell. Physiol.* **2012**, *227*, 1081–1089. [CrossRef] [PubMed]

89. Bungard, D.; Fuerth, B.J.; Zeng, P.Y.; Faubert, B.; Maas, N.L.; Viollet, B.; Carling, D.; Thompson, C.B.; Jones, R.G.; Berger, S.L. Signaling kinase AMPK activates stress-promoted transcription via histone H2B phosphorylation. *Science* **2010**, *329*, 1201–1205. [CrossRef] [PubMed]

90. Lo, W.S.; Duggan, L.; Emre, N.C.; Belotserkovskya, R.; Lane, W.S.; Shiekhattar, R.; Berger, S.L. Snf1-a histone kinase that works in concert with the histone acetyltransferase Gcn5 to regulate transcription. *Science* **2001**, *293*, 1142–1146. [CrossRef] [PubMed]

91. Lan, F.; Cacicedo, J.M.; Ruderman, N.; Ido, Y. SIRT1 modulation of the acetylation status, cytosolic localization, and activity of LKB1. Possible role in AMP-activated protein kinase activation. *J. Biol. Chem.* **2008**, *283*, 27628–27635. [CrossRef] [PubMed]

92. Hou, X.; Xu, S.; Maitland-Toolan, K.A.; Sato, K.; Jiang, B.; Ido, Y.; Lan, F.; Walsh, K.; Wierzbicki, M.; Verbeuren, T.J.; et al. SIRT1 regulates hepatocyte lipid metabolism through activating AMP-activated protein kinase. *J. Biol. Chem.* **2008**, *283*, 20015–20026. [CrossRef] [PubMed]

93. Jiang, R.; Carlson, M. The Snf1 protein kinase and its activating subunit, Snf4, interact with distinct domains of the Sip1/Sip2/Gal83 component in the kinase complex. *Mol. Cell. Biol.* **1997**, *17*, 2099–2106. [CrossRef] [PubMed]

94. Lu, J.-Y.; Lin, Y.-Y.; Sheu, J.-C.; Wu, J.-T.; Lee, F.-J.; Chen, Y.; Lin, M.-I.; Chiang, F.-T.; Tai, T.-Y.; Berger, S.L.; et al. Acetylation of yeast AMPK controls intrinsic aging idependently of caloric restriction. *Cell* **2011**, *146*, 969–979. [CrossRef] [PubMed]

Permissions

List of Contributors

Gyöngyi Kudlik, Tamás Takács, Anita Kurilla, Bálint Szeder, Kitti Koprivanacz, Balázs L. Merő and Virag Vas
Institute of Enzymology, Research Centre for Natural Sciences, 1117 Budapest, Hungary

László Radnai
Institute of Enzymology, Research Centre for Natural Sciences, 1117 Budapest, Hungary
Department of Molecular Medicine, The Scripps Research Institute, Jupiter, FL 33458, USA
Department of Neuroscience, The Scripps Research Institute, Jupiter, FL 33458, USA

László Buday
Institute of Enzymology, Research Centre for Natural Sciences, 1117 Budapest, Hungary
Department of Medical Chemistry, Semmelweis University Medical School, 1085 Budapest, Hungary

Yasmine Ould Amer and Etienne Hebert-Chatelain
Department of Biology, University of Moncton, Moncton, NB E1A 3E9, Canada
Canada Research Chair in Mitochondrial Signaling and Physiopathology, University of Moncton, Moncton, NB E1A 3E9, Canada

Natalia A. Vilchinskaya and Boris S. Shenkman
Myology Laboratory, Institute of Biomedical Problems RAS, Moscow 123007, Russia

Igor I. Krivoi
Department of General Physiology, St. Petersburg State University, St. Petersburg 199034, Russia

Nina Zippel and Annemarieke E. Loot
Institute for Vascular Signalling, Centre for Molecular Medicine, Johann Wolfgang Goethe University, 60590 Frankfurt, Germany

Heike Stingl, Voahanginirina Randriamboavonjy, Ingrid Fleming and Beate Fisslthaler
Institute for Vascular Signalling, Centre for Molecular Medicine, Johann Wolfgang Goethe University, 60590 Frankfurt, Germany
DZHK (German Centre for Cardiovascular Research) partner site RhineMain, Theodor Stern Kai 7, 60590 Frankfurt, Germany

Prashanta Silwal
Department of Microbiology, Chungnam National University School of Medicine, Daejeon 35015, Korea
Infection Control Convergence Research Center, Chungnam National University School of Medicine, Daejeon 35015, Korea

Jin Kyung Kim and Eun-Kyeong Jo
Department of Microbiology, Chungnam National University School of Medicine, Daejeon 35015, Korea
Infection Control Convergence Research Center, Chungnam National University School of Medicine, Daejeon 35015, Korea
Department of Medical Science, Chungnam National University School of Medicine, Daejeon 35015, Korea

Jae-Min Yuk
Department of Infection Biology, Chungnam National University School of Medicine, Daejeon 35015, Korea

Isaac Tamargo-Gómez and Guillermo Mariño
Instituto de Investigación Sanitaria del Principado de Asturias, 33011 Oviedo, Spain
Departamento de Biología Funcional, Universidad de Oviedo, 33011 Oviedo, Spain

Baile Wang
State Key Laboratory of Pharmaceutical Biotechnology, The University of Hong Kong, Hong Kong, China
Department of Medicine, The University of Hong Kong, Hong Kong, China

Kenneth King-Yip Cheng
Department of Health Technology and Informatics, The Hong Kong Polytechnic University, Hong Kong, China

Yan Yan and H. Eric Xu
Center for Cancer and Cell Biology, Van Andel Research Institute, 333 Bostwick Ave. N.E., Grand Rapids, MI 49503, USA
VARI/SIMM Center, Center for Structure and Function of Drug Targets, CAS-Key Laboratory of Receptor Research, Shanghai Institute of Materia Medica, Chinese Academy of Sciences, Shanghai 201203, China

X. Edward Zhou and Karsten Melcher
Center for Cancer and Cell Biology, Van Andel Research Institute, 333 Bostwick Ave. N.E., Grand Rapids, MI 49503, USA

Marc Foretz and Benoit Viollet
INSERM, U1016, Institut Cochin, Département d'Endocrinologie Métabolisme et Diabète, 24, rue du Faubourg Saint Jacques, 75014 Paris, France
CNRS, UMR8104, 75014 Paris, France
Université Paris Descartes, Sorbonne Paris Cité, 75014 Paris, France

Patrick C. Even
UMR PNCA, AgroParisTech, INRA, Université Paris-Saclay, 75005 Paris, France

Sébastien Didier, Florent Sauvé, Manon Domise, Luc Buée, Claudia Marinangeli and Valérie Vingtdeux
Université de Lille, Inserm, Centre Hospitalo-Universitaire de Lille, UMR-S1172—JPArc—Centre de Recherche Jean-Pierre AUBERT, F-59000 Lille, France

Natalie R. Janzen, Jamie Whitfield and Nolan J. Hoffman
Exercise and Nutrition Research Program, Mary MacKillop Institute for Health Research, Australian Catholic University, Level 5, 215 Spring Street, Melbourne, Victoria 3000, Australia

Philipp Glosse
Institute of Agricultural and Nutritional Sciences, Martin Luther University Halle-Wittenberg, D-06120 Halle (Saale), Germany

Michael Föller
Institute of Physiology, University of Hohenheim, D-70599 Stuttgart, Germany

Ales Vancura, Shreya Nagar, Pritpal Kaur, Pengli Bu, Madhura Bhagwat and Ivana Vancurova
Department of Biological Sciences, St. John's University, New York, NY 11439, USA

Index

Printed in the USA
CPSIA information can be obtained
at www.ICGtesting.com
JSHW051407091023
49903JS00006B/311

9 781639 877331